THE RANDOM HOUSE
Geographical
DICTIONARY

THE RANDOM HOUSE
Geographical
DICTIONARY

RANDOM HOUSE
NEW YORK

Copyright © 1992 by Random House, Inc.

All rights reserved under International and Pan-American Copyright Conventions. No part of this publication may be reproduced in any form or by any means, electronic or mechanical, including photocopying, without permission in writing from the publisher. All inquiries should be addressed to: Reference and Electronic Publishing Division, Random House, Inc., 201 East 50th Street, New York, NY 10022. Published in the United States by Random House, Inc., New York, and simultaneously in Canada by Random House of Canada, Limited, Toronto.

Library of Congress Cataloging-in-Publication Data

The Random House geographical dictionary.
       p.   cm.—(Random House pocket dictionaries and guides)
    ISBN 0-679-41570-X
    1. Gazetteers.   I. Series.
GA103.5.R36   1992
910′.3—dc20   92-25713
                                                    CIP

Manufactured in the United States of America

1 2 3 4 5 6 7 8 9

First Edition

Contents

Staff

Enid Pearsons, *Editor*
Sol Steinmetz, *Executive Editor*
Mia McCroskey, *Managing Editor*

Alice Kovac Somoroff, *Project Editor*
Constance Baboukis Padakis, *Associate Editor*
Judy Kaplan Johnson, *Supervising Copyeditor*
Sasan Doroudian, Lani Mysak, Maria Padilla,
  Editorial Assistants

Patricia W. Ehresmann, *Production Director*
Rita E. Rubin, *Production Associate*
Charlotte Staub, *Designer*

Publisher, Michael Mellin
Associate Publishers, Catherine Fowler,
  John L. Hornor III

Preface

The Random House Geographical Dictionary is designed to provide a convenient source of authoritative, essential information about the most important places in the world, past and present. This handy portable gazetteer includes more than 8,000 entries identifying countries and their capitals, states, provinces, cities, national parks, mountain ranges, deserts, rivers, seas, and other geographic features around the globe.

Based on Random House's constantly updated "Living Dictionary" database, the book contains hundreds of entries that reflect the recent political and geographic changes in Germany, Eastern Europe, the former Soviet Union, and Yugoslavia, making this the most up-to-date portable geographical reference book available.

All the entries are listed in convenient alphabetical order. Each boldface entry is divided into syllables to show end-of-line hyphenation, although such hyphenation for proper names should be avoided whenever possible. Each place name is followed by a parenthesized pronunciation. (See the Pronunciation Key on page ix.) Entries include population figures as well as facts about location, length, height, and area (both in miles and meters). The foreign names of many places are often indicated both alongside the English name (**Ire·land** . . . Latin, **Hibernia** . . . Irish, **Eire**) and in a separate entry (**Hi·ber·ni·a** . . . Latin name of IRELAND). Many entries also include historical, political, and other useful information (**Ith·a·ca** . . . legendary home of Ulysses; **I′vory Coast′** . . . gained independence 1960; **K2** . . . second highest peak in the world). In addition, derived forms appear at the end of many entries. These are shown with part-of-speech labels. The label *n.* (noun) normally indicates a term designating a "native or inhabitant" of a place, and an *adj.* (adjective) label designates a common descriptive term meaning "of or pertaining to" that place. The word **Canadian,** for example, shown at the entry for **Canada,** has both of these meanings. The noun senses of **French, English,** and similar terms represent instances where the derived noun form refers to "natives or inhabitants" in the plural.

Here, then, in convenient pocket-size format, is a reference book for students, travelers, and all others interested in obtaining accurate information on places and place names throughout the world.

Pronunciation Key

STRESS

Pronunciations are marked for stress to reveal the relative differences in emphasis between syllables. In words of two or more syllables, a primary stress mark (′), as in **mother** (muth′ər), follows the syllable having greatest stress. A secondary stress mark (′), as in **grandmother** (grand′ muth′ər), follows a syllable having slightly less stress than primary but more stress than an unmarked syllable.

ENGLISH SOUNDS

a	act, bat, marry
ā	age, paid, say
â(r)	air, dare, Mary
ä	ah, part, balm
b	back, cabin, cab
ch	child, beach
d	do, madder, bed
e	edge, set, merry
ē	equal, bee, pretty
ēr	ear, mere
f	fit, differ, puff
g	give, trigger, beg
h	hit, behave
hw	which, nowhere
i	if, big, mirror
ī	ice, bite, deny
j	just, tragic, fudge
k	keep, token, make
l	low, mellow, bottle (bot′l)
m	my, summer, him
n	now, sinner, button (but′n)
ng	sing, Washington
o	ox, bomb, wasp
ō	over, boat, no
ô	order, ball, raw

oi	oil, joint, joy
o͝o	book, tour
o͞o	ooze, fool, too
ou	out, loud, cow
p	pot, supper, stop
r	read, hurry, near
s	see, passing, miss
sh	shoe, fashion, push
t	ten, matter, bit
th	thin, ether, path
ᵺ	that, either, smooth
u	up, sun
û(r)	urge, burn, cur
v	voice, river, live
w	witch, away
y	yes, onion
z	zoo, lazy, those
zh	treasure, mirage
ə	used in unaccented syllables to indicate the sound of the reduced vowel in *a*lone, syst*e*m, eas*i*ly, gall*o*p, circ*u*s
°	used between i and r and between ou and r to show triphthongal quality, as in fire(fī°r), hour (ou°r)

NON-ENGLISH SOUNDS

A	as in French *ami* (A mē′)
KH	as in Scottish *loch* (lôKH)
N	as in French *bon* (bôN) [used to indicate that the preceding vowel is nasalized.]
Œ	as in French *feu* (fŒ)
R	[a symbol for any non-English *r*, including a trill or flap in Italian and Spanish and a sound in French and German similar to KH but pronounced with voice]
Y	as in French *tu* (tY)
°	as in French *Bastogne* (bA stôn′y°)

Abbreviations
Used in this Book

ab.	about
adj.	adjective
Arab.	Arabic
Brit.	British
cap., caps	capital, capitals
Ch.	Chinese
Dan.	Danish
def., defs.	definition, definitions
Du.	Dutch
esp.	especially
Fr.	French
Ger.	German
Gk.	Greek
Heb.	Hebrew
It.	Italian
Japn.	Japanese
km	kilometers
mi.	miles
n.	noun
n.pl.	plural noun
Norw.	Norwegian
pl.	plural
Pol.	Polish
Port.	Portuguese

Rom.	Romanian
Russ.	Russian
Sp.	Spanish
sq.	square
Sw.	Swedish
usu.	usually

A

Aa·chen (ä′kən, ä′ĸнən), a city in W Germany, near the Belgian and Dutch borders. 232,000. French, **Aix-la-Chapelle.**

Aal·borg (ôl′bôrg), ÅLBORG.

Aalst (älst), a city in central Belgium. 76,441. French, **Alost.**

Aar·au (är′ou), the capital of Aargau, in N Switzerland. 15,927.

Aar·e (är′ə) also **Aar** (är), a river in central Switzerland, flowing N to the Rhine. 175 mi. (280 km) long.

Aar·gau (är′gou), a canton in N Switzerland. 490,400; 542 sq. mi. (1400 sq. km). *Cap.:* Aarau. French, **Argovie.**

Aar·hus (ôr′hōōs′), ÅRHUS.

AB, Alberta.

A·ba (ä bä′), a town in SE Nigeria. 216,000.

Ab·a·co (ab′ə kō′), two islands **(Great Abaco** and **Little Abaco)** in the N Bahamas. 7271; 776 sq. mi. (2010 sq. km).

A·ba·dan (ä′bə dän′, ab′ə-), a city in SW Iran, on the Shatt-al-Arab: oil refineries. 306,000.

A·ba·kan (ä′bə kän′), the capital of the Khakass Autonomous Region in the Russian Federation in Asia, on the Yenisei River. 154,000.

Ab·be·ville (ab′ē vil′, ab vēl′), a town in N France, on the Somme River: site of Paleolithic artifacts. 26,581.

A·be·o·ku·ta (ä′bā ō′kōō tä) a city in SW Nigeria. 308,800.

Ab·er·dare (ab′ər dâr′), a city in Mid Glamorgan, in S Wales. 65,400.

Ab·er·deen (ab′ər dēn′ *for 1, 2;* ab′ər dēn′ *for 3*), **1.** Also called **Ab′er·deen′shire** (-shēr, -shər) a historic county in NE Scotland. **2.** a seaport in NE Scotland, on the North Sea: administrative center of the Grampian region. 213,228. **3.** a city in NE South Dakota. 25,973. —**Ab′er·do′ni·an** (-dō′nē ən), *adj., n.*

Ab·i·djan (ab′i jän′), a seaport in and the capital of the Ivory Coast. 1,850,000.

Ab·i·lene (ab′ə lēn′), a city in central Texas. 106,654.

Ab·ing·ton (ab'ing tən), a town in SE Pennsylvania. 59,084.

Ab·i·tib·i (ab'i tib'ē), **1.** a lake in E Ontario and W Quebec, Canada. 369 sq. mi. (956 sq. km). **2.** a river flowing N from this lake. 340 mi. (547 km) long.

Ab·kha·zi·a or **Ab·kha·si·a** (ab kä'zhə, -zē ə, -kä'-), an autonomous republic in the Georgian Republic, on the E coast of the Black Sea. 537,000; 3320 sq. mi. (8600 sq. km). *Cap.:* Sukhumi. Formerly, **Ab·khaz'** **Auton'omous So'viet So'cialist Repub'lic** (äb käz'). —**Ab·kha'zi·an** (-kä'zē ən, -zhən), *adj., n.*

Å·bo (ô'bŏŏ), Swedish name of TURKU.

Ab·o·mey (ab'ə mā', ə bŏ'mē), a city in SW Benin. 54,418.

A·bou·kir (ä'bŏŏ kēr', ab'ŏŏ-), ABUKIR.

A·bruz·zi (ə brŏŏt'sē), a region in central Italy, on the Adriatic. 1,257,988; 4168 sq. mi. (10,794 sq. km).

Ab·sa'ro·ka Range' (ab sär'ə kə), a mountain range in S Montana and NW Wyoming: part of the Rocky Mountains. Highest peak, Franks Peak, 13,140 ft. (4005 m).

A·bu Dha·bi (ä'bŏŏ dä'bē), **1.** a sheikdom in the N United Arab Emirates, on the S coast of the Persian Gulf. 670,125. **2.** the capital of this sheikdom and the provisional capital of the United Arab Emirates. 242,975.

A·bu·ja (ə bŏŏ'jə), the capital of Nigeria, in the central part. 378,671.

A·bu·kir or **A·bou·kir** (ä'bŏŏ kēr', ab'ŏŏ-), a bay in N Egypt, between Alexandria and the Rosetta mouth of the Nile: French fleet defeated here by British fleet 1798.

A·bur·y (ā'bə rē), AVEBURY.

A·bu Sim·bel (ä'bŏŏ sim'bel, -bəl), a former village in S Egypt, on the Nile: inundated by Lake Nasser, created by the Aswan High Dam; site of two temples of Ramses II, now moved to higher ground.

A·by·dos (ə bī'dəs), **1.** an ancient ruined city in central Egypt, near Thebes: temples and necropolis. **2.** an ancient town in NW Asia Minor, at the narrowest part of the Hellespont.

Ab·y·la (ab'ə lə), ancient name of JEBEL MUSA.

Ab·ys·sin·i·a (ab'ə sin'ē ə), **1.** former name of

ETHIOPIA (def. 1). **2.** ETHIOPIA (def. 2). —**Ab'ys·sin'i·an,** *adj., n.*

A·ca·di·a (ə kā/dē ə), a region and former French colony on the N Atlantic coast of North America, including the present Canadian provinces of Nova Scotia, New Brunswick, and Prince Edward Island, and part of Maine: ceded to the British 1713. —**A·ca/di·an,** *adj., n.*

Aca/dia Na/tional Park', a national park in Maine, on Mount Desert Island. 44 sq. mi. (114 sq. km).

A·ca·pul·co (ak/ə pŏŏl/kō, ä/kə-), a seaport and resort in SW Mexico, on the Pacific. 456,700.

Ac·ar·na·ni·a (ak/ər nā/nē ə), a coastal region in W central Greece, on the Ionian Sea. —**Ac'ar·na/ni·an,** *adj., n.*

Ac·cra (ak/rə, ə krä/), a seaport in and the capital of Ghana, on the Gulf of Guinea. 867,459.

A·chae·a (ə kē/ə), an ancient district in S Greece, on the Gulf of Corinth. —**A·chae/an,** *adj., n.*

Ach·ill (ak/əl), an island off the coast of NW Ireland. 14 mi. (23 km) long; 11 mi. (18 km) wide.

Ac·o·ma (ak/ə mô/, -mə, ä/kə-), a Pueblo Indian village near Albuquerque, N. M.: oldest continuously inhabited location in the U.S.

A·con·ca·gua (ä/kông kä/gwə), a mountain in W Argentina, in the Andes: highest peak in the Western Hemisphere. 22,834 ft. (6960 m).

A·cre (ä/krə *for 1;* ä/kər, ā/kər *for 2*), **1.** a state in W Brazil. 386,200; 58,900 sq. mi. (152,550 sq. km). *Cap.:* Rio Branco. **2.** a seaport in NW Israel: besieged and captured by Crusaders 1191. 38,700.

Ac·ti·um (ak/tē əm, -shē əm), a promontory in NW ancient Greece: Antony and Cleopatra were defeated by Octavian and Agrippa in a naval battle near here in 31 B.C.

Ad·ams (ad/əmz), **1. Mount,** a mountain in SW Washington, in the Cascade Range. 12,307 ft. (3751 m). **2. Mount,** a mountain in N New Hampshire, in the White Mountains. 5798 ft. (1767 m).

Ad/am's Bridge', an island chain in the Gulf of Mannar between NW Sri Lanka and SE India; ownership divided between Sri Lanka and India. 30 mi. (48 km) long.

A·da·na (ä′dä nä′), a city in S Turkey, on the Seyhan River. 776,000. Also called **Seyhan.**

A·da·pa·za·ri (ä′də pä′zə rē′), a city in NW Turkey, SE of Istanbul. 130,977.

Ad·dis A·ba·ba (ad′is ab′ə bə), the capital of Ethiopia, in the central part. 1,412,575.

Ad·e·laide (ad′l ād′), a city in and the capital of South Australia, in Australia. 993,100.

A·dé′lie Coast′ (ə dā′lē; Fr. A dā lē′), a coastal region of Antarctica, south of Australia: claimed by France.

A·den (äd′n, ād′n), **1.** the economic capital of the Republic of Yemen, a seaport on the Gulf of Aden. 318,000. **2.** a former British colony and protectorate on the Gulf of Aden, in SW Arabia: became People's Democratic Republic of Yemen in 1967; since 1990 part of the Republic of Yemen. **3. Gulf of,** an arm of the Arabian Sea between the E tip of Africa and the S coast of Arabia.

A·di·ge (ä′di jā′), a river in N Italy, flowing SE to the Adriatic Sea. 220 mi. (354 km) long.

Ad′i·ron′dack Moun′tains (ad′ə ron′dak, ad′-), a mountain range in NE New York: a part of the Appalachian Mountains. Highest peak, Mt. Marcy, 5344 ft. (1629 m). Also called **Ad′i·ron′-dacks.**

Ad′miralty Is′lands, a group of islands in the SW Pacific, N of New Guinea: part of Papua New Guinea. 30,160; ab. 800 sq. mi. (2070 sq. km).

Ad′miralty Range′, a mountain range in Antarctica, NW of the Ross Sea.

A·do-E·ki·ti (ä′dō ek′i tē′, -ā′ki-), a town in SE Nigeria. 265,800.

A·do·wa (ä′dōō wä′), ADUWA.

A·dri·an·o·ple (ā′drē ə nō′pəl), EDIRNE.

A′dri·at′ic Sea′ (ā′drē at′ik, ā′drē-), an arm of the Mediterranean between Italy and the Balkan Peninsula. Also called **Adriatic.**

A·du·wa or **A·do·wa** (ä′dōō wä′), a town in N Ethiopia: Italians defeated 1896. 26,782.

A′dy·gei Auton′omous Re′gion (ä′də gā′, ä′də gā′), an autonomous region in the Russian Federation, part of the Krasnodar territory, in the NW Caucasus Mountains. 432,000; 2,934 sq. mi. (7,600 sq. km). *Cap.:* Maikop.

A·dzhar·i·stan (ə jär′ə stan′, -stän′), an autonomous republic in the Georgian Republic, in Transcaucasia. 393,000; 1160 sq. mi. (3000 sq. km). *Cap.:* Batumi. Formerly, **A·dzhar/ Auton/omous So/viet So/cialist Repub/lic** (ə jär/).

Ae·ga/di·an (or **Ae·ga/de·an**) **Is/lands** (i gā/dē ən), EGADI.

Ae·ga·tes (i gā/tēz), ancient name of EGADI.

Ae·ge/an Is/lands (i jē/ən), the islands of the Aegean Sea, including the Dodecanese, Cyclades, and Sporades.

Aege/an Sea/, an arm of the Mediterranean Sea between Greece and Turkey. Also called **Aegean.**

Ae·gi·na (ē jī/nə, i jē/-), **1.** an island in the Saronic Gulf. 32 sq. mi. (83 sq. km). **2.** a seaport on this island. 6333. Modern Greek, **Aiyina.**

Ae·gos·pot·a·mi (ē/gəs pot/ə mi/), a river in ancient Thrace, flowing into the Hellespont: near its mouth the Athenian fleet was defeated by Lysander, 405 B.C., in the last battle of the Peloponnesian War.

Ae·o·lis (ē/ə lis) also **Ae·o·li·a** (ē ō/lē ə), an ancient coastal region and Greek colony in NW Asia Minor. —**Ae·o/li·an,** *adj., n.*

Aet·na (et/nə), **Mount,** ETNA, Mount.

Ae·to·li·a (ē tō/lē ə), a region in W central Greece. —**Ae·to/li·an,** *adj., n.*

A·fars/ and Is/sas (ä färz/ ənd ē/säz), **French Territory of the,** a former name of DJIBOUTI (def. 1).

Af·ghan·i·stan (af gan/ə stan/), a republic in central Asia, NW of Pakistan and E of Iran. 15,810,000; 251,773 sq. mi. (652,090 sq. km). *Cap.:* Kabul.

Af·ri·ca (af/ri kə), a continent S of Europe and between the Atlantic and Indian oceans. ab. 11,700,000 sq. mi. (30,303,000 sq. km). —**Af/ri·can,** *adj., n.*

A·ga·dir (ä/gä dēr/), a seaport in SW Morocco. 110,479.

A·ga·ña (ä gä/nyä), the capital of Guam. 2119.

A·gar·ta·la (ug/ər tul/ə), the capital of Tripura state, in NE India. 132,186.

Ag·as·siz (ag/ə sē), **Lake,** a lake existing in the prehistoric Pleistocene Epoch in central North America. 700 mi. (1127 km) long.

Ag·in·court (aj/in kôrt/, -kōrt/, azh/in kŏŏr/), a

village in N France, near Calais: victory of the English over the French 1415. 276.

A·gra (ä′grə), a city in SW Uttar Pradesh, in N India: site of the Taj Mahal. 770,000.

A·gri·gen·to (ä′grē jen′tō), a city in S Italy. 51,931. Formerly, **Girgenti.**

A·gua·dil·la (ä′gwə dē′ə), a seaport in NW Puerto Rico. 56,600.

A·guas·ca·lien·tes (ä′gwäs kä lyen′läs), **1.** a state in central Mexico. 684,247; 2499 sq. mi. (6470 sq. km). **2.** the capital of this state. 359,454.

A·gul·has (ə gul′əs), **Cape,** the southernmost point of Africa.

Ah·med·a·bad or **Ah·mad·a·bad** (ä′məd äbäd′), a city in E Gujarat state, in W India, N of Bombay. 2,515,000.

Ah·vaz (ä väz′) also **Ah·waz** (ä wäz′), a city in SW Iran. 579,826.

Ah·ve·nan·maa (äʜ′ve nän mä′), Finnish name of the Åland Islands.

Ain·tab (ïn täb′), former name of Gaziantep.

Aisne (ān), a river in N France, flowing NW and W to the Oise. 175 mi. (280 km) long.

Aix-en-Pro·vence (eks′än prə väns′, -väns′), a city in SE France, N of Marseilles. 124,550. Also called **Aix.**

Aix-la-Cha·pelle (eks′lä shä pel′), French name of Aachen.

Aix-les-Bains (eks′lä bänz′, -baɴ′), a town in SE France, N of Chambéry. 22,293.

Ai·yi·na (e′yē nä), Modern Greek name of Aegina.

Ai·zawl (ï zōl′), the capital of Mizoram state, in NE India. 74,493.

A·jac·cio (ä yät′chō), a seaport on the W coast of Corsica: the birthplace of Napoleon I. 54,089.

A·jan·ta (ə jun′tə), a village in N Maharashtra, in W central India: caves and shrines containing Buddhist frescoes and sculptures.

Aj·mer (uj mēr′), a city in central Rajasthan, in NW India. 374,000.

Ajmer′-Mer·wa′ra (-mer wär′ə), a former province in NW India. 2400 sq. mi. (6216 sq. km).

A·jodh·ya (ə yōd′yə), a city in E Uttar Pradesh, in N India: one of the seven sacred Hindu centers.

AK, Alaska.

A·ka·ba (ä′kə bə, ak′ə-), AQABA.

A·ka·shi (ä kä′shē), a city on W Honshu, in Japan. 261,000.

Ak·hi·sar (äk′hi sär′), a town in W Turkey, NE of Izmir. 61,491. Ancient, **Thyatira.**

A·ki·ta (ə kē′tə), a seaport on N Honshu, in N Japan, on the Sea of Japan. 294,000.

Ak·kad or **Ac·cad** (ak′ad, ä′käd), **1.** an ancient region in Mesopotamia, the N division of Babylonia. **2.** an ancient city in this region. —**Ak·ka·di·an** (ə kā′dē ən), *adj., n.*

Ak·ron (ak′rən), a city in NE Ohio. 223,019.

Ak·sum or **Ax·um** (äk′sŏŏm), a town in N Ethiopia: the capital of an ancient kingdom 1st to c7th centuries B.C.

Ak·tyu·binsk (äk tyŏŏ′binsk), a city in NW Kazakhstan. 248,000.

Ak·yab (ak yab′, ak′yab), former name of SIT-TWE.

AL, Alabama.

Ala., Alabama.

Al·a·bam·a (al′ə bam′ə), **1.** a state in the SE United States. 4,040,587; 51,609 sq. mi. (133,670 sq. km). *Cap.:* Montgomery. *Abbr.:* AL, Ala. **2.** a river flowing SW from central Alabama to the Mobile River. 315 mi. (505 km) long. —**Al′a·bam′i·an, Al′a·bam′an,** *adj., n.*

A·la·go·as (ä′lə gō′əs), a state in NE Brazil. 2,302,800; 10,674 sq. mi. (27,650 sq. km). *Cap.:* Maceió.

A·lai′ Moun′tains (ə lī′), a mountain range in SW Kyrgyzstan: part of the Tien Shan Mountains; highest peak, ab. 19,000 ft. (5790 m).

Al·a·me·da (al′ə mē′də, -mā′-), a city in W California. 76,459.

A·la·mein (al′ə mān′), EL ALAMEIN.

Al·a·mo·gor·do (al′ə mə gôr′dō), a city in S New Mexico: first atomic bomb exploded in the desert ab. 50 mi. (80 km) NW of here, July 16, 1945. 24,024.

Å′land Is′lands (ä′lənd, ô′lənd), a group of Finnish islands in the Baltic Sea, between Sweden and Finland. 23,761; 572 sq. mi. (1480 sq. km). Finnish, **Ahvenanmaa.**

Alas., Alaska.

A·las·ka (ə las′kə), **1.** a state of the United States in NW North America. 550,043; 586,400 sq. mi. (1,519,000 sq. km). *Cap.*: Juneau. *Abbr.*: AK, Alas. **2. Gulf of,** a gulf of the Pacific Ocean on the coast of S Alaska. —**A·las/kan,** *adj.*, *n.*

Alas/ka Penin/sula, a peninsula in SW Alaska. 500 mi. (800 km) long.

Alas/ka Range/, a mountain range in S Alaska. Highest peak, Mt. McKinley, 20,320 ft. (6194 m).

Al·a·va (al/ə və), **Cape,** a cape in NW Washington: westernmost point in the contiguous U.S.

Al·ba·ce·te (äl/vä thā/tā, -sā/-), a city in SE Spain. 127,169.

Al·ba Lon·ga (al/bə lông/gə, long/-), a city of ancient Latium, SE of Rome: legendary birthplace of Romulus and Remus.

Al·ba·ni·a (al bā/nē ə), a republic in S Europe, in the Balkan Peninsula, W of Macedonia and NW of Greece. 3,080,000; 10,632 sq. mi. (27,535 sq. km). *Cap.*: Tiranë. —**Al·ba/ni·an,** *adj.*, *n.*

Al·ba·ny (ôl/bə nē), **1.** the capital of New York, in the E part, on the Hudson. 101,082. **2.** a city in SW Georgia. 78,122. **3.** a river in central Canada, flowing E from W Ontario to James Bay. 610 mi. (980 km) long.

Al/be·marle Sound/ (al/bə märl/), an inlet of the Atlantic Ocean, in NE North Carolina. 60 mi. (97 km) long.

Al·bert (al/bərt), **Lake,** a lake in central Africa, between Uganda and Zaire: a source of the Nile. 100 mi. (160 km) long; 2061 sq. mi. (5338 sq. km); 2030 ft. (619 m) above sea level.

Al·ber·ta (al bûr/tə), a province in W Canada. 2,365,825; 255,285 sq. mi. (661,190 sq. km). *Cap.*: Edmonton. *Abbr.*: AB, Alta. —**Al·ber/tan,** *adj.*, *n.*

Al·bert·ville (al/bərt vil/, -bərt-; *Fr.* al ber vēl/), former name of KALEMIE.

Al·bi (al bē/), a city in S France: center of the Albigenses. 49,456.

Ål·borg or **Aal·borg** (ôl/bôrg), a seaport in NE Jutland, in Denmark. 154,582.

Al·bu·quer·que (al/bə kûr/kē), a city in central New Mexico. 384,736.

Al·ca·traz (al/kə traz/), an island in San Francisco Bay: site of a U.S. penitentiary 1933–63.

Al·dan (ul dän/), a river in the Russian Federation in Asia, flowing NE from the Stanovoi Mountains to the Lena. ab. 1500 mi. (2415 km) long.

Al·der·ney (ôl′dər nē), one of the Channel Islands in the English Channel. 1785; 3 sq. mi. (8 sq. km).

A·len·çon (A län sôɴ′; *Eng.* ə len′sən, -son), a city in NW France: lace manufacture. 34,666.

A·lep·po (ə lep′ō), a city in NW Syria. 1,878,701. French, **A·lep** (A lep′).

A·les·san·dri·a (al′ə san′drē ə), a city in NW Italy, in Piedmont. 102,774.

A·leu′tian Is′lands (ə lōō′shən), an archipelago extending SW from the Alaska Peninsula: part of Alaska. Also called **Aleutians.**

Aleu′tian Range′, a mountain range extending along the E coast of the Alaska Peninsula. Highest peak, Mt. Katmai, 6715 ft. (2047 m).

Al·ex·an′der Archipel′ago (al′ig zan′dər, -zän′-), an archipelago off the SE coast of Alaska.

Al′exander Is′land, an island off the coast of Antarctica, in the Bellingshausen Sea. Formerly, **Alexander I Island.**

Al·ex·an·dret·ta (al′ig zan dret′ə, -zän-), former name of Iskenderun.

Al·ex·an·dri·a (al′ig zan′drē ə, -zän′-), **1.** a seaport in N Egypt, in the Nile delta: founded in 332 b.c. by Alexander the Great; ancient center of learning. 2,893,000. **2.** a city in NE Virginia, S of the District of Columbia. 111,183. **3.** a city in central Louisiana, on the Red River. 49,188.

Al·ge·ci·ras (al′ji sir′əs), a seaport in S Spain, in Andalusia, on the Strait of Gibraltar. 97,213.

Al·ge·ri·a (al jēr′ē ə), a republic in NW Africa: gained independence from France 1962. 23,850,000; 919,352 sq. mi. (2,381,122 sq. km). *Cap.:* Algiers. —**Al·ger′i·an,** *adj., n.*

Al·giers (al jērz′), **1.** the capital of Algeria, in the N part. 1,839,000. **2.** one of the former Barbary States in N Africa: now Algeria.

Al·ham·bra (al ham′brə), a city in SW California, near Los Angeles. 82,106.

Al·i·can·te (al′ə kan′tē), a seaport in SE Spain, on the Mediterranean. 265,543.

Al′ice Springs′ (al′is), a town in Northern Territory, in central Australia. 13,400. Also called **the Alice.** Formerly, **Stuart.**

A·li·garh (ä′lē gur′, al′ə gär′), a city in W Uttar Pradesh. 320,000.

Alk·maar (älk′mär), a city in the W Netherlands: cheese market. 88,085.

Al Ku·fa (al kōō′fǝ, -fa), KUFA.

Al·lah·a·bad (al′ǝ hǝ bad′, ä′lǝ hä bäd′), a city in SE Uttar Pradesh, in N India, on the Ganges. 642,000.

Al·le·ghe·ny (al′i gā′nē) a river flowing NW from Pennsylvania into SW New York and then S through W Pennsylvania, joining the Monongahela at Pittsburgh to form the Ohio River. 325 mi. (525 km) long. —**Al′le·ghe′ni·an, Al′le·gha′ni·an,** *adj.*

Al′leghe·ny Moun′tains, a mountain range in Pennsylvania, Maryland, West Virginia, and Virginia: a part of the Appalachian Mountains. Also called **Al′le·ghe′nies.**

Al·len·town (al′ǝn toun′), a city in E Pennsylvania. 105,090.

Al·lep·pey (ǝ lep′ē), a port in SW Kerala, in S India, on the Arabian Sea. 169,934.

Al·lier (A lyā′), a river flowing N from S France to the Loire. ab. 250 mi. (400 km) long.

Al·lo·way (al′ǝ wā′), a hamlet in SW Scotland, near Ayr: birthplace of Robert Burns.

Al·ma-A·ta (al′mǝ ä tä′), the capital of Kazakhstan, in the SE part. 1,108,000.

Al·ma·dén (al′mǝ den′, -dān′, äl′-), a town in central Spain: mercury mines. 10,774.

Al·me·rí·a (al′mǝ rē′ǝ, äl′-), a seaport in S Spain, on the Mediterranean. 156,838. —**Al′me·ri′an,** *adj., n.*

A·lost (A lôst′), French name of AALST.

Alps (alps), a mountain range in S Europe, extending from France through Switzerland and Italy into Austria, Slovenia, and Croatia. Highest peak, Mont Blanc, 15,781 ft. (4810 m).

Al·sace (al sas′, -sās′, al′sas, -sās), **1.** a historic region and former province of NE France: between the Vosges mountains and the Rhine. **2.** a metropolitan region in NE France. 1,599,800; 3196 sq. mi. (8280 sq. km). —**Al·sa·tian** (-sā′shǝn), *adj., n.*

Al′sace-Lor·raine′, a region in NE France, including the former provinces of Alsace and Lorraine: part of Germany 1871–1919, 1940–45.

Alta., Alberta.

Al·tai or **Al·tay** (al′tī), a territory of the Russian

Federation in central Asia. 2,675,000; 101,000 sq. mi. (261,700 sq. km). *Cap.*: Barnaul.

Al·tai Moun'tains, a mountain range in central Asia, mostly in Mongolia, China, Kazakhstan, and the S Russian Federation. Highest peak, Belukha, 15,157 ft. (4506 m).

Al·ta·mi·ra (al'tə mēr'ə), a cave in N Spain, near Santander, noted for its Upper Paleolithic polychrome paintings of bison, deer, and pigs.

Alt·dorf (ält'dôrf'), a town in and the capital of Uri, in central Switzerland, near Lucerne: legendary home of William Tell. 8600.

Al·ti·pla·no (al'tə plä'nō, äl'-), a plateau region in South America, situated in the Andes of Argentina, Bolivia, and Peru.

Al·too·na (al tōō'nə), a city in central Pennsylvania. 51,881.

Al·U·bay·yid (al'ōō bā'id), EL OBEID.

A·ma·ga·sa·ki (ä'mə gə sä'kē, am'ə-), a city on SW Honshu, in S Japan. 523,657.

A·ma·pá (ä'mä pä'), a federal territory in N Brazil. 232,400; 54,160 sq. mi. (140,276 sq. km). *Cap.*: Macapá.

Am·a·ril·lo (am'ə ril'ō), a city in NW Texas. 157,615.

Am·a·zon (am'ə zon', -zən), a river in N South America, flowing E from the Peruvian Andes through N Brazil to the Atlantic Ocean: the largest river in the world in volume of water carried. 3900 mi. (6280 km) long. —**Am'a·zo'ni·an** (-zō'nē ən), *adj.*

A·ma·zo·nas (am'ə zō'nəs), a state in NW Brazil. 1,842,800; 601,769 sq. mi. (1,558,582 sq. km). *Cap.*: Manaus.

Am·a·zo·ni·a (am'ə zō'nē ə), the region around the Amazon, in N South America.

Am·ba·to (äm bä'tō), a city in central Ecuador, ab. 8500 ft. (2590 m) above sea level. 100,454.

Am·bon (am'bon) also **Am·boi·na** (-boi'nə), **1.** an island in the central Moluccas, in E Indonesia. 72,679; 314 sq. mi. (813 sq. km). **2.** a seaport on this island. 56,037. —**Am'bo·nese'** (-bə nēz', -nēs'), *n., pl.* **-nese,** *adj.*

Am'brose Chan'nel (am'brōz), a ship channel at the entrance to New York harbor, near Sandy Hook. 7½ mi. (12 km) long.

Am·chit·ka (am chit'kə), an island off the coast

of SW Alaska, in the W part of the Aleutian Islands.

A·mer·i·ca (ə mer/i kə), **1.** UNITED STATES. **2.** NORTH AMERICA. **3.** SOUTH AMERICA. **4.** Also called **the Americas.** North and South America, considered together. —**A·mer/i·can,** *adj., n.*

Amer/ican Samo/a, the part of Samoa belonging to the U.S., comprising mainly Tutuila and the Manua Islands. 32,297; 76 sq. mi. (197 sq. km). *Cap.:* Pago Pago. *Abbr.:* AS Compare SAMOA, WESTERN SAMOA.

A·mers·foort (ä/mərz fôrt/, -fōrt/, -mərs-), a city in the central Netherlands. 93,516.

Am·i·ens (A myan/), a city in N France, on the Somme. 135,992.

A/min·di/vi Is/lands (ä/min dē/vē, ä/min-), a group of islands in the NE Laccadive Islands, off the SW coast of India. 3.75 sq. mi. (9.71 sq. km).

Am·man (ä män/, ä/män), the capital of Jordan, in the W part. 777,500. Also called **Rabbah, Rabbath.**

Am·mon (am/ən), an ancient country east of the Jordan River, inhabited by a Semitic people, the Ammonites. —**Am/mon·ite/** (-ə nit/), *n.*

A·moy (ä moi/, am/oi), XIAMEN.

Am·rit·sar (əm rit/sər), a city in NW Punjab, in NW India: site of the Golden Temple. 589,000.

Am·ster·dam (am/stər dam/), the official capital of the Netherlands. 712,294. Compare HAGUE, The.

A·mu Dar·ya (ä/mōō där/yə), a river in central Asia, flowing NW from the Pamirs to the Aral Sea. ab. 1400 mi. (2250 km) long. Also called **Oxus.**

A/mund·sen Sea/ (ä/mənd sən, ä/mən-), an arm of the S Pacific Ocean off Marie Byrd Land, Antarctica.

A·mur (ä mōōr/), a river in E Asia, forming most of the boundary between N Manchuria and the SE Russian Federation, flowing into the Sea of Okhotsk. ab. 2700 mi. (4350 km) long. Chinese, **Heilong Jiang.**

A·na·dyr/ Range/ (ä/nə dēr/, an/ə-), a mountain range in NE Siberia: a part of the Kolyma Range.

An·a·heim (an/ə him/), a city in SW California, SE of Los Angeles. 266,406.

A·ná·huac (ə nä/wäk), the central plateau of

Mexico, between the Sierra Madre Occidental and the Sierra Madre Oriental ranges (3700 to 9000 ft.; 1128 to 2743 m): center of Aztec civilization.

A·na·pur·na (an'ə pŏŏr'nə, -pûr'-), ANNAPURNA.

An·a·to·li·a (an'ə tō/lē ə), a vast plateau between the Black and the Mediterranean seas: in ancient usage, synonymous with Asia Minor; in modern usage, applied to Turkey in Asia. Compare ASIA MINOR. —**An'a·to/li·an,** *adj., n.*

An·ch'ing (än'ching'), ANQING.

An·chor·age (ang'kər ij), a seaport in S Alaska. 226,338.

An·co·hu·ma (äng'kō ōō'mə), a peak of Mount Sorata, in W Bolivia.

An·co·na (ang kō'nə, an-), a seaport in E Italy, on the Adriatic Sea. 104,255.

An·da·lu·sia (an'dl ōō'zhə, -shē ə), a region in S Spain, bordering on the Atlantic Ocean and the Mediterranean Sea. 33,712 sq. mi. (87,314 sq. km). Spanish, **An·da·lu·cí·a** (än'dä lōō thē'ä, -sē'ä). —**An'da·lu'sian,** *adj., n.*

An'da·man and Nic'obar Is'lands (an'də-mən), a union territory of India, comprising the Andaman and Nicobar island groups in the E part of the Bay of Bengal, SW of Burma. 188,254; 3143 sq. mi. (8140 sq. km). *Cap.:* Port Blair.

An'daman Is'lands, a group of islands of India in the E part of the Bay of Bengal, W of the Malay Peninsula, part of Andaman and Nicobar Islands. 157,821; 2508 sq. mi. (6496 sq. km).

An'daman Sea', a part of the Bay of Bengal, E of the Andaman and Nicobar Islands. 300,000 sq. mi. (777,000 sq. km).

An·der·lecht (an'dər leкнt'), a city in central Belgium, near Brussels. 103,796.

An·der·son (an'dər sən), a city in central Indiana. 59,459.

An·der·son·ville (an'dər sən vil'), a village in SW Georgia: site of a Confederate military prison. 267.

An·des (an'dēz), a mountain range in W South America, extending ab. 4500 mi. (7250 km) from N Colombia and Venezuela south to Cape Horn. Highest peak, Aconcagua, 22,834 ft. (6960 m). —**An·de·an** (an'dē ən, an dē'-), *adj.*

An·dhra Pra·desh (än'drə prə däsh'), a state in SE India, formed from portions of Madras and Hyderabad states 1956. 53,593,000; 106,204 sq. mi. (275,068 sq. km). *Cap.:* Hyderabad.

An·di·zhan (än/di zhän/), a city in E Uzbekistan, SE of Tashkent. 288,000.

An·dor·ra (an dôr/ə, -dor/ə), **1.** a republic in the E Pyrenees between France and Spain. 51,400; 181 sq. mi. (468 sq. km). **2.** Also called **An·dor·ra la Ve·lla** (*Catalan.* än dôr/rä lä ve/lyä). the capital of this republic. 15,639. —**An·dor/ran,** *adj., n.*

An·dre·a/nof Is/lands (an/drē an/ôf, -of, än/drē ä/nôf, -nof), a group of islands in the W part of the Aleutian Islands. 1432 sq. mi. (3710 sq. km).

An·dro·pov (an drō/pôf, -pof), a city in the W Russian Federation, NE of Moscow, on the Volga. 254,000. Formerly, **Rybinsk** (1958–84), **Shcherbakov** (1946–57).

An·dros (an/drəs), the largest island in the Bahamas, in the W part of the group. 8845; 1600 sq. mi. (4144 sq. km).

An·dros·cog·gin (an/drə skog/in), a river flowing from NE New Hampshire through SW Maine into the Kennebec River. 171 mi. (275 km) long.

A·ne·to (ä nā/tō), **Pi·co de** (pē/kō dä) a mountain in NE Spain: highest peak of the Pyrenees. 11,165 ft. (3400 m). French, **Pic de Néthou.**

An·ga·ra (äng/gə rä/), a river in the S Russian Federation in Asia, flowing NW from Lake Baikal to the Yenisei River: called Upper Tunguska in its lower course. 1151 mi. (1855 km) long.

An·garsk (äng gärsk/), a city in the S Russian Federation in Asia, near Lake Baikal. 262,000.

An·ge·les (än/hä läs/), a city in the Philippines, on S central Luzon. 236,00.

An/gel Falls/, a waterfall in SE Venezuela: world's highest. 3212 ft. (979 m) high.

An·gers (än zhā/), a city in W France. 163,191.

Ang·kor (ang/kôr, -kōr), a vast assemblage of ruins of the Khmer empire, in NW Cambodia: elaborately carved and decorated temples, statues, gateways, and towers.

An·gle·sey (ang/gəl sē), an island and historic county in Gwynedd, in NW Wales.

An·gli·a (ang/glē ə), Latin name of ENGLAND.

An/glo-E·gyp/tian Sudan/ (ang/glō-), former name of SUDAN.

An·go·la (ang gō′lə), a republic in SW Africa: formerly an overseas province of Portugal; gained independence 1975. 9,390,000; 481,226 sq. mi. (1,246,375 sq. km). *Cap.:* Luanda. Formerly, **Portuguese West Africa.** —**An·go′lan,** *adj., n.*

An·go·ra (ang gôr′ə, -gōr′ə, an-), former name of Ankara.

An·gou·mois (äng′gōōm wä′), a region and former province of W France: famous as source of cognac.

An·guil·la (ang gwil′ə), an island in the N Leeward Islands, in the E West Indies; a British dependency. 6500; 34 sq. mi. (88 sq. km). Compare St. Kitts-Nevis-Anguilla.

An·gus (ang′gəs), a historic county in E Scotland. Formerly, Forfar.

An·halt (än′hält), a former state in central Germany: now part of Saxony-Anhalt.

An·hui (än′hwē′) also **An·hwei** (-hwā′), a province in E China. 52,170,000; 54,015 sq. mi. (139,899 sq. km). *Cap.:* Hefei.

An·i·ak′chak Cra′ter (an′ē ak′chak, an′-), an active volcanic crater on the Alaska Peninsula, with a diameter of 6 mi. (10 km).

An·jou (an′jōō; *Fr.* än zhōō′), a region and former province in W France, in the Loire Valley.

An·ka·ra (ang′kər ə), the capital of Turkey, in the central part. 2,251,533. Formerly, Angora.

An·king (än′king′), Anqing.

An·na·ba (an nä′bə), a seaport in NE Algeria: site of Hippo Regius. 348,322. Formerly, Bône.

An Na·fud (an′ na fōōd′), Nefud Desert.

An-Na·jaf (an naj′af), Najaf.

An·nam (ə nam′), a former kingdom and French protectorate along the E coast of French Indochina: now part of Vietnam.

An·nap·o·lis (ə nap′ə lis), the capital of Maryland, in the central part, on Chesapeake Bay: U.S. Naval Academy. 33,360.

An·na·pur·na or **A·na·pur·na** (an′ə pŏŏr′nə, -pûr′-), a mountain in N Nepal, in the Himalayas. 26,503 ft. (8078 m).

Ann Ar·bor (an är′bər), a city in SE Michigan. 109,592.

An·ne·cy (An′ sē′), a city in SE France. 54,954.

An Nhon (än′ nôn′), a city in S central Vietnam. 117,000. Formerly, Binh Dinh.

An·qing or **An·ch'ing** (än'ching'), a city in S Anhui province, in E China, on the Chang Jiang. 160,000.

An·shan (än'shän'), a city in E Liaoning province, in NE China. 1,500,000.

An·ta·ki·ya (än'tä kē'yä), Arabic name of ANTIOCH.

An·ta·kya (än tä'kyä), Turkish name of ANTIOCH.

An·tal·ya (än täl'yä), a seaport in SW Turkey. 258,139.

An·ta·na·na·ri·vo (än'tə nä'nə rē'vō, an'tə-nan'ə-), the capital of Madagascar, in the central part. 703,000. Formerly, **Tananarive**.

Ant·arc·tic (ant ärk'tik, -är'tik), **the,** the Antarctic Ocean and Antarctica.

Ant·arc·ti·ca (ant ärk'ti kə, -är'ti-), the continent surrounding the South Pole: almost entirely covered by an ice sheet. ab. 5,000,000 sq. mi. (12,950,000 sq. km). Also called **Antarc'tic Con'tinent.**

Antarc'tic Cir'cle, an imaginary line drawn parallel to the equator, at 23° 28' N of the South Pole.

Antarc'tic O'cean, the waters surrounding Antarctica, comprising the southernmost parts of the Pacific, Atlantic, and Indian oceans.

Antarc'tic Penin'sula, a peninsula in Antarctica, S of South America. Formerly, PALMER PENINSULA. Compare GRAHAM LAND.

An·tibes (än tēb'), a seaport in SE France, SW of Nice: preserved ruins of 4th-century B.C. Roman town. 56,309.

An·ti·cos·ti (an'tə kô'stē, -kos'tē), an island at the head of the Gulf of St. Lawrence in E Canada, in E Quebec province. 135 mi. (217 km) long; 3043 sq. mi. (7880 sq. km).

An·tie·tam (an tē'təm), a creek flowing from S Pennsylvania through NW Maryland into the Potomac: Civil War battle fought near here at Sharpsburg, Maryland, in 1862.

An·ti·gua (an tē'gə), one of the Leeward Islands, in the E West Indies. 80,000; 108 sq. mi. (280 sq. km). —**An·ti'guan,** *adj., n.*

Anti'gua and Barbu'da, an island state in the E West Indies, comprising Antigua, Barbuda, and a smaller island: formerly a British crown colony;

gained independence 1981. 81,500; 171 sq. mi. (442 sq. km). *Cap.:* St. John's.

An·ti-Leb·a·non (an/tē leb/ə nən), a mountain range in SW Asia, between Syria and Lebanon, E of the Lebanon Mountains.

An·til·les (an til/ēz), a chain of islands in the West Indies, divided into two parts, one including Cuba, Hispaniola, Jamaica, and Puerto Rico **(Greater Antilles)**, the other including a large arch of smaller islands to the SE and S **(Lesser Antilles or Caribees).** —**An·til/le·an,** *adj., n.*

An·ti·och (an/tē ok/), **1.** Arabic, **Antakiya.** Turkish, **Antakya.** a city in S Turkey: capital of the ancient kingdom of Syria 300–64 B.C. 94,942. **2.** a city in W California. 62,195. —**An/ti·o/chi·an** (-ō/kē ən), *n., adj.*

An·tip·o·des (an tip/ə dēz/), a group of islands SE of and belonging to New Zealand. 24 sq. mi. (62 sq. km).

An·ti·sa·na (än/tē sä/nə), **Mount,** an active volcano in N central Ecuador, near Quito. 18,885 ft. (5756 m).

An·to·fa·gas·ta (än/tō fə gä/stə), a seaport in N Chile. 149,720.

An·trim (an/trim), a county in NE Northern Ireland. 642,300; 1093 sq. mi. (2831 sq. km).

An·tung (än/dŏong/), former name of DANDONG.

Ant·werp (an/twərp), **1.** a province in N Belgium. 1,587,450; 1104 sq. mi. (2860 sq. km). **2.** the capital of this province and a seaport, on the Scheldt. 476,044. French, **An·vers** (än vɛʀ/). Flemish, **Ant·wer·pen** (änt/vɛʀ pən).

A·nu·ra·dha·pu·ra (ə nŏŏr/ə də pŏŏr/ə, un/ŏŏ-rä/də-), a city in N central Sri Lanka: ruins of ancient Buddhist temples. 30,000.

An·yang (än/yäng/), a city in N Henan province, in E China: site of the ancient city of Yin, the center of the Shang dynasty. 225,000.

An·zhe·ro-Su·dzhensk (un zhâr/ə sŏŏ/jinsk), a city in the S Russian Federation in central Asia. 105,000.

An·zi·o (an/zē ō/), a port in Italy, S of Rome on the Tyrrhenian coast: site of Allied beachhead in World War II. 27,094.

A·o·mo·ri (ä/ō môr/ē), a seaport on N Honshu, in N Japan. 293,000.

A·o·ran·gi (ä/ō räng/gē), COOK, Mount.

Ap·a·lach·ee Bay' (ap/ə lach/ē, ap/-), a bay of the Gulf of Mexico, on the coast of N Florida.

Ap·a·lach·i·co·la (ap/ə lach/ə kō/lə), a river flowing S from NW Florida into the Gulf of Mexico. 90 mi. (145 km) long.

A·pel·doorn (ä/pəl dōrn′, -dôrn′), a city in central Netherlands. 146,337.

Ap·en·nines (ap/ə ninz/), a mountain range in Italy, extending across the length of the entire peninsula from NW to SW. Highest peak, Monte Corno, 9585 ft. (2922 m). Also called **Ap/ennine Moun/tains.**

A·pi·a (ä pē/ə, ä/pē ä/), the capital of Western Samoa, on N Upolu. 33,170.

A·po (ä/pō), an active volcano in the S Philippines, on S Mindanao: highest peak in the Philippines. 9690 ft. (2954 m).

Ap·pa·la·chi·a (ap/ə lā/chē ə, -chə, -lach/ē ə, -lach/ə), a region in the E United States, in the area of the S Appalachian Mountains, usu. including NE Alabama to SW Pennsylvania. —**Ap/pa·la/chi·an,** *adj., n.*

Appala/chian Moun/tains, a mountain range in E North America, extending from S Quebec province to N Alabama. Highest peak, Mt. Mitchell, 6684 ft. (2037 m). Also called **Appalachians.**

Appala/chian Trail/, a hiking trail extending through the Appalachian Mountains from central Maine to N Georgia. 2050 mi. (3300 km) long.

Ap·pen·zell (ä/pən tsel/), a canton in NE Switzerland, divided into two independent areas.

Ap/pen·zell Aus/ser Rho/den (ou/sər rōd/n), a demicanton in NE Switzerland. 49,600; 94 sq. mi. (245 sq. km).

Ap/pen·zell In/ner Rho/den (in/ər rōd/n), a demicanton in NE Switzerland. 13,400; 66 sq. mi. (170 sq. km).

Ap/pi·an Way/ (ap/ē ən), an ancient Roman highway extending from Rome to Brindisi. ab. 350 mi. (565 km) long.

Ap·ple·ton (ap/əl tən), a city in E Wisconsin. 65,695.

A·pu·lia (ə pyōōl/yə), a region in SE Italy. 3,828,322; 7442 sq. mi. (19,275 sq. km). *Cap.:* Bari. Italian, **Puglia.** —**A·pu/lian,** *adj., n.*

A·pu·re (ä pōōr/ä), a river flowing E from W Venezuela to the Orinoco. ab. 500 mi. (805 km) long.

A·pu·rí·mac (ä/pōō rē/mäk), a river flowing NW from S Peru to the Ucayali River. ab. 550 mi. (885 km) long.

A·qa·ba or **A·ka·ba** (ä/kə bə, ak/ə-), **1.** a seaport in SW Jordan at the N end of the Gulf of Aqaba. 10,000. **2. Gulf of,** an arm of the Red Sea between Saudi Arabia and Egypt. 100 mi. (160 km) long.

Aq·ui·la (ak/wə lə, ä/kwə-), a city in central Italy. 300,950. Also called **L'Aquila, A/quila de/gli A·bruz/zi** (del/yē ä brōōt/sē).

Aq·ui·taine (ak/wi tān/), **1.** Latin, **Aq/ui·ta/ni·a.** a historic region in SW France, formerly an ancient Roman province and medieval duchy. **2.** a metropolitan region in SW France. 2,718,200; 15,949 sq. mi. (41,308 sq. km).

AR, Arkansas.

A·ra·bi·a (ə rā/bē ə), a peninsula in SW Asia including Saudi Arabia, Yemen, Oman, the United Arab Emirates, Qatar, and Kuwait. ab. 1,000,000 sq. mi. (2,600,000 sq. km). Also called **Ara/bian Penin/sula. —A·ra/bi·an,** adj., n.

Ara/bian Gulf/, PERSIAN GULF.

Ara/bian Sea/, the NW arm of the Indian Ocean between India and Arabia.

Ar/ab Repub/lic of E/gypt (ar/əb), official name of EGYPT.

Ar·a·by (ar/ə bē), Chiefly Literary. Arabia.

A·ra·ca·jú (är/ə kə zhōō/), the capital of Sergipe, in NE Brazil. 299,622.

A·rad (ä räd/), a city in W Romania on the Mureș River. 185,892.

A/ra·fu/ra Sea/ (är/ə fŏŏr/ə, är/ə-), a part of the Pacific between N Australia and SW New Guinea.

Ar·a·gon (ar/ə gon/), a region in NE Spain: formerly a kingdom; later a province. 18,181 sq. mi. (47,089 sq. km). Spanish, **A·ra·gón** (ä/rä gōn/).

A·ra·gua·ya (är/ə gwä/yə), a river flowing N from central Brazil to the Tocantins River. ab. 1100 mi. (1770 km) long.

A·ra·kan Yo·ma (är/ə kän/ yō/mə, ar/ə kan/), a mountain range on the W border of Burma. Highest peak, Saramati, 12,633 ft. (3851 m).

Ar/al Sea/ (ar/əl), an inland sea between Kazakhstan and Uzbekistan, E of the Caspian Sea. 26,166 sq. mi. (67,770 sq. km). Russian, **A·ral·sko·ye Mo·re** (u räl/skə yə mô/ryə).

Ar′an Is′lands (ar′ən), a group of three islands off the W central coast of Ireland. ab. 18 sq. mi. (47 sq. km).

Ar·a·rat (ar′ə rat′), a mountain in E Turkey, near the borders of Iran and Armenia: traditionally considered the landing place of Noah's Ark. 16,945 ft. (5165 m).

A·ras (ä räs′), a river in SW Asia, flowing from E Turkey along part of the boundary between NW Iran and Armenia and Azerbaijan into the Kura River. ab. 660 mi. (1065 km) long. Ancient, **Araxes.**

Ar·au·ca·ni·a (ar′ô kā′nē ə), a region in central Chile.

A·rax·es (ə rak′sēz), ancient name of ARAS.

Ar·be·la (är bē′lə), an ancient city of Assyria, E of the Tigris, on the site of modern Erbil. Compare GAUGAMELA.

Ar·bil (är′bil), ERBIL.

Ar·ca·di·a (är kā′dē ə), a mountainous region of ancient Greece in the central Peloponnesus. —**Ar·ca′di·an,** *adj., n.*

Arch·an·gel (ärk′ān′jəl), **1.** Russian, **Arkhangelsk.** a seaport in the NW Russian Federation in Europe, on Dvina Bay. 416,000. **2. Gulf of,** former name of DVINA BAY.

Arch′es Na′tional Park′, a national park in E Utah: natural arch formations. 114 sq. mi. (295 sq. km).

Arc·tic (ärk′tik, är′-), **the,** the region lying north of the Arctic Circle.

Arc′tic Cir′cle, an imaginary line drawn parallel to the equator, at 23° 28′ S of the North Pole.

Arc′tic O′cean, an ocean N of North America, Asia, and the Arctic Circle.

Ar·den (är′dn), **Forest of,** a forest district in central England, in N Warwickshire.

Ar·dennes (är den′), **Forest of,** a wooded plateau region in NE France, SE Belgium, and Luxembourg.

A·re·ci·bo (är′ə sē′bō), a seaport in N Puerto Rico. 90,960.

A·re·qui·pa (ar′ə kē′pə, är′-), a city in S Peru. 591,700.

A·rez·zo (ä ret′sō), a city in central Italy. 91,535.

Ar·gen·teuil (AR zhän tœ/y°, är′zhən-), a city in N France, on the Seine near Paris. 103,141.

Ar·gen·ti·na (är′jən tē′nə), a republic in S South America. 31,060,000; 1,084,120 sq. mi. (2,807,870 sq. km). *Cap.:* Buenos Aires. Also called **the Ar′gen·tine′** (-tēn′, -tin′). Official name, **Ar′gentine Repub′lic.** —**Ar′gen·tine′** (-tēn′), **Ar′gen·tin/e·an** (-tin/ē ən), *adj., n.*

Ar·go·lis (är′gə lis), **1.** an ancient district in SE Greece. **2. Gulf of,** a gulf of the Aegean, in SE Greece. ab. 30 mi. (48 km) long.

Ar′gonne For′est (är′gon, är gon′), a wooded region in NE France: battles, World War I, 1918; World War II, 1944. Also called **Ar·gonne.**

Ar·gos (är′gos, -gəs), an ancient city in SE Greece, on the Gulf of Argolis: a powerful rival of Sparta, Athens, and Corinth.

Ar·go·vie (AR gô vē′), French name of AARGAU.

Ar·gun (är gōōn′), a river in NE Asia, forming part of the boundary between the Russian Federation and China. ab. 450 mi. (725 km) long.

Ar·gyll (är gīl′), a historic county in W Scotland. Also called **Ar·gyll′shire** (-shēr, -shər).

År·hus or **Aar·hus** (ôr′hōōs), a seaport in E Jutland, in Denmark. 258,028.

A·ri·ca (ə rē′kə), a seaport in N Chile. 169,774.

Ar·i·ma·thae·a or **Ar·i·ma·the·a** (ar′ə mə thē′ə), a town in ancient Palestine.

A·rim·i·num (ə rim′ə nəm), ancient name of RIMINI.

Ariz., Arizona.

Ar·i·zo·na (ar′ə zō′nə), a state in SW United States. 3,665,228; 113,909 sq. mi. (295,025 sq. km). *Cap.:* Phoenix. *Abbr.:* AZ, Ariz. —**Ar/i·zo′nan, Ar/i·zo′ni·an,** *adj., n.*

Ark., Arkansas.

Ar·kan·sas (är′kən sô′; *also for 2* är kan′zəs), **1.** a state in S central United States; 2,350,725; 53,103 sq. mi. (137,537 sq. km). *Cap.:* Little Rock. *Abbr.:* AR, Ark. **2.** a river flowing E and SE from central Colorado into the Mississippi in SE Arkansas. 1450 mi. (2335 km) long. —**Ar·kan′san,** *n., adj.*

Ar·khan·gelsk (UR КНän′gyilsk), Russian name of ARCHANGEL.

Arles (ärl), a city in SE France, on the Rhone River. 50,345.

Ar·ling·ton (är′ling tən), **1.** a county in NE Virginia, opposite Washington, D.C.: site of national cemetery **(Ar′lington Na′tional Cem′etery)**. 152,599. **2.** a city in N Texas. 261,721.

Ar′lington Heights′, a city in NE Illinois, near Chicago. 75,460.

Ar·magh (är mä′), a county in S Northern Ireland. 118,800; 484 sq. mi. (1254 sq. km).

Ar·ma·vir (är′mə vēr′), a city in the SW Russian Federation, E of Krasnodar. 172,000.

Ar·me·ni·a (är mē′nē ə), **1.** an ancient country in W Asia: now divided between Armenia, Turkey, and Iran. **2.** Also called **Arme′nian Repub′lic.** a republic in Transcaucasia, S of Georgia and W of Azerbaijan. 3,283,000; ab. 11,490 sq. mi. (29,800 sq. km). *Cap.:* Yerevan. —**Ar·me′ni·an,** *n., adj.*

Ar·men·tières (är′mən tyâr′, -tērz′), a city in N France. 27,473.

Ar·mor·i·ca (är môr′i kə, -mor′-), an ancient region in NW France, corresponding generally to Brittany.

Arn·hem (ärn′hem, är′nəm), a city in the central Netherlands, on the Rhine River: World War II battle 1944. 128,717.

Arn′hem Land′ or **Arn′hem·land′,** a region in N Northern Territory, Australia: site of Aborigine reservation.

Ar·no (är′nō), a river flowing W from central Italy to the Ligurian Sea. 140 mi. (225 km) long.

A·roos·took (ə rōōs′tŏŏk, -tik), a river flowing NE from N Maine to the St. John River. 140 mi. (225 km) long.

Ar·ran (ar′ən), an island in SW Scotland, in the Firth of Clyde. 3705; 166 sq. mi. (430 sq. km).

Ar·tois (är twä′), a former province in N France.

A·ru·ba (ə rōō′bə), a self-governing Dutch island in the S Caribbean, off the NW coast of Venezuela: formerly (1845–1986) a part of the Netherlands Antilles. 62,500; 75 sq. mi. (193 sq. km).

A′ru Is′lands (är′ōō), an island group in Indonesia, SW of New Guinea. 3306 sq. mi. (8565 sq. km).

A·ru·na·chal Pra·desh (är′ə nä′chəl prə-dāsh′), a state in NE India. 631,839; 34,262 sq. mi. (88,743 sq. km).

A·ru·wi·mi (är′ōō wē′mē), a river in Zaire, flowing SW and W into the Zaire River. ab. 800 mi. (1300 km) long.

Ar·vad·a (är vad′ə), ʸa city in central Colorado, near Denver. 89,235.

AS, American Samoa.

A·sa·hi·ga·wa (ä′sä hē′gä wä) also **A·sa·hi·ka·wa** (-kä wä), a city on W central Hokkaido, in N Japan. 363,000.

A·san·sol (ä′sən sōl′), a city in NW West Bengal, in E India. 365,000.

As·bur·y Park′ (az′ber′ē, -bə rē), a city in E New Jersey: seashore resort. 17,015.

As·cen·sion (ə sen′shən), a British island in the S Atlantic Ocean: constituent part of St. Helena. 1130; 34 sq. mi. (88 sq. km).

As·cot (as′kət), a village in SE Berkshire, in S England: annual horse races.

A·shan·ti (ə shan′tē, ə shän′-), a region in S central Ghana, formerly a kingdom.

Ash·dod (ash′dod), a town in W Israel. 68,000.

Ashe·ville (ash′vil), a city in W North Carolina. 61,607.

Ash·kha·bad (äsh′kə bäd′), the capital of Turkmenistan, in the S part. 398,000. Formerly, **Poltoratsk.**

A·sia (ā′zhə, ā′shə), a continent bounded by Europe and the Arctic, Pacific, and Indian oceans. ab. 16,000,000 sq. mi. (41,440,000 sq. km). —**A′sian,** *adj., n.*

A′sia Mi′nor, a peninsula in W Asia between the Black and Mediterranean seas, including most of Asian Turkey. Compare ANATOLIA.

A·sir (ä sēr′), a region in SW Saudi Arabia.

As·ma·ra (äs mär′ə), the capital of Eritrea, in N Ethiopia. 276,355.

A·so·san (ä′sō sän′), a volcano in SW Japan, in central Kyushu. 5225 ft. (1593 m); crater 12 mi. (19 km) across.

As·pen (as′pən), a town in central Colorado: ski resort. 3678.

As·sam (a sam′), a state in NE India. 19,902,826; 30,283 sq. mi. (78,438 sq. km).

As′sa·teague Is′land (as′ə tēg′), an island in SE Maryland and E Virginia on Chincoteague Bay: wild pony roundup.

As·sin·i·boine (ə sin′ə boin′), a river in S Canada, flowing S and E from SE Saskatchewan into the Red River in S Manitoba. 450 mi. (725 km) long.

As·si·si (ə sē′zē), a town in E Umbria, in central Italy: birthplace of St. Francis of Assisi. 24,002.

As·syr·i·a (ə sēr′ē ə), an ancient kingdom and empire of SW Asia, centered in N Mesopotamia: greatest extent from c750 to 612 B.C. —**As·syr′i·an,** *adj., n.*

As·ti (ä′stē, as′tē), a city in the Piedmont region of Italy, S of Turin: center of wine-producing region. 76,950.

As·tra·khan (as′trə kən, -kän′), a city in the S Russian Federation in Europe, at the mouth of the Volga. 509,000.

As·tu·ri·as (a stŏŏr′ē əs, a styŏŏr′-), a former kingdom and province in NW Spain.

A·sun·ción (ä′sŏŏn syŏn′), the capital of Paraguay, in the S part. 457,210.

As·wan (as′wän), **1.** Ancient, **Syene.** a city in SE Egypt, on the Nile. 258,600. **2.** a dam near this city, extending across the Nile. 6400 ft. (1950 m) long.

A·syut (ä syŏŏt′), a city in central Egypt, on the Nile. 291,300.

A′ta·ca′ma Des′ert (at′ə käm′ə, at′-, ä′tə kä′mə, ä′tə-), an arid region in N Chile: nitrate deposits. ab. 70,000 sq. mi. (181,300 sq. km).

At·ba·ra (ät′bə rə, at′-), a river in NE Africa, flowing NW from NW Ethiopia to the Nile in E Sudan. ab. 500 mi. (800 km) long.

Ath·a·bas·ka (ath′ə bas′kə), **1. Lake,** a lake in W Canada, in NW Saskatchewan and NE Alberta. ab. 200 mi. (320 km) long; ab. 3000 sq. mi. (7800 sq. km). **2.** a river in W Canada flowing NE from W Alberta to Lake Athabaska. 765 mi. (1230 km) long.

Ath·ens (ath′inz), **1.** Greek, **A·the·nai** (ä thē′ne). the capital of Greece, in the SE part: ancient city-state. 885,136. **2.** a city in N Georgia. 42,549. —**A·the′ni·an,** *adj., n.*

Ath·os (ath′ōs, ā′thos), **Mount,** the easternmost of three prongs of the peninsula of Chalcidice, in NE Greece: site of an autonomous theocracy comprising 20 monasteries. 1713; 131 sq. mi. (340 sq. km); ab. 35 mi. (56 km) long.

A·ti·tlán (ä′tē tlän′), **Lake,** a crater lake in SW Guatemala, 4700 ft. (1433 m) above sea level. ab. 53 sq. mi. (137 sq. km).

At·lan·ta (at lan′tə), the capital of Georgia, in the N part. 394,017.

At·lan′tic Cit′y (at lan′tik), a city in SE New Jersey: seashore resort. 40,199.

Atlan′tic O′cean, an ocean bounded by North America and South America in the Western Hemisphere and by Europe and Africa in the Eastern Hemisphere. ab. 31,530,000 sq. mi. (81,663,000 sq. km); greatest known depth, 30,246 ft. (9219 m). Also called **Atlantic.**

Atlan′tic Prov′inces, the Canadian provinces bordering the Atlantic Ocean, comprising New Brunswick, Newfoundland, Nova Scotia, and Prince Edward Island.

At′las Moun′tains (at′ləs), a mountain range in NW Africa, extending through Morocco, Algeria, and Tunisia. Highest peak, Mt. Tizi, 14,764 ft. (4500 m).

A·trek (ə trek′, ä trek′) also **A·trak** (ə trak′, ä trak′), a river arising in NE Iran, flowing W, then along the Iran–Turkmenistan border, and then through Turkmenistan into the Caspian Sea. ab. 300 mi. (485 km) long.

At·ro·pa·te·ne (a′trə pə tē′nə), MEDIA ATROPATENE.

At·ti·ca (at′i kə), a region in SE Greece, surrounding Athens: under Athenian rule in ancient times. —**At′tic,** *adj.*

At·tu (at′tōō′), the westernmost of the Aleutian Islands.

Aube (ōb), a river in N France, flowing NW to the Seine. 125 mi. (200 km) long.

Au·ber·vil·liers (ō bâr vēl yā′), a town in N France, a suburb of Paris. 72,997.

Auck·land (ôk′lənd), a seaport on N North Island, in New Zealand. 841,700.

Augs·burg (ôgz′bûrg), a city in Bavaria, in S Germany. 247,700.

Au·gus·ta (ô gus′tə, ə gus′-), **1.** a city in E Georgia, on the Savannah River. 47,532. **2.** the capital of Maine, in the SW part, on the Kennebec River. 21,819.

Au·ri·gnac (ô rē nyak′), a village in S France: many prehistoric artifacts found in area. 1149.

Au·ro·ra (ə rôr′ə, ə rōr′ə), **1.** a city in central Colorado, near Denver. 222,103. **2.** a city in NE Illinois. 99,581.

Au·sa·ble (ô sā/bəl), a river in NE New York, flowing NE through a gorge **(Ausa/ble Chasm/)** into Lake Champlain. 20 mi. (32 km) long.

Ausch·witz (oush/vits), a town in SW Poland: site of Nazi death camp during World War II. 39,600. Polish, **Oświęcim.**

Aus·ter·litz (ô/stər lits, ou/stər-), a town in S Moravia, in central Czechoslovakia: Russian and Austrian armies defeated by Napoleon I 1805. Czech, **Slavkov.**

Aus·tin (ô/stən), the capital of Texas, in the central part, on the Colorado River. 465,622.

Austral., Australia.

Aus·tral·a·sia (ô/strə lā/zhə, -shə), Australia, New Zealand, and neighboring islands in the S Pacific Ocean. —**Aus/tral·a/sian,** adj., n.

Aus·tral·ia (ô strāl/yə), **1.** a continent SE of Asia, between the Indian and Pacific oceans. 2,948,366 sq. mi. (7,636,270 sq. km). **2. Commonwealth of,** a nation consisting of the continent of Australia and the island of Tasmania: a member of the Commonwealth of Nations. 16,250,000; 2,974,581 sq. mi. (7,704,165 sq. km). Cap.: Canberra. —**Aus/tral/ian,** adj., n.

Austral/ian Alps/, a mountain range in SE Australia. Highest peak, Mt. Kosciusko, 7316 ft. (2230 m).

Austral/ian Cap/ital Ter/ritory, a federal territory on the continent of Australia in the SE part: includes Canberra, capital of the Commonwealth of Australia. 260,700; 939 sq. mi. (2430 sq. km). Formerly, **Federal Capital Territory.**

Aus·tra·sia (ô strā/zhə, -shə), the E part of the kingdom of the Franks of the 6th–8th centuries, composed of what is now NE France, W Germany, and Belgium.

Aus·tri·a (ô/strē ə), a republic in central Europe. 7,555,333; 32,381 sq. mi. (83,865 sq. km). Cap.: Vienna. German, **Österreich.** —**Aus/tri·an,** adj., n.

Aus/tria-Hun/gary, a former monarchy (1867–1918) in central Europe that included what is now Austria, Hungary, Czechoslovakia, Croatia, Slovenia, and parts of Romania, Poland, and Italy. —**Aus/tro-Hungar/ian,** (ô/strō-), adj., n.

Aus·tro·ne·sia (ô/strō nē/zhə, -shə), the islands of the central and S Pacific. —**Aus/tro·ne/sian,** adj.

Au·teuil (ō tœ′/yᵉ), a former town, now a district in W Paris, France.

Au·vergne (ō vârn′, ō vûrn′), **1.** a historic region in S central France. **2.** a metropolitan region in S central France. 1,334,400; 10,044 sq. mi. (26,013 sq. km). **3.** a mountain range in S central France. Highest peak, 6188 ft. (1886 m).

Av′a·lon Penin′sula (av′ə lon′), a peninsula in SE Newfoundland, in E Canada.

Ave·bur·y (āv′bə rē), a village in Wiltshire, England: site of one of the largest ceremonial megalithic structures in Europe. Often, **Abury.**

A·ve·lla·ne·da (ä vā/yä nä/thä), a city in E Argentina, near Buenos Aires. 330,654.

Av·en·tine (av′ən tīn′, -tin), one of the seven hills on which ancient Rome was built.

A·ver·nus (ə vûr′nəs), a lake in the caldera of a volcano near Naples, Italy, regarded in ancient times as the entrance to the underworld.

A·vi·gnon (A vē nyôN′), a city in SE France, on the Rhone River: papal residence 1309–77. 93,024.

Av·lo·na (av lō′nə), former name of VLORË.

A·von (ā′vən, av′ən), **1.** a river in central England, flowing SE past Stratford-on-Avon to the Severn. 96 mi. (155 km) long. **2.** a river in S England, flowing W to the mouth of the Severn. ab. 75 mi. (120 km) long. **3.** a river in S England, flowing S to the English Channel. ab. 60 mi. (100 km) long. **4.** a county in SW England. 951,200; 520 sq. mi. (1346 sq. km).

A·wa·ji (ä wä′jē), an island in Japan, S of Honshu and N of Shikoku. 230 sq. mi. (596 sq. km).

Ax·el Hei·berg (ak′səl hī′bûrg), the largest island belonging to the Sverdrup group in the Canadian Northwest Territories. 15,779 sq. mi. (40,868 sq. km).

Ax·um (äk′sōōm), AKSUM.

A·ya·cu·cho (ä/yä kōō/chō), a city in SW Peru: decisive victory of Bolívar over Spanish troops 1824. 94,200.

Ayers′ Rock′ (ârz), a conspicuous red monadnock in central Australia, in the SW Northern Territory. 1143 ft. (348 m) high.

Ayr (âr), **1.** a seaport in SW Scotland. 49,481. **2.** AYRSHIRE.

Ayr·shire (âr′shēr, -shər), a historic county in SW Scotland. Also called **Ayr.**

A·yut·tha·ya (ä yōō′tä yä), a city in central Thailand, on the Chao Phraya: former national capital. 47,189.

AZ, Arizona.

A·za·ni·a (ə zā′nē ə), the Republic of South Africa: a designation used by black liberationists. —**A·za′ni·an,** *n., adj.*

Az·ca·po·tzal·co (äs′kä pō tsäl′kō), a city in central Mexico: suburb of Mexico City; a cultural center during the pre-Columbian period. 545,513.

Az·er·bai·jan (az′ər bī jän′, ä′zər-), **1.** Also, **Az′er·bai·dzhan′.** Formerly, **Azerbaijan′ So′viet So′cialist Repub′lic.** a republic in Transcaucasia, N of Iran and W of the Caspian Sea. 7,029,000; 33,430 sq. mi. (86,600 sq. km). *Cap.:* Baku. **2.** a region of NW Iran. —**Az′er·bai·ja′ni** (-jä′nē), *n., adj.*

A·zores (ə zôrz′, ə zōrz′, ā′zôrz, ā′zōrz), a group of islands in the N Atlantic, W of Portugal: politically part of Portugal. 253,500; 890 sq. mi. (2305 sq. km). —**A·zo′re·an, A·zo′ri·an,** *adj., n.*

A·zov (az′ôf, -of, ā′zôf, ā′zof), **Sea of,** a northern arm of the Black Sea, connected with the Black Sea by Kerch Strait. ab. 14,500 sq. mi. (37,555 sq. km).

B

Baal·bek (bäl′bek, bā′əl-), a town in E Lebanon: ruins of ancient city. 16,000. Ancient Greek, **Heliopolis.**

Bab el Man·deb (bäb′ el män′deb), a strait between NE Africa and the SW tip of the Arabian peninsula, connecting the Red Sea and the Gulf of Aden. 20 mi. (32 km) wide.

Ba·bu·yan′ Is′lands (bä′bōō yän′), a group of islands in the Philippines, N of Luzon. 225 sq. mi. (580 sq. km).

Bab·y·lon (bab′ə lən, -lon′), an ancient city in SW Asia, on the Euphrates River: capital of Babylonia and later of the Chaldean empire.

Bab·y·lo·ni·a (bab′ə lō′nē ə, -lōn′yə), any of a succession of states, having Babylon as their principal city, that existed in S Mesopotamia between c1900 B.C. and 539 B.C. —**Bab′y·lo′ni·an,** *adj., n.*

Ba·că·u (bə kou′), a city in E Romania. 175,299.

Back′ Bay′, a prosperous residential and commercial area of Boston, Mass.

Ba·co·lod (bä kō′lod), a seaport on N Negros, in the central Philippines. 364,000.

Bac·tra (bak′trə), ancient name of **BALKH.**

Bac·tri·a (bak′trē ə), an ancient country in W Asia, between the Oxus River and the Hindu Kush Mountains. *Cap.:* Bactra. —**Bac′tri·an,** *adj., n.*

Ba·da·joz (bä′ŧ͟Hä hōth′, -hōs′), a city in SW Spain. 126,340.

Ba·da·lo·na (bä′ŧ͟Hä lō′nä), a seaport in NE Spain, near Barcelona. 223,444.

Ba·den (bäd′n), **1.** a region in SW Germany, formerly a state, now incorporated in Baden-Württemberg. **2.** BADEN-BADEN.

Ba′den-Ba′den, a city in W Baden-Württemberg, in SW Germany: spa. 48,680.

Ba′den-Würt′temberg, a state in SW Germany. 9,390,000; 13,800 sq. mi. (35,751 sq. km). *Cap.:* Stuttgart.

Bad′ Go′desberg (bät), official name of GODESBERG.

Bad′lands Na′tional Park′, a national park in SW South Dakota: rock formations and animal fossils. 380 sq. mi. (985 sq. km).

Baf′fin Bay′ (baf′in), a part of the Arctic Ocean between W Greenland and E Baffin Island.

Baf′fin Is′land, a Canadian island in the Arctic Ocean, between Greenland and N Canada. ab. 1000 mi. (1600 km) long; 190,000 sq. mi. (492,000 sq. km). Also called **Baf′fin Land′.**

Bagh·dad or **Bag·dad** (bag′dad, bəg dad′), the capital of Iraq, in the central part, on the Tigris. 4,648,609.

Ba·gui·o (bä′gē ō′), a city on W Luzon, in the N Philippines: summer capital. 119,009; 4961 ft. (1512 m) high.

Ba·ha·mas (bə hä′məz, -hä′-), an independent country comprising a group of islands (**Baha′ma Is′lands**) in the W Atlantic Ocean, SE of Florida: formerly a British colony; gained independence 1973. 236,000; 5353 sq. mi. (13,864 sq. km). *Cap.:* Nassau. Official name, **Com′monwealth of the Baha′mas.** —**Ba·ha′mi·an** (-hā′-, -hä′-), *n., adj.*

Ba·ha·wal·pur (bə hä′wəl pŏŏr′, bä′wəl-), a state in E Pakistan. 4,652,000; 32,443 sq. mi. (83,000 sq. km).

Ba·hi·a (bä ē′ə, bə-), **1.** a coastal state of E Brazil. 11,086,600; 216,130 sq. mi. (559,700 sq. km). *Cap.:* Salvador. **2.** a former name of SALVADOR (def. 2).

Ba·hi·a de Co·chi·nos (bä ē′ä łħä kō chē′nōs), Spanish name of BAY OF PIGS.

Bah·rain or **Bah·rein** (bä rān′, -rīn′, bə-), **1.** a sheikdom in the Persian Gulf, consisting of a group of islands. 486,000; 266 sq. mi. (688 sq. km). *Cap.:* Manama. **2.** the largest island in this group: oil fields. 213 sq. mi. (552 sq. km). —**Bah·rain′i,** *n., pl.* **-is,** *adj.*

Bai·kal (bī käl′), **Lake,** a lake in the Russian Federation, in S Siberia: the deepest lake in the world. 13,200 sq. mi. (34,188 sq. km); 5714 ft. (1742 m) deep.

Bai·le Átha Cli·ath (blä klē′ə), Irish name of DUBLIN.

Ba′ja Califor′nia (bä′hä), a peninsula in NW Mexico between the Gulf of California and the Pacific. Also called **Ba′ja, Lower California.**

Ba′ja Califor′nia Nor′te (nôr′tē, -tä), a state in NW Mexico, in the N part of Baja California. 1,388,500; 26,997 sq. mi. (69,921 sq. km). *Cap.:* Mexicali.

Ba/ja Califor/nia Sur/ (sŏŏr), a state in NW Mexico, in the S part of Baja California. 315,100; 28,369 sq. mi. (73,475 sq. km). *Cap.:* La Paz.

Ba/ker Lake/ (bā/kər), a lake in the Northwest Territories, in N Canada. 975 sq. mi. (2525 sq. km).

Ba·kers·field (bā/kərz fēld/), a city in S California. 174,820.

Bakh·ta·ran (bäкн/tä rän/), a city in W Iran. 560,514. Formerly, **Kermanshah.**

Ba·ku (bu kōō/), the capital of Azerbaijan, in the E part, on the Caspian Sea. 1,757,000.

Ba·kwan·ga (bə kwäng/gə), former name of Mbuji-Mayi.

Bal·a·kla·va (bal/ə klä/və), a seaport in S Crimea, in S Ukraine, on the Black Sea.

Ba·la·ton (bal/ə ton/), a lake in W Hungary: the largest lake in central Europe. ab. 50 mi. (80 km) long; 230 sq. mi. (596 sq. km).

Bald/win Park/ (bôld/win), a city in SW California, near Los Angeles. 69,330.

Bâle (bäl), French name of Basel.

Bal·e·ar/ic Is/lands (bal/ē ar/ik), a group of islands including Ibiza, Majorca, and Minorca, and constituting a province of Spain in the W Mediterranean Sea. 754,777; 1936 sq. mi. (5015 sq. km). *Cap.:* Palma. Spanish, **Ba·le·a·res** (bä/le ä/res).

Ba·li (bä/lē, bal/ē), an island in Indonesia, E of Java. 2,469,930; 2147 sq. mi. (5561 sq. km).

Ba·lik·pa·pan (bä/lik pä/pän), a seaport on E Borneo, in central Indonesia. 280,875.

Bal/kan Moun/tains, a mountain range extending from W Bulgaria to the Black Sea: highest peak, 7794 ft. (2370 m).

Bal/kan Penin/sula, a peninsula in S Europe, S of the Danube River and bordered by the Adriatic, Ionian, Aegean, and Black seas.

Bal·kans (bôl/kənz), **the,** the countries in the Balkan Peninsula: Yugoslavia, Bosnia and Herzegovina, Croatia, Macedonia, Slovenia, Romania, Bulgaria, Albania, Greece, and the European part of Turkey. Also called **the Bal/kan States/.** —**Bal/kan,** *adj.*

Balkh (bälкн), a town in N Afghanistan: capital of ancient Bactria. Ancient, **Bactra.**

Bal·khash (bal kash′, bäl käsh′), a salt lake in SE Kazakhstan. ab. 7115 sq. mi. (18,430 sq. km).

Bal·la·rat (bal′ə rat′), a city in S Victoria, in SE Australia. 79,000.

Bal′tic Sea′ (bôl′tik), a sea in N Europe, bounded by Denmark, Sweden, Finland, Estonia, Latvia, Lithuania, Poland, and Germany. ab. 160,000 sq. mi. (414,000 sq. km).

Bal′tic States′, Estonia, Latvia, Lithuania, and sometimes Finland.

Bal·ti·more (bôl′tə môr′, -mōr′), a seaport in N Maryland, on an estuary near the Chesapeake Bay. 736,014.

Ba·lu·chi·stan (bə lōō′chə stän′, -stan′), 1. an arid mountainous region in S Asia, in SE Iran and SW Pakistan, bordering on the Arabian Sea. 2. a province in SW Pakistan. 4,908,000; 134,050 sq. mi. (347,190 sq. km). *Cap.:* Quetta.

Ba·ma·ko (bam′ə kō′, bä′mə kō′), the capital of Mali: inland port on the Niger River. 404,022.

Ban·at (ban′it, bä′nit), a fertile low-lying region extending through Hungary, Romania, and Yugoslavia.

Ban·dar Se·ri Be·ga·wan (bun′dər ser′ē bə gä′wən), the capital of the sultanate of Brunei, on the NW coast of Borneo, in the Malay Archipelago. 63,868.

Ban′da Sea′ (bän′də, ban′-), a sea between Sulawesi (Celebes) and New Guinea, S of the Moluccas and N of Timor.

Ban·djar·ma·sin (bän′jər mä′sin), BANJARMASIN.

Ban·dung (bän′dŏŏng, -dŏŏng, ban′-), a city in W Java, in Indonesia. 1,462,637. Dutch, **Ban′doeng** (-dŏŏng).

Banff (bamf), a historic county in NE Scotland. Also called **Banff·shire** (bamf′shēr, -shər).

Banff′ Na′tional Park′, a national reserve, 2585 sq. mi. (6695 sq. km), in the Rocky Mountains, in SW Alberta, Canada.

Ban·ga·lore (bang′gə lôr′, -lōr′), the capital of Karnataka, in SW India. 2,914,000.

Bang·ka or **Ban·ka** (bang′kə), an island in Indonesia, E of Sumatra: tin mines. 4611 sq. mi. (11,942 sq. km).

Bang·kok (bang′kok, bang kok′), the capital of Thailand, in the S central part, on the Chao Phraya. 5,609,352.

Ban·gla·desh (bäng/glə desh/, bang/-), a republic in S Asia, N of the Bay of Bengal: a member of the Commonwealth of Nations; a former province of Pakistan. 104,100,000; 54,501 sq. mi. (141,158 sq. km). *Cap.:* Dhaka. Compare EAST PAKISTAN. —**Ban/gla·desh/i,** *n., pl.* **-is,** *adj.*

Ban·gor (bang/gôr, -gər), a seaport in S Maine, on the Penobscot River. 31,643.

Ban·gui (bän gē/), the capital of the Central African Republic, in the SW part. 596,776.

Bang·we·u·lu (bang/wē ōō/lōō), a shallow lake and swamp in NE Zambia. ab. 150 mi. (240 km) long.

Ban·jar·ma·sin or **Ban·djar·ma·sin** (bän/jərmä/sin), a seaport on the S coast of Borneo, in Indonesia. 381,286.

Ban·jul (bän/jōōl), the capital of the Gambia. 44,188.

Ban·ka (bang/kə), BANGKA.

Banks/ Is/land, an island in the W Northwest Territories, in NW Canada. 24,600 sq. mi. (63,700 sq. km).

Ban·nock·burn (ban/ək bûrn/, ban/ək bûrn/), a village in central Scotland: site of the victory (1314) of the Scots under Robert the Bruce over the English, which assured the independence of Scotland.

Ban·tam (ban/təm), a village in W Java, in S Indonesia: first Dutch settlement in the East Indies.

Bao·ding or **Pao·ting** (bou/ding/), a city in central Hebei province, in NE China. 502,394. Formerly, **Tsingyuan.**

Bao·ji or **Pao·chi** (bou/jē/), a city in W Shaanxi province, in central China. 338,754.

Ba·ra·cal·do (bär/ə käl/dō), a city in N Spain. 112,854.

Ba·ra·co·a (bär/ə kō/ə), a seaport in E Cuba: oldest Spanish town in Cuba; settled 1512. 35,538.

Ba·ra·no·vi·chi (bə rä/nə vich/ē), a city in central Belarus, SW of Minsk. 146,000.

Bar·ba·dos (bär bā/dōz, -dōs, -dəs), an island in the E West Indies constituting an independent state in the Commonwealth of Nations: formerly a British colony. 253,881; 166 sq. mi. (430 sq. km). *Cap.:* Bridgetown. —**Bar·ba/di·an,** *adj., n.*

Bar·ba·ry (bär/bə rē), a region in N Africa, extending from W of Egypt to the Atlantic Ocean and including the former Barbary States.

Bar′bary Coast′, the Mediterranean coastline of the former Barbary States.

Bar′bary States′, Morocco, Algiers, Tunis, and Tripoli, c1520–1830, when they were the refuge of pirates.

Bar·bu·da (bär bōō′də), one of the NE Leeward Islands, in the E West Indies: part of Antigua and Barbuda. 62 sq. mi. (161 sq. km).

Bar·ca (bär′kə), CYRENAICA. —**Bar′can,** *adj.*

Bar·ce·lo·na (bär′sə lō′nə), a seaport in NE Spain, on the Mediterranean. 2,000,000.

Ba·reil·ly or **Ba·re·li** (bə rā′lē), a city in N central Uttar Pradesh, in N India. 438,000.

Bar′ents Sea′ (bar′ənts, bär′-), a part of the Arctic Ocean between NE Europe and the islands of Spitsbergen, Franz Josef Land, and Novaya Zemlya.

Bar′ Har′bor, a town on Mount Desert Island, in S Maine: summer resort. 4124.

Ba·ri (bär′ē), a seaport in SE Italy, on the Adriatic. 358,906.

Ba·ri·sal (bur′ə säl′, bar′ə sôl′), a port in S Bangladesh, on the Ganges River. 142,098.

Bar·king (bär′king), a borough of Greater London, England. 154,200.

Bar·na·ul (bär′nə ōōl′), the capital of the Altai territory in the Russian Federation, on the Ob River, S of Novosibirsk. 602,000.

Bar·net (bär′nit), a borough of Greater London, England. 305,900.

Barns·ley (bärnz′lē), a city in South Yorkshire, in N England. 224,000.

Ba·ro·da (bə rō′də), former name of VADODARA.

Ba·rot·se·land (bə rot′sə land′), a region in W Zambia. 410,087; 44,920 sq. mi. (116,343 sq. km).

Bar·qui·si·me·to (bär′kē sē mā′tō), a city in N Venezuela. 523,101

Bar·ran·qui·lla (bär′än kē′yə), a seaport in N Colombia, on the Magdalena River. 899,781.

Bar′ren Grounds′, a sparsely inhabited region of tundra in N Canada, esp. in the area W of Hudson Bay. Also called **Bar′ren Lands′.**

Bar′rier Reef′, GREAT BARRIER REEF.

Bar·row (bar′ō), **1.** Also called **Bar′row-in-Fur′-**

ness (fûr′nis). a seaport in Cumbria, in NW England. 73,900. **2. Point,** the N tip of Alaska: the northernmost point of the U.S.

Ba·sel (bä′zəl) also **Basle, 1.** a city in NW Switzerland, on the Rhine River. 192,800. **2.** a canton in N Switzerland, divided into two independent areas. French, **Bâle.**

Ba·shan (bā′shən), a region in ancient Palestine, E of the Jordan River.

Bash·kir/ Auton′omous Repub′lic (bäsh kēr′, bash-), an autonomous republic in the Russian Federation in Europe. 3,952,000; 55,430 sq. mi. (143,600 sq. km). *Cap.:* Ufa.

Ba·si·lan (bə sē′län), an island in the Philippines, SW of Mindanao. 495 sq. mi. (1282 sq. km).

Ba·sil·don (bā′zəl dən, baz′əl-), a town in S Essex, in SE England. 157,400.

Ba·si·li·ca·ta (bä zē/lē kä′tä), Italian name of LUCANIA.

Basle (bäl), BASEL.

Basque′ Prov′inces (bask), a region in N Spain, on the Bay of Biscay.

Bas·ra (bus′rə, bäs′rä) also **Busra, Busrah,** a port in SE Iraq, N of the Persian Gulf. 616,700.

Bas·sein (bə sān′), a city in SW Burma, on the Irrawaddy River. 144,092.

Basse-Nor·man·die (bäs nôr män dē′), a metropolitan region in NW France, including the historic region of Normandy. 1,373,400; 6,791 sq. mi. (17,589 sq. km).

Basse·terre (bäs târ′), the capital of St. Kitts-Nevis. 15,897.

Basse-Terre (bäs târ′), **1.** the capital of Guadeloupe, in the French West Indies. 15,690. **2.** See under GUADELOUPE.

Bass′ Strait′ (bas), a strait between Australia and Tasmania. 80–150 mi. (130–240 km) wide.

Bas·togne (ba stōn′; *Fr.* ᴅa stôn′yᵉ), a town in SE Belgium: U.S. forces besieged here during German counteroffensive in 1944. 6816.

Ba·su·to·land (bə sōō′tōō land′, -tō-), former name of LESOTHO.

Ba·taan or **Ba·taán** (bə tan′, -tän′; *locally* bä′tä än′), a peninsula on W Luzon, in the Philippines: U.S. troops surrendered to Japanese April 9, 1942.

Ba·tan·gas (bä täng′gäs), a seaport on SW Luzon, in the N central Philippines. 184,000.

Ba·ta·vi·a (bə tā′vē ə), former name of JAKARTA.

Bath (bath, bäth), a city in Avon, in SW England: mineral springs. 84,300.

Bath·urst (bath′ərst), former name of BANJUL.

Bat·on Rouge (bat′n rōōzh′), the capital of Louisiana, in the SE part: a port on the Mississippi. 219,531.

Bat·ter·sea (bat′ər sē), a former borough of London, England, now part of Wandsworth, on the Thames.

Bat′tle Creek′, a city in S Michigan. 53,540.

Ba·tu·mi (bä tōō′mē), the capital of Adzharistan, in the SW Georgian Republic, on the Black Sea. 136,000. Formerly, **Ba·tum** (bä tōōm′).

Bat Yam (bät′ yäm′), a city on the coast of the Mediterranean Sea in W central Israel, S of Tel Aviv. 132,100.

Ba·var·i·a (bə vâr′ē ə), a state in SE Germany. 11,082,600; 27,240 sq. mi. (70,550 sq. km). *Cap.:* Munich. German, **Bay′ern** (bī′ərn). —**Ba·var′i·an,** *adj., n.*

Ba·ya·món (bä′yä mōn′), a city in N Puerto Rico. 211,616.

Bay′ of Pigs′, a bay of the Caribbean Sea in SW Cuba: site of attempted invasion of Cuba by anti-Castro forces April 1961. Spanish, **Bahía de Cochinos.**

Ba·yonne (bā yōn′ *for 1;* bä yôn′ *for 2),* **1.** a seaport in NE New Jersey. 61,444. **2.** a seaport in SW France, near the Bay of Biscay. 44,706.

Bay·reuth (bī′roit, bī roit′), a city in NE Bavaria, in SE Germany: annual music festivals founded by Richard Wagner. 71,848.

Bay·town (bā′toun′), a city in SE Texas, on Galveston Bay. 63,850.

Bear′ Riv′er, a river in N Utah, SW Wyoming, and SE Idaho, flowing into the Great Salt Lake. 350 mi. (565 km) long.

Beau′fort Sea′ (bō′fərt), a part of the Arctic Ocean, NE of Alaska.

Beau·jo·lais (bō′zhə lā′), a winegrowing region in E France.

Beau·mont (bō′mont), a city in SE Texas. 114,323.

Beau·port (bō′pôrt′), a city in E Quebec, in E

Canada: suburb of Quebec, on the St. Lawrence River. 62,869.

Beau·vais (bō vāʹ), a city in NW France. 56,725.

Bea·ver·ton (bēʹvər tən), a city in NW Oregon. 53,310.

Bech·u·a·na·land (bechʹōō äʹnə landʹ, bekʹyōō-), former name of Botswana.

Bed·ford·shire (bedʹfərd shērʹ, -shər), a county in central England. 525,900; 477 sq. mi. (1235 sq. km). Also called **Bedʹford.**

Bedʹloe's Isʹland (bedʹlōz) also **Bedʹloe Isʹland,** former name of Liberty Island.

Beer·she·ba (bēr shēʹbə, bērʹshə-), a city in Israel, near the N limit of the Negev desert: the southernmost city of ancient Palestine. 114,600.

Bei·jing (beiʹjingʹ) also **Peking,** the capital of the People's Republic of China, in the NE part, in central Hebei province. 6,800,000. Formerly (1928–49), **Peiping.**

Bei·ra (bāʹrə), a seaport in central Mozambique. 269,700.

Bei·rut (bā rōōtʹ), the capital of Lebanon, a seaport. 702,000.

Be·la·rus (byelʹə rōōsʹ, belʹ-), n. a republic in E Europe, N of Ukraine. 10,200,000; 80,154 sq. mi. (207,600 sq. km). Cap.: Minsk. Also called **Belorussia.** Formerly, **Belorussian Soviet Socialist Republic.**

Be·lém (bə lemʹ), the capital of Pará state, in N Brazil on the Pará River. 755,984.

Bel·fast (belʹfast, -fäst, bel fastʹ, -fästʹ), a seaport in and the capital of Northern Ireland, on the E coast. 296,900.

Bel·fort (bel fôrʹ, bā-), a fortress city in E France on a mountain pass between the Vosges and Jura mountains. 57,317.

Belʹgian Conʹgo, a former name (1908–60) of Zaire (def. 1).

Bel·gium (belʹjəm), a kingdom in W Europe, bordering the North Sea, N of France. 9,813,152; 11,779 sq. mi. (30,508 sq. km). Cap.: Brussels. French, **Belgique** (bel zhēkʹ); Flemish, **Bel·gi·ë** (belʹкнē ə).—**Belʹgian** (-jən), n., adj.

Bel·go·rod (belʹgə rodʹ), a city in the W Russian Federation, N of Kharkov. 293,000.

Bel·grade (belʹgrād, -gräd, -grad, bel grädʹ, -grädʹ, -gradʹ), the capital of Yugoslavia and the republic of Serbia, at the confluence of the Danube and Sava rivers. 1,470,073. Serbo-Croatian, **Beograd.**

Bel·gra·vi·a (bel grā'vē ə), a fashionable district in London, England, adjoining Hyde Park. —**Bel·gra'vi·an,** *adj.*

Be·li·tung (be lē'tong) also **Billiton,** an island in Indonesia, between Borneo and Sumatra. 1866 sq. mi. (4833 sq. km).

Be·lize (bə lēz'), **1.** Formerly, **British Honduras.** a parliamentary democracy in N Central America: a former British crown colony; gained independence 1981. 193,000; 8867 sq. mi. (22,966 sq. km). *Cap.:* Belmopan. **2.** Also called **Belize/ City/y.** a seaport in and the main city of Belize. 43,600. —**Be·li·ze·an** (bə lē'zē ən), *adj., n.*

Bel/leau Wood/ (bel/ō, be lō'), a forest in N France, NW of Château-Thierry: a memorial to the U.S. Marines who won a battle there 1918.

Belle/ Isle/, Strait of, a strait between Newfoundland and Labrador, Canada. 10–15 mi. (16–24 km) wide.

Belle·vue (bel'vyōo), a city in W Washington. 86,874.

Bell·flow·er (bel'flou'ər), a city in SW California, near Los Angeles. 61,815.

Bel·ling·ham (bel'ing ham'), a seaport in NW Washington. 52,179.

Bel'lings·haus·en Sea/ (bel'ingz hou'zən), an arm of the S Pacific Ocean, W of Antarctic Peninsula.

Bel·lin·zo·na (bel'in zō'nə), the capital of Ticino, in S Switzerland. 17,000.

Bel·mo·pan (bel'mō pan'), the capital of Belize, in the central part. 5280.

Be·lo Ho·ri·zon·te (bā'lō ôr'ə zon'tē), the capital of Minas Gerais, in SE Brazil. 1,814,990.

Be·lo·rus·sia or **Bye·lo·rus·sia** (byel'ə rush'ə, bel'ə-), BELARUS. —**Be'lo·rus'sian,** *adj., n.*

Belorus'sian So'viet So'cialist Repub'lic, former name of BELARUS.

Be·lo·stok (*Russ.* byi lu stôk'), BIALYSTOK.

Bel·sen (bel'sən, -zən), locality in NW Germany: site of Nazi concentration camp **(Bergen-Belsen)** during World War II.

Belt·way (belt'wā'), **the,** the Washington, D.C. area.

Be·na·res (bə när′is, -ēz), former name of VARA-
NASI.

Ben·de·ry (ben der′ē), a city in E central Moldova, SE of Kishinev. 130,000.

Ben·di·go (ben′di gō′), a city in central Victoria
in SE Australia: gold mining. 64,790.

Ben·e·lux (ben′l uks′), Belgium, the Netherlands,
and Luxembourg considered together.

Ben·gal (ben gôl′, -gäl′, beng-), **1.** a former
province in NE India, now divided between India
and Bangladesh. Compare EAST BENGAL, WEST
BENGAL. **2. Bay of,** a part of the Indian Ocean between India and Burma. —**Ben·ga′li,** *n., adj.*

Beng·bu (bä/nĕ) also **Pengpu,** a city in N
Anhui province, in E China. 550,000.

Ben·gha·zi or **Ben·ga·si** (ben gä′zē, beng-), a
seaport in N Libya: former capital. 485,386.

Be·ni (bā/nē), a river flowing NE from W Bolivia to
the Brazilian border, where with the Mamoré it
forms the Madeira River. ab. 600 mi. (965 km)
long.

Be·nin (be nēn′), **1.** Formerly, **Dahomey.** a republic in W Africa: formerly part of French West
Africa; gained independence in 1960. 4,440,000;
44,290 sq. mi. (114,711 sq. km). *Cap.:* Porto
Novo. **2. Bight of,** a bay in N Gulf of Guinea in W
Africa. **3.** a historic kingdom of W Africa centered
in Edo-speaking regions W of the Niger River. **4.** a
river in S Nigeria flowing into the Bight of Benin.

Benin′ Cit′y, a city in S Nigeria. 165,900.

Be·ni-Suef (ben/ē swāf′), a city in NW Egypt on
the Nile River. 107,100.

Ben Lo·mond (ben lō′mənd), a mountain in
central Scotland in Stirlingshire on the E shore of
Loch Lomond. 3192 ft. (975 m).

Ben Ne·vis (ben nē′vis, nev′is), a mountain in
NW Scotland in the Grampians: highest peak in
Great Britain. 4406 ft. (1343 m).

Ben·ning·ton (ben′ing tən), a town in SW Vermont: defeat of British by the Green Mountain
Boys 1777. 15,815.

Be·no·ni (bə nō′ni, -nē), a city in the NE of the
Republic of South Africa near Johannesburg: gold
mines. 167,000.

Be·nue (bā′nwā), a river in W Africa flowing W
from Cameroon to the Niger River in Nigeria. 870
mi. (1400 km).

Ben·xi (bun/shē/) also **Penchi, Penki,** a city in E Liaoning province, in NE China. 792,401.

Be·o·grad (bā/ō gräd), Serbo-Croatian name of BELGRADE.

Ber·be·ra (bûr/bər ə), a seaport in Somalia on the Gulf of Aden: former capital of British Somaliland. 65,000.

Ber·dyansk (bər dyansk/), a city in S Ukraine, on the Sea of Azov. 129,000.

Be·re·zi·na (bi rā/zə nə), a river in central Belarus, flowing SE into the Dnieper River. 350 mi. (565 km) long.

Be·rez·ni·ki (bə rez/ni kē), a city in the Russian Federation, on the Kama river, near the Ural Mountains. 200,000.

Ber·ga·ma (bər gä/mə), a town in W Turkey in Asia: site of ancient Pergamum. 34,716.

Ber·gen (bûr/gən, bâr/-), a city in SW Norway on the Atlantic Ocean. 213,594.

Ber·gen-Bel·sen (bûr/gən bel/sən, -zən, bâr/-), See under BELSEN.

Ber/ing Sea/ (bēr/ing, bâr/-), a part of the N Pacific N of the Aleutian Islands. 878,000 sq. mi. (2,274,000 sq. km).

Ber/ing Strait/, a strait between Alaska and the Russian Federation in Asia connecting the Bering Sea and the Arctic Ocean. 36 mi. (58 km) wide.

Berke·ley (bûrk/lē), a city in W California on San Francisco Bay. 102,724.

Berk·shire (bûrk/shēr, -shər; *Brit.* bärk/-), a county in S England: constitutes a state. 740,600; 485 sq. mi. (1255 sq. km). Also called **Berks** (bûrks; *Brit.* bärks).

Berk/shire Hills/ (bûrk/shēr, -shər), a range of low mountains in W Massachusetts. Highest peak, 3505 ft. (1070 m). Also called **Berk/shires.**

Ber·lin (bər lin/), the capital of Germany, in the NE part: constitutes a state. 3,121,000; 341 sq. mi. (883 sq. km). Formerly (1948–90) divided into a western zone **(West Berlin),** a part of West Germany; and an eastern zone **(East Berlin),** the capital of East Germany. —**Ber·lin/er,** *n.*

Ber·me·jo (bər mā/hō), a river in N Argentina flowing SE to the Paraguay River. 1000 mi. (1600 km) long.

Ber·mu·da (bər myōō/də), a group of islands in the Atlantic, 580 mi. (935 km) E of North Carolina: a British colony; resort. 58,080; 19 sq. mi.

(49 sq. km). *Cap.:* Hamilton. —**Ber·mu′dan, Ber· mu′di·an,** *adj., n.*

Bern or **Berne** (bûrn, bârn), **1.** the capital of Switzerland, in the W part: capital of Bern canton. 136,300. **2.** a canton in W Switzerland. 928,800; 2658 sq. mi. (6885 sq. km). *Cap.:* Bern. —**Ber· nese** (bûr nēz′, -nēs′), *adj., n., pl.* **-nese.**

Ber′nese Alps′, a mountain range in SW Switzerland, part of the Alps. Highest peak, Finsteraarhorn, 14,026 ft. (4275 m).

Ber·ni·na (bər nē′nə), a mountain in SE Switzerland: highest peak of the Rhaetian Alps, 13,295 ft. (4052 m).

Ber·ry or **Ber·ri** (ber′ē, be rē′), a former province in central France.

Ber·wick (ber′ik), a historic county in SE Scotland. Also called **Ber′wick·shire′** (-shēr′, -shər).

Be·san·çon (bə zän sôn′), a city in E France: Roman ruins. 119,687.

Bes·sa·ra·bi·a (bes′ə rā′bē ə), a region in Moldova, on the W shore of the Black Sea: formerly part of Romania. —**Bes′sa·ra′bi·an,** *adj., n.*

Beth·a·ny (beth′ə nē), a village in W Jordan, near Jerusalem, at the foot of the Mount of Olives: home of Lazarus in the Bible.

Beth·el (beth′əl, -el, beth′el′), a village in W Jordan, near Jerusalem.

Be·thes·da (bə thez′də), a city in central Maryland: residential suburb of Washington, D.C. 62,936.

Beth·le·hem (beth′li hem′, -lē əm), **1.** a town in NW Jordan, near Jerusalem, occupied by Israel 1967: birthplace of Jesus and David. 16,313. **2.** a city in E Pennsylvania. 71,428.

Bev′erly Hills′ (bev′ər lē), a city in SW California, surrounded by the city of Los Angeles. 32,367.

Bex·ley (beks′lē), a borough of Greater London, England. 220,600.

Bey·oğ·lu (bā′ə lōō′, bā′ə lōō′), a modern section of Istanbul, Turkey, N of the Golden Horn. Formerly, **Pera.**

Bé·ziers (bā zyā′), a city in S France, SW of Montpellier. 78,477.

Bez·wa·da (bez wä′də), former name of Vijaya-wada.

Bhat·pa·ra (bät′pär ə), a city in SW West Bengal in E India. 204,750.

Bhau·na·gar (bou nug′ər) also **Bhav·na·gar** (bäv-), a seaport in S Gujarat in W India. 308,000.

Bho·pal (bō päl′), **1.** a former state in central India: now part of Madhya Pradesh. **2.** the capital of Madhya Pradesh. 672,000.

Bhu·ba·nes·war (bub′ə nesh′wər), the capital of Orissa state, in E India. 219,419.

Bhu·tan (bōō tän′), a kingdom in the Himalayas, NE of India: foreign affairs under Indian jurisdiction. 1,400,000; ab. 19,300 sq. mi. (50,000 sq. km). *Cap.:* Thimphu. —**Bhu·tan·ese** (bōōt′n ēz′, -ēs′), *n., pl.* **-ese**, *adj.*

Bi·a·fra (bē ä′frə), **1.** a former secessionist state (1967–70) in SE Nigeria, in W Africa. *Cap.:* Enugu. **2. Bight of,** a wide bay in the E part of the Gulf of Guinea off the W coast of Africa. —**Bi·a′fran**, *adj., n.*

Bi·ak (bē yäk′), an island in Indonesia, N of Irian Jaya. 948 sq. mi. (2455 sq. km).

Bia·ly·stok (byä′li stôk′, -wi-), a city in E Poland. 245,000. Russian, **Belostok, Byelostok.**

Biar·ritz (bē′ə rits′, bē′ə rits′), a city in SW France on the Bay of Biscay: resort. 27,653.

Biel (bēl), **Lake,** Bienne, Lake of.

Bie·le·feld (bē′lə felt′), a city in NW Germany. 315,000.

Biel·sko-Bia·ła (byel′skô byä′lä, -byä′wä), a city in S Poland. 174,000.

Bienne (byen), **Lake of,** a lake in NW Switzerland: traces of prehistoric lake dwellings. 16 sq. mi. (41 sq. km). Also called **Lake Biel.**

Big′ Bend′ Na′tional Park′, a national park in W Texas on the Rio Grande. 1080 sq. mi. (2800 sq. km).

Big′ Di′omede, See under Diomede Islands.

Big·horn (big′hôrn′), a river flowing from central Wyoming to the Yellowstone River in S Montana. 336 mi. (540 km) long.

Big′horn Moun′tains, a mountain range in N Wyoming, part of the Rocky Mountains. Highest peak, 13,165 ft. (4013 m). Also called **Big′horns′.**

Big′ Mud′dy Riv′er, a river in SW Illinois, flowing SW into the Mississippi. ab. 120 mi. (195 km) long.

Big/ Sur/ (sûr), a rugged coastal region in W California, S of Carmel.

Bi·har (bi här/), **1.** a state in NE India. 69,823,154; 67,164 sq. mi. (173,955 sq. km). *Cap.*: Patna. **2.** a city in the central part of this state. 151,308.

Bi·ka·ner (bē/kə nēr/, -ner/, bik/ə-), a city in NW Rajasthan, India. 280,000.

Bi·ki·ni (bi kē/nē), an atoll in the N Pacific, in the Marshall Islands: atomic bomb tests 1946. 3 sq. mi. (8 sq. km).

Bil·ba·o (bil bou/), a seaport in N Spain, near the Bay of Biscay. 433,000.

Bil·lings (bil/ingz), a city in S Montana. 81,151.

Bil·li·ton (bi lē/ton), BELITUNG.

Bi·lox·i (bi lok/sē; *locally* -luk/-), a city in SE Mississippi, on the Gulf of Mexico. 49,311.

Bing·ham·ton (bing/əm tən), a city in S New York, on the Susquehanna River. 53,008.

Binh Dinh or **Binh·dinh** (bin/ din/), former name of AN NHON.

Bí·o-Bí·o (bē/ō bē/ō), a river in central Chile, flowing NW from the Andes to the Pacific at Concepción. ab. 240 mi. (384 km) long.

Bi·o·ko (bē ō/kō), an island in the Bight of Biafra, near the W coast of Africa: a province of Equatorial Guinea. 80,000; ab. 800 sq. mi. (2072 sq. km). Formerly, **Fernando Po, Macías Nguema Biyogo.**

Bir·ken·head (bûr/kən hed/), a seaport in Merseyside metropolitan county, in W England, on the Mersey River. 336,500.

Bir·ming·ham (bûr/ming əm *for 1;* -ham/ *for 2*), **1.** a city in West Midlands, in central England. 1,084,600. **2.** a city in central Alabama. 265,968.

Bi·ro·bi·dzhan or **Bi·ro·bi·jan** (bir/ō bi jän/), the capital of the Jewish Autonomous Region, in E Siberia, in the SE Russian Federation in Asia, W of Khabarovsk. 82,000.

Bi·sa·yas (bē sä/yäs), Spanish name of the VISAYAN ISLANDS.

Bis·cay (bis/kā, -kē), **Bay of,** a bay of the Atlantic between W France and N Spain.

Bis/cayne Bay/ (bis/kān, bis kān/), an inlet of the Atlantic Ocean, on the SE coast of Florida.

Bish·kek (bish kek/), the capital of Kyrgyzstan, in

the N part. 616,000. Formerly, **Pishpek** (until 1926), **Frunze** (1926–91).

Bisho (bē/shō), the capital of Ciskei, in SE South Africa.

Bisk or **Biysk** (byēsk), a city in the S Russian Federation in Asia, near the Ob River, SE of Barnaul. 231,000.

Bis·marck (biz/märk), the capital of North Dakota, in the central part. 44,485.

Bis/marck Archipel/ago, a group of islands in Papua New Guinea, in the SW Pacific Ocean, including the Admiralty Islands, New Britain, New Ireland, and adjacent islands. ab. 23,000 sq. mi. (59,570 sq. km).

Bis·sau (bi sou/) also **Bis·são** (bē souN/), a seaport in and the capital of Guinea-Bissau, in the W part. 109,214.

Bi·thyn·i·a (bi thin/ē ə), an ancient state in NW Asia Minor. —**Bi·thyn/i·an,** *adj., n.*

Bi·to·la (bē/tō/lä), a city in S Macedonia. 137,636. Serbo-Croatian, **Bi·tolj** (-tōl, -tōl/yə). Turkish, **Monastir.**

Bit/ter Lakes/, two lakes in NE Egypt, forming part of the Suez Canal.

Bit/ter·root Range/ (bit/ər rōōt/, -rŏŏt/), a mountain range between Idaho and Montana, a part of the Rocky Mountains: highest peak, 11,393 ft. (3473 m).

Bi·wa (bē/wä), **Lake,** the largest lake in Japan, on Honshu, near Kyoto. 260 sq. mi. (673 sq. km).

Biysk (byēsk), **Bisk.**

Bi·zer·te (bi zûr/tə, -tē, -zârt/) also **Bi·zer·ta** (-tə), a seaport in N Tunisia. 94,509.

Black·burn (blak/bərn), **1.** a city in central Lancashire, in NW England. 142,200. **2. Mount,** a mountain in SE Alaska, in the Wrangell Mountains. 16,140 ft. (4920 m).

Black/ Can/yon, a canyon of the Colorado River between Arizona and Nevada: site of Boulder Dam.

Black/ For/est, a wooded mountain region in SW Germany. German, **Schwarzwald.**

Black/ Hills/, a group of mountains in W South Dakota and NE Wyoming. Highest peak, Harney Peak, 7242 ft. (2207 m).

Black/ Moun/tains, a mountain range in W North Carolina, part of the Appalachian Mountains. Highest peak, Mount Mitchell, 6684 ft. (2037 m).

Black·pool (blak′pōōl′), a seaport in W Lancashire, in NW England: resort. 147,000.

Black′ Sea′, a sea between Europe and Asia, bordered by Turkey, Romania, Bulgaria, Ukraine, Georgia, and the Russian Federation. 164,000 sq. mi. (424,760 sq. km). Also called **Euxine Sea.** Ancient, **Pontus Euxinus.**

Black′ Vol′ta, a river in W Africa, in Ghana: the upper branch of the Volta River. ab. 500 mi. (800 km) long.

Bla·go·ve·shchensk (blä′gə vesh′ensk, -chensk), a city in the SE Russian Federation in Asia, on the Amur River. 202,000.

Blanc (blän), **Mont,** Mont Blanc.

Blan′ca Peak′ (blang′kə), a mountain in S Colorado: highest peak in the Sangre de Cristo Range. 14,390 ft. (4385 m).

Blan·tyre (blan tī°r′), a city in S Malawi. 355,200.

Blen·heim (blen′əm), a village in S Germany, on the Danube: victory of the Duke of Marlborough over the French, 1704. German, **Blindheim.**

Bli·da (blē′dä), a city in N Algeria. 191,314.

Blind·heim (blint′him′), German name of Blen-heim.

Block′ Is′land, an island off the coast of and a part of Rhode Island, at the E entrance to Long Island Sound.

Bloem·fon·tein (blōōm′fon tān′), the capital of the Orange Free State, in the central Republic of South Africa. 232,984.

Blois (blwa), a city in central France, on the Loire River. 51,950.

Bloo·ming·ton (blōō′ming tən), **1.** a city in SE Minnesota. 86,335. **2.** a city in S Indiana. 60,633. **3.** a city in central Illinois. 51,972.

Blooms·bur·y (blōōmz′bə rē, -brē), a district in central London, N of the Thames: a literary, artistic, and intellectual center in the early 20th century.

Blue′grass Re′gion, a region in central Kentucky, famous for its horse farms and fields of bluegrass. Also called **Blue′grass Coun′try.**

Blue′ Moun′tains, a range of low mountains in NE Oregon and SE Washington.

Blue′ Nile′, a river in E Africa, flowing NNW from

Lake Tana in Ethiopia into the Nile at Khartoum: a tributary of the Nile. ab. 950 mi. (1530 km) long. Compare NILE.

Blue′ Ridge′, a mountain range extending SW from N Virginia to N Georgia: part of the Appalachian Mountains. Also called **Blue′ Ridge′ Moun′tains.**

Bo·bruisk (bə broo̅′isk), a city in SE Belarus, SE of Minsk. 232,000.

Bo·ca Ra·ton (bō′kə rə tōn′), a city in SE Florida. 61,492.

Bo·chum (bō′кнŏŏm), a city in central North Rhine-Westphalia, in W Germany. 413,400.

Bo·den·see (bōd′n zā′), German name of Lake CONSTANCE.

Boe·o·tia (bē ō′shə), a district in ancient Greece, NW of Athens. *Cap.:* Thebes.

Bo·gaz·koy or **Bo·ğaz·köy** (bō′äz kœ′ē, -koi′), a village in N central Turkey: site of the ancient Hittite capital.

Bo·gor (bō′gôr), a city on W Java, in Indonesia. 247,409.

Bo·go·tá (bō′gə tä′), the capital of Colombia, in the central part. 3,982,941.

Bo·hai or **Po·hai** (bô′hī′), an arm of the Yellow Sea in NE China.

Bo·he·mi·a (bō hē′mē ə), a region in W Czechoslovakia: formerly a kingdom in central Europe; under Hapsburg rule 1526–1918. Czech, **Čechy.** —**Bo·he′mi·an,** *n., adj.*

Bohe′mia-Mora′via, a former German protectorate including Bohemia and Moravia, 1939–45.

Bo·hol (bō hôl′), an island in the S central Philippines. 1492 sq. mi. (3864 sq. km).

Bois′ de Boulogne′ (bwä′), a park W of Paris, France.

Boi·se (boi′zē *or, esp. locally,* -sē), the capital of Idaho, in the SW part. 125,738.

Bo·kha·ra (bō kär′ə), BUKHARA.

Boks·burg (boks′bûrg), a city in Transvaal, NE South Africa. 108,850.

Bo·liv·i·a (bə liv′ē ə, bō-), a republic in W South America. 7,000,000; 404,388 sq. mi. (1,047,370 sq. km). *Caps.:* La Paz and Sucre. —**Bo·liv′i·an,** *adj., n.*

Bo·lo·gna (bə lōn′yə), a city in N Italy. 459,080.

Bol·ton (bōl′tn), a borough in Greater Manchester, in NW England. 263,300.

Bol·za·no (bōl zä′nō, bōlt sä′-), a city in NE Italy. 101,230.

Bom·bay (bom bā′), a seaport in and the capital of Maharashtra, in W India, on the Arabian Sea. 8,227,000.

Bo·mu (bō′mōō) also **Mbomu,** a river in central Africa, forming part of the boundary between Zaire and the Central African Republic, flowing N and W into the Uele River to form the Ubangi River. ab. 500 mi. (805 km) long.

Bon·aire (bô nâr′), an island in the E Netherlands Antilles, in the S West Indies. 9137; 95 sq. mi. (245 sq. km).

Bône (bōn), former name of ANNABA.

Bo′nin Is′lands (bō′nin), a group of islands in the N Pacific, SE of and belonging to Japan: under U.S. administration 1945–68. 40 sq. mi. (104 sq. km).

Bonn (bon, bôn), a city in W Germany, on the Rhine: seat of the government. 291,400.

Boo·thi·a (bōō′thē ə), **1.** a peninsula in N Canada: the northernmost part of the mainland of North America; former location of the north magnetic pole. **2. Gulf of,** a gulf between this peninsula and Baffin Island.

Bo·phu·that·swa·na (bō′pōō tät swä′nə), a self-governing black homeland in South Africa, consisting of several noncontiguous enclaves in the N central part: granted independence in 1977. 1,660,000; 16,988 sq. mi. (44,000 sq. km). *Cap.:* Mmabatho.

Bo·ra Bo·ra (bôr′ə bôr′ə; bōr′ə bōr′ə), an island in the Society Islands, in the S Pacific, NW of Tahiti. ab. 2000; 15 sq. mi. (39 sq. km).

Bo·rås (bōō rôs′), a city in S Sweden, near Göteborg. 102,129.

Bor·deaux (bôr dō′), a seaport in SW France, on the Garonne River. 226,281.

Bor·ders (bôr′dərz), a region in SE Scotland. 102,141; 1804 sq. mi. (4671 sq. km).

Bor·ne·o (bôr′nē ō′), an island in the Malay Archipelago, politically divided among Indonesia, Malaysia, and Brunei. 290,000 sq. mi. (750,000 sq. km). —**Bor′ne·an,** *adj., n.*

Born·holm (bôrn′hōm, -hōlm), a Danish island

in the Baltic Sea, S of Sweden. 47,126; 227 sq. mi. (588 sq. km).

Bo·ro·di·no (bôr′ə dē′nō, bor′-), a village in the W Russian Federation, 70 mi. (113 km) W of Moscow: Napoleon's victory here made possible the capture of Moscow, 1812.

Bos·ni·a (boz′nē ə), a historic region in SE Europe: a former Turkish province; a part of Austria 1879–1918; now part of Bosnia and Herzegovina. —**Bos′ni·an,** *adj., n.*

Bos′nia and Herzegovi′na, a republic in SE Europe: formerly part of Yugoslavia. 4,360,000; 19,741 sq. mi. (51,129 sq. km). *Cap.:* Sarajevo.

Bos·po·rus (bos′pər əs) also **Bos·pho·rus** (-fər-), a strait connecting the Black Sea and the Sea of Marmara. 18 mi. (29 km) long.

Bos′sier Cit′y (bō′zhər), a city in NW Louisiana. 52,721.

Bos·ton (bô′stən, bos′tən), the capital of Massachusetts, in the E part. 574,283. —**Bos·to′ni·an** (-stō′nē ən), *adj., n.*

Bot′any Bay′, a bay on the SE coast of Australia, near Sydney: site of early British penal colony.

Both·ni·a (both′nē ə), **Gulf of,** an arm of the Baltic Sea, extending N between Sweden and Finland. ab. 400 mi. (645 km) long.

Bot·swa·na (bot swä′nə), a republic in S Africa: formerly a British protectorate; gained independence 1966; member of the Commonwealth of Nations. 1,210,000; 275,000 sq. mi. (712,250 sq. km). *Cap.:* Gaborone. Formerly, **Bechuanaland.**

Bot·trop (bot′rop), a city in W Germany, in the Ruhr region. 112,300.

Boua·ké (bwä kā′, bwä′kā), a city in central Ivory Coast. 200,000.

Bou·gain·ville (boo′gən vil′, bō′-), the largest of the Solomon Islands, in the W Pacific Ocean: part of Papua New Guinea. 4080 sq. mi. (10,567 sq. km).

Boul·der (bōl′dər), a city in N central Colorado. 83,312.

Boul′der Can′yon, a canyon of the Colorado River between Arizona and Nevada, above Boulder Dam.

Boul′der Dam′, a dam on the Colorado River, on the boundary between SE Nevada and NW Arizona. 726 ft. (221 m) high; 1244 ft. (379 m) long. Official name, **Hoover Dam.**

Bou·logne (bŏŏ lōn′, -loin′; *Fr.* -lôn/yᵉ), a seaport in N France, on the English Channel. 49,284. Also called **Boulogne/-sur-Mer/** (-ѕʏʀ мᴇʀ′).

Boulogne/ Bil·lan·court/ (bē yän kŏŏR′), a suburb of Paris, in N France. 103,948. Also called **Boulogne/-sur-Seine/** (-ѕʏʀ sen′).

Bourges (bŏŏrzh), a city in central France: cathedral. 80,379.

Bour·gogne (bŏŏr gôn/yᵉ), French name of Bᴜʀ-ɢᴜɴᴅʏ.

Bourne·mouth (bŏrn′məth, bôrn′-, bōrn′-), a city in Dorset in S England: seashore resort. 154,200.

Bow·er·y (bou/ə rē, bou/rē), **the,** a street and area in New York City, noted for its cheap bars and flophouses.

Bowl/ing Green/, a city in S Kentucky. 40,450.

Boyne (boin), a river in E Ireland: William III defeated James II near here 1690. 70 mi. (110 km) long.

Bo·yo/ma Falls/ (bô yō′mə), seven cataracts of the Lualaba River where it becomes the Zaire (Congo) River, in NE Zaire, S of Kisangani. Formerly, **Stanley Falls.**

Boz·ca·a·da (bōz/jä ä dä′, -jä dä′), an island belonging to Turkey in the NE Aegean, near the entrance to the Dardanelles. Ancient, **Tenedos.**

Bra·bant (brə bant′, -bän′, brä′bənt), **1.** a former duchy in W Europe, now divided between the Netherlands and Belgium. **2.** a province in central Belgium. 2,220,088; 1268 sq. mi. (3285 sq. km). *Cap.:* Brussels. —**Bra·bant/ine** (-ban′tin, -tin), *adj.*

Brack·nell (brak′nl), a town in E Berkshire, in S England. 95,000.

Brad·ford (brad′fərd), a city in West Yorkshire, in N England. 460,600.

Bra·ga (brä/gə), a city in N Portugal: an ecclesiastical center. 63,033.

Brah·ma·pu·tra (brä/mə pŏŏ′trə), a river in S Asia, flowing from S Tibet through NE India and joining the Ganges River in Bangladesh. ab. 1700 mi. (2700 km) long.

Brä·i·la (brə ē/lä), a port in E Romania, on the Danube River. 234,600.

Brak·pan (brak′pan′), a city in the NE Republic of South Africa, near Johannesburg. 85,044.

Bramp·ton (bramp′tən), a city in SE Ontario, in S Canada, near Toronto. 188,498.

Bran·den·burg (bran′dən bûrg′), **1.** a state in NE central Germany. 2,700,000; 10,039 sq. mi. (26,000 sq. km). *Cap.:* Potsdam. **2.** a city in NE Germany. 95,203. —**Bran′den·burg′er,** *n.*

Bran·don (bran′dən), a city in SW Manitoba, in S central Canada. 38,708.

Bran·dy·wine (bran′dē win′), a creek in SE Pennsylvania and N Delaware: British defeat of Americans 1777.

Brant·ford (brant′fərd), a city in S Ontario, in SE Canada, near Lake Erie. 76,146.

Bra·sil (*Port.* brə zēl′), Brazil.

Bra·síl·ia (brə zil′yə), the capital of Brazil, on the central plateau. 411,505.

Bra·şov (brä shôv′), a city in central Romania. 346,640.

Bra·ti·sla·va (brat′ə slä′və, brä′tə-), the capital of Slovakia, in S central Czechoslovakia, on the Danube River. 435,000.

Bratsk (brätsk), a city in the S central Russian Federation in Asia, on the Angara River. 249,000.

Braun·schweig (broun′shvīk′), German name of BRUNSWICK.

Bra·zil (brə zil′), a federal republic in South America. 155,560,000; 3,286,170 sq. mi. (8,511,180 sq. km). *Cap.:* Brasília. Official name, **Fed′erative Repub′lic of Brazil′.** Portuguese, **Brasil.** —**Bra·zil′ian,** *adj., n.*

Bra·zos (braz′əs, brä′zəs), a river flowing SE from N Texas to the Gulf of Mexico. 870 mi. (1400 km) long.

Braz·za·ville (braz′ə vil′, brä′zə-), the capital of the People's Republic of the Congo, in the S part, on the Congo (Zaire) River. 585,812.

Breck·nock·shire (brek′nək shēr′, -shər, -nok-), a historic county in S Wales, now part of Powys, Gwent, and Mid Glamorgan.

Bre·da (brā dä′), a city in the S Netherlands. 156,173.

Breed's′ Hill′ (brēdz), a hill adjoining Bunker Hill, where the Battle of Bunker Hill was actually fought.

Bre·genz (brā′gents), a city in W Austria, on Lake Constance. 24,683.

Brem·en (brem′ən, brā′mən), **1.** a state in NW Germany. 654,000; 156 sq. mi. (405 sq. km). **2.** the capital of this state, on the Weser River. 522,000.

Brem·er·ha·ven (brem′ər hā′vən, brā′-), a seaport in NW Germany, at the mouth of the Weser River. 132,200.

Brem·er·ton (brem′ər tən), a city in W Washington, on Puget Sound: navy yard. 36,208.

Bren′ner Pass′ (bren′ər), a mountain pass in the Alps, on the border between Italy and Austria. 4494 ft. (1370 m) high.

Brent (brent), a borough of Greater London, England. 262,800.

Bre·scia (bre′shə), a city in central Lombardy, in N Italy. 198,839. —**Bre′scian,** *adj.*

Bres·lau (brez′lou, bres′-), German name of Wroclaw.

Brest (brest), **1.** a seaport in the W extremity of France. 160,355. **2.** Formerly, **Brest Litovsk.** a city in SW Belarus, on the Bug River: formerly in Poland; German-Russian peace treaty 1918. 238,000.

Brest′ Li·tovsk′ (li tôfsk′), former name (until 1921) of Brest (def. 2).

Bre·tagne (brə taN′y°), French name of Brittany.

Bridge·port (brij′pôrt′, -pōrt′), a seaport in SW Connecticut, on Long Island Sound. 141,686.

Bridge·town (brij′toun′), the capital of Barbados, on the SW coast. 7466.

Brie (brē), a region in NE France, between the Seine and the Marne.

Bri·enz (brē ents′), **Lake of,** a lake in SE Bern canton in Switzerland. 11.5 sq. mi. (30 sq. km). German, **Bri·enz·er See** (brē en′tsər zā′).

Brigh·ton (brīt′n), a city in East Sussex, in SE England: seashore resort. 140,900.

Brin·di·si (brin′də zē′, brēn′-), an Adriatic seaport in SE Apulia, in SE Italy. 87,420.

Bris·bane (briz′bān, -bən), the capital of Queensland, in E Australia. 1,171,300.

Bris·tol (bris′tl), **1.** a seaport in Avon, in SW England, on the Avon River near its confluence with the Severn estuary. 420,100. **2.** a city in central Connecticut. 60,640.

Bris/tol Chan/nel, an inlet of the Atlantic, between S Wales and SW England, extending to the mouth of the Severn estuary. 85 mi. (137 km) long.

Brit·ain (brit/n), **1.** GREAT BRITAIN. **2.** BRITANNIA (def. 1). —**Brit/ish,** *adj., n.* —**Brit·on,** *n.*

Bri·tan·ni·a (bri tan/ē ə, -tan/yə), **1.** the ancient Roman name of the island of Great Britain, esp. the S part where the early Roman provinces were. **2.** Great Britain or the British Empire.

Brit/ish Antarc/tic Ter/ritory, a British colony in the S Atlantic, comprising the South Shetland Islands, the South Orkney Islands, and Graham Land: formerly dependencies of the Falkland Islands.

Brit/ish Cameroons/, CAMEROONS (def. 2).

Brit/ish Colum/bia, a province in W Canada on the Pacific coast. 2,883,367; 366,255 sq. mi. (948,600 sq. km). *Cap.:* Victoria. —**Brit/ish Colum/bian,** *n., adj.*

Brit/ish Com/monwealth of Na/tions, former name of the COMMONWEALTH OF NATIONS. Also called **Brit/ish Com/monwealth.**

Brit/ish East/ Af/rica, the former British territories of Kenya, Uganda, and Tanzania.

Brit/ish Em/pire, (formerly) the United Kingdom and the territories under the leadership or control of the British crown.

Brit/ish Guia/na, former name of GUYANA.

Brit/ish Hondu/ras, former name of BELIZE (def. 1). —**Brit/ish Hondu/ran,** *adj., n.*

Brit/ish In/dia, a part of India, comprising 17 provinces, that prior to 1947 was subject to British law: now divided among India, Pakistan, and Bangladesh.

Brit/ish In/dian O/cean Ter/ritory, a British colony in the Indian Ocean, consisting of the Chagos Archipelago. 76 sq. mi. (177 sq. km).

Brit/ish Isles/, a group of islands in W Europe: Great Britain, Ireland, the Isle of Man, and adjacent small islands. 120,592 sq. mi. (312,300 sq. km).

Brit/ish Malay/a, the former British possessions on the Malay Peninsula and the Malay Archipelago: now part of Malaysia.

Brit′ish Soma′liland, a former British protectorate in E Africa, on the Gulf of Aden: now the N part of Somalia.

Brit′ish Vir′gin Is′lands, a British colony comprising several small islands in the West Indies, E of Puerto Rico. 13,246; 67 sq. mi. (174 sq. km). *Cap.:* Road Town.

Brit′ish West′ In′dies, (formerly) the possessions of Great Britain in the West Indies. Compare WEST INDIES (def. 2).

Brit·ta·ny (brit′n ē), **1.** a historic region in NW France, on a peninsula between the English Channel and the Bay of Biscay: a former duchy and province. **2.** a metropolitan region in NW France. 2,764,200; 10,505 sq. mi. (27,208 sq. km). French, **Bretagne.**

Br·no (bûr′nō), a city in central Czechoslovakia. 390,000.

Broads (brôdz), **the,** a low-lying region in E England, in Norfolk and Suffolk: bogs and marshy lakes.

Broad·way (brôd′wā′), a street in New York City, famous for its theaters, restaurants, and bright lights.

Brock·en (brok′ən), a mountain in N central Germany: the highest peak in the Harz Mountains. 3745 ft. (1141 m).

Brock·ton (brok′tən), a city in E Massachusetts. 92,788.

Bro′ken Ar′row, a town in NE Oklahoma. 58,043.

Bro′ken Hill′, former name of KABWE.

Brom·ley (brom′lē, brum′-), a borough of Greater London, England. 294,900.

Bronx (brongks), **the,** a borough of New York City, N of Manhattan. 1,203,789; 43.4 sq. mi. (112 sq. km). **—Bronx′ite,** *n.*

Brook·line (brŏok′lin′), a town in E Massachusetts, near Boston. 55,062.

Brook·lyn (brŏok′lin), a borough of New York City, on W Long Island. 2,300,664; 76.4 sq. mi. (198 sq. km). **—Brook·lyn·ite′,** *n.*

Brook′lyn Park′, a town in central Minnesota. 56,381.

Brooks′ Range′ (brŏoks), a mountain range in N Alaska, forming a watershed between the Yukon River and the Arctic Ocean: highest peak, 9239 ft. (2815 m).

Bros·sard (brô särd′, -sär′), a town in S Quebec, in E Canada: suburb of Montreal. 57,441.

Browns·ville (brounz′vil), a seaport in S Texas, near the mouth of the Rio Grande. 98,962.

Bru·ges (brōō′jiz, brōōzh), a city in NW Belgium: connected by canal with its seaport, Zeebrugge. 119,718. Flemish, **Brug·ge** (brœкн′ə).

Bru·nei (brōō nī′, -nā′), an independent sultanate on the NW coast of Borneo: a former British protectorate (1889–1983). 241,400; 2226 sq. mi. (5765 sq. km). *Cap.:* Bandar Seri Begawan. Official name, **Brunei′ Da·rus·sa·lam′** (dä′rōō sä-läm′). —**Bru·nei′an,** *adj., n.*

Bruns·wick (brunz′wik), **1.** a former state of Germany: now part of Lower Saxony in Germany. **2.** a city in Lower Saxony, in N central Germany. 247,800. German, **Braunschweig.**

Brus·sels (brus′əlz), the capital of Belgium, in the central part. 1,050,787 (with suburbs). Flemish, **Brus·sel** (brʏs′əl); French, **Brux·elles** (brʏ-sel′, brʏk-).

Bry·an (brī′ən), a city in E Texas. 55,002.

Bry·ansk (brē änsk′), a city in the W Russian Federation, on the Desna river, SW of Moscow. 445,000.

Bryce′ Can′yon Na′tional Park′ (bris), a national park in SW Utah: colorful rock formations. 56 sq. mi. (145 sq. km).

Bu·ca·ra·man·ga (bōō′kär ə mäng′gä), a city in N Colombia. 352,326.

Bu·cha·rest (bōō′kə rest′, byōō′-), the capital of Romania, in the S part. 1,975,808. Romanian, **Bucureşti.**

Bu·chen·wald (bōō′kən wôld′, -vält′, -кнən-), the site of a former Nazi concentration camp in central Germany, near Weimar.

Buck·ing·ham·shire (buk′ing əm shēr′, -shər), a county in S England. 621,300; 294 sq. mi. (761 sq. km). Also called **Buck′ing·ham, Bucks** (buks).

Bu·co·vi·na or **Bu·ko·vi·na** (bōō′kə vē′nə), a region in E central Europe, formerly a district in N Romania: now divided between Romania and Ukraine. 4031 sq. mi. (10,440 sq. km).

Bu·cu·reşti (bōō kŏō resht′), Romanian name of BUCHAREST.

Bu·da·pest (bōō′də pest′, -pesht′, bŏŏd′ə-), the capital of Hungary, in the central part, on the Danube. 2,104,000.

Buddh Ga·ya (bŏŏd′ gə yä′), a village in central Bihar, in NE India: site of tree under which Siddhartha became the Buddha.

Bud·weis (bŏŏt′vis), German name of ČESKÉ BUDĚJOVICE.

Bue′na Park′ (byōō′nə), a city in SW California. 68,784.

Bue·na·ven·tu·ra (bwä′nə ven tŏŏr′ə, -tyŏŏr′ə), a seaport in W Colombia. 193,185.

Bue·nos Ai·res (bwā′nəs i°r′iz, bō′nəs), the capital of Argentina, in the E part, on the Río de la Plata. 9,927,404.

Buf·fa·lo (buf′ə lō′), a port in W New York, on Lake Erie. 328,123.

Bug (bŏŏg, bŏŏk), **1.** a river in E central Europe, rising in W Ukraine and forming part of the boundary between Poland and Ukraine, flowing NW to the Vistula in Poland. 450 mi. (725 km) long. **2.** a river in SW Ukraine, flowing SE to the Dnieper estuary. ab. 530 mi. (850 km) long.

Bu·gan·da (bŏŏ gan′də, byŏŏ-), a historic kingdom of East Africa, located N of Lake Victoria and W of the Nile in Uganda.

Bu·jum·bu·ra (bōō′jŏŏm bŏŏr′ə), the capital of Burundi, in the W part, on Lake Tanganyika. 272,600. Formerly, **Usumbura.**

Bu·ka·vu (bŏŏ kä′vōō), a city in E Zaire. 180,633. Formerly, **Costermansville.**

Bu·kha·ra (bŏŏ kär′ə, bōō-) also **Bokhara, 1.** a city in S central Uzbekistan, W of Samarkand. 220,000. **2.** a former khanate in SW Asia: now incorporated into Uzbekistan.

Bu·ko·vi·na (bōō′kə vē′nə), BUCOVINA.

Bu·la·wa·yo (bōō′lə wä′yō, -wä′-), a city in SW Zimbabwe: mining center. 414,800.

Bul·gar·i·a (bul gâr′ē ə, bŏŏl-), a republic in SE Europe. 8,761,000; 42,800 sq. mi. (110,850 sq. km). *Cap.:* Sofia. —**Bul·gar′i·an,** *n., adj.*

Bull′ Run′, a creek in NE Virginia: Union forces defeated near here in major Civil War battles 1861, 1862.

Bun′ker Hill′, a hill in Charlestown, Mass.: the

first major battle of the American Revolution, known as the Battle of Bunker Hill, was fought on adjoining Breed's Hill on June 17, 1775.

Bur·bank (bûr/bangk/), a city in SW California. 93,643.

Bur·gas (bŏŏr gäs/), a seaport in E Bulgaria, on the Black Sea. 197,555.

Bur·gen·land (bŏŏr/gən länt/, bûr/gən land/), a province in E Austria, bordering Hungary. 272,274; 1530 sq. mi. (3960 sq. km).

Bur·gos (bŏŏr/gōs), a city in N Spain. 163,910.

Bur·gun·dy (bûr/gən dē), **1.** a historic region in central France: a former kingdom, duchy, and province. **2.** a metropolitan region in central France. 1,607,200; 12,194 sq. mi. (31,582 sq. km). French, **Bourgogne.** —**Bur·gun·di·an** (bər-gun/dē ən), *adj., n.*

Bur·ki·na Fa·so (bər kē/nə fä/sō), a republic in W Africa: formerly part of French West Africa. 8,530,000; 106,111 sq. mi. (274,827 sq. km). *Cap.:* Ouagadougou. Formerly, **Upper Volta.**

Bur·ling·ton (bûr/ling tən), **1.** a city in S Ontario, in S Canada, on Lake Ontario. 116,675. **2.** a city in NW Vermont, on Lake Champlain. 37,712.

Bur·ma (bûr/mə), a republic in SE Asia, on the Bay of Bengal. 39,840,000; 261,789 sq. mi. (678,034 sq. km). *Cap.:* Yangon. Official name, **Union of Myanmar.** —**Bur·mese/**, *n., pl.* -**mese**, *adj.*

Burn·ley (bûrn/lē), a city in E Lancashire, in NW England. 92,700.

Burns·ville (bûrnz/vil), a city in SE Minnesota. 51,288.

Bur·sa (bŏŏr sä/) a city in NW Turkey in Asia: a former capital of the Ottoman Empire. 614,133.

Bu·run·di (bŏŏ rŏŏn/dē), a republic in central Africa, E of Zaire: formerly the S part of the Belgian trust territory of Ruanda-Urundi; gained independence 1962. 5,130,000; 10,747 sq. mi. (27,834 sq. km). *Cap.:* Bujumbura. —**Bu·run/di·an,** *adj., n.*

Bur·yat/ Auton/omous Repub/lic (bŏŏr yät/, bŏŏr/ē ät/), an automomous republic in the Russian Federation in Asia, E of Lake Baikal. 1,042,000; ab. 135,650 sq. mi. (351,300 sq. km). *Cap.:* Ulan Ude.

Bur·y St. Ed·munds (ber/ē sänt ed/məndz,

-sənt-), a city in W Suffolk, in E England: medieval shrine. 25,629.

Bus·ra or **Bus·rah** (bus/rə), BASRA.

Bute (byōōt), **1.** Also, **Bute·shire** (byōōt/shēr, -shər). a historic county in SW Scotland, composed of three islands in the Firth of Clyde. **2.** an island in the Firth of Clyde, in SW Scotland: part of the county Bute. 50 sq. mi. (130 sq. km).

Butte (byōōt), a city in SW Montana: mining center. 33,380.

Bu·tu·an (bə tōō/än), a city in the Philippines, on NE Mindanao. 228,000.

Bu·tung (bōō/tŏŏng), an island of Indonesia, SE of Sulawesi Island. 100 mi. (161 km) long.

Buz/zard's Bay/, an inlet of the Atlantic, in SE Massachusetts. 30 mi. (48 km) long.

Byd·goszcz (bid/gôshch), a city in N Poland. 361,000.

Bye·lo·rus·sia (byel/ə rush/ə, bel/ə-), BELARUS. —**Bye/lo·rus/sian,** *adj., n.*

Bye·lo·stok (*Russ.* byi lu stôk/), BIALYSTOK.

By·tom (bē/tôm), a city in S Poland. 239,000.

Byz/an·tine Em/pire (biz/ən tēn/), the Eastern Roman Empire after the fall of the Western Empire in A.D. 476: became extinct after the fall of Constantinople, its capital, in 1453.

By·zan·ti·um (bi zan/shē əm, -tē əm), an ancient Greek city on the Bosporus and the Sea of Marmara: rebuilt by Constantine I and renamed Constantinople A.D. 330. Compare ISTANBUL.

C

CA, California.

Ca·ba·na·tuan (kä′vä nä twän′), a city on central Luzon, in the N Philippines. 173,000.

Ca·bi·mas (kə bē′məs), a city in NW Venezuela, on the E coast of Lake Maracaibo. 135,529.

Ca·bin·da (kə bēn′də), an exclave of Angola, on the W coast of Africa. 114,000; 2807 sq. mi. (7270 sq. km).

Ca·diz (kä′dēs), a city in the Philippines, on N Negros. 120,000.

Cá·diz (kə diz′, kā′diz; *Sp.* kä′thēth, -thēs), a seaport in SW Spain, on a bay of the Atlantic **(Gulf′ of Cá·diz′).** 154,051.

Cae·li·an (sē′lē ən), the southeasternmost of the seven hills on which ancient Rome was built.

Caen (kän, kän), a city in NW France, SW of Le Havre. 117,119.

Caer·nar·von·shire (kär när′vən shēr′, -shēr), a historic county in Gwynedd, in NW Wales. Also called **Caer·nar′von.**

Caes·a·re·a (sē′zə rē′ə, ses′ə-, sez′ə-), **1.** an ancient seaport in NW Israel: Roman capital of Palestine. **2.** ancient name of KAYSERI.

Ca·ga·yan de O·ro (kä′gä yän′ dā ôr′ō), a city in the Philippines, on NW Mindanao. 340,000.

Ca·glia·ri (käl′yə rē), a seaport in S Sardinia. 221,790.

Ca·guas (kä′gwäs), a city in E central Puerto Rico. 126,298.

Ca·ho′ki·a Mounds′ (kə hō′kē ə), a group of very large prehistoric Indian earthworks in SW Illinois.

Cai′cos Is′lands (kī′kōs, kā′-), TURKS AND CAICOS ISLANDS.

Cai·ro (kī′rō), the capital of Egypt, in the N part on the E bank of the Nile. 6,325,000.

Caith·ness (kāth′nes), a historic county in NE Scotland. Also called **Caith′ness·shire′** (-shēr′, -shər).

Cal., California.

Cal·a·bar (kal′ə bär′, kal′ə bär′), a seaport in SE Nigeria. 126,000.

Ca·la·bri·a (kə lä′brē ə, -lä′-), **1.** a region in S Italy. 2,145,724. 5828 sq. mi. (15,100 sq. km).

Cap.: Reggio Calabria. **2.** an ancient district at the extreme SE of the Italian peninsula. —**Ca·la′-bri·an,** *n., adj.*

Cal·ais (kal′ā, ka lā′), a seaport in N France, on the Strait of Dover: the French port nearest England. 76,935.

Cal·cut·ta (kal kut′ə), the capital of West Bengal state, in E India, on the Hooghly River: former capital of British India. 9,166,000.

Cal·e·do·ni·a (kal′i dō′nē ə), *Chiefly Literary.* Scotland. —**Cal′e·do′ni·an,** *n., adj.*

Caledo′nian Canal′, a canal in N Scotland, extending NE from the Atlantic to the North Sea. 60½ mi. (97 km) long.

Cal·ga·ry (kal′gə rē), a city in S Alberta, in SW Canada. 636,104.

Ca·li (kä′lē), a city in SW Colombia. 1,350,565.

Cal·i·cut (kal′i kut′), a seaport in W Kerala, in SW India. 546,000. Formerly, **Kozhikode.**

Calif., California.

Cal·i·for·nia (kal′ə fôrn′yə, -fôr′nē ə), **1.** a state in W United States, on the Pacific coast. 29,760,021; 158,693 sq. mi. (411,015 sq. km). *Cap.:* Sacramento. *Abbr.:* CA, Cal., Calif. **2. Gulf of,** an arm of the Pacific Ocean, extending NW between the coast of W Mexico and the peninsula of Baja California. ab. 750 mi. (1207 km) long; 62,600 sq. mi. (162,100 sq. km). —**Cal′i·for′-nian,** *adj., n.*

Ca·llao (kä you′), a seaport in W Peru, near Lima. 560,000.

Ca·lo·o·can (kal′ə ō′kän, kä′lə-), a city in the Philippines, on SW Luzon. 746,000.

Cal·pe (kal′pē), ancient name of the Rock of GIBRALTAR.

Cal·va·ry (kal′və rē), the place where Jesus was crucified, near Jerusalem.

Cal·y·don (kal′i don′), an ancient city in W Greece, in Aetolia. —**Cal′y·do′ni·an** (-dō′nē ən), *adj.*

Cam (kam), a river in E England flowing NE by Cambridge, into the Ouse River. 40 mi. (64 km) long. Also called **Granta.**

Ca·ma·güey (kam′ə gwā′), a city in central Cuba. 260,800.

Cam·a·ril·lo (kam′ə ril′ō), a city in SW California. 52,303.

Cam·bo·di·a (kam bōʹdē ə), **State of**, a republic in SE Asia: formerly part of French Indochina. 6,230,000; 69,866 sq. mi. (180,953 sq. km). *Cap.:* Phnom Penh. Formerly, **People's Republic of Kampuchea, Khmer Republic.** —**Cam·boʹdi·an**, *adj.*, *n.*

Cam·bri·a (kamʹbrē ə), medieval name of WALES.

Cam·bridge (kāmʹbrij), **1.** a city in Cambridgeshire, in E England: famous university founded in 12th century. 98,400. **2.** a city in E Massachusetts, near Boston. 95,802. **3.** CAMBRIDGESHIRE. **4.** a city in SE Ontario, in S Canada. 79,920. —**Can·ta·brig·i·an** (kanʹtə brijʹē ən), *adj.*, *n.*

Cam·bridge·shire (-sherʹ, -shər), a county in E England. 642,400; 1316 sq. mi. (3410 sq. km). Also called **Cambridge.**

Cam·den (kamʹdən), **1.** a borough of Greater London, England. 184,900. **2.** a port in SW New Jersey, on the Delaware River opposite Philadelphia. 87,492.

Cam·e·roon (kamʹə roonʹ), **1.** a republic in W equatorial Africa: formed in 1960 by the French trusteeship of Cameroun; joined in 1961 by the S part of the British trusteeship of Cameroons. 11,000,000; 179,558 sq. mi. (465,054 sq. km). *Cap.:* Yaoundé. **2. Mount,** an active volcano in W Cameroon: highest peak on the coast of W Africa. 13,370 ft. (4075 m). French, **Cameroun.**

Cam·e·roons (kamʹə roonzʹ), (*used with a sing. v.*) **1.** a region in W Africa: a German protectorate 1884–1919; divided in 1919 into British and French mandates. **2.** Also called **British Cameroons.** a former British mandate (1919–46) and trusteeship (1946–60) in W Africa: by a 1961 plebiscite the S part joined Cameroon and the N part joined Nigeria. —**Cam·e·roonʹi·an**, *adj.*, *n.*

Came·roun (kam roonʹ; *Fr.* kam RŌŌNʹ), **1.** CAMEROON. **2.** Also called **French Cameroons.** a former French mandate (1919–46) and trusteeship (1946–60) in W Africa: gained independence 1960; now part of Cameroon.

Cam·pa·gna (kam pänʹyə, kəm-), a low plain surrounding the city of Rome, Italy.

Cam·pa·ni·a (kam päʹnē ə, -pänʹyə, käm päʹnyə), a region in SW Italy. 5,463,134; 5214 sq. mi. (13,505 sq. km). *Cap.:* Naples. —**Cam·paʹni·an**, *adj.*, *n.*

Cam·pe·che (käm pe/che), **1.** a state in SE Mexico, on the peninsula of Yucatán. 592,933; 19,672 sq. mi. (50,950 sq. km). **2.** the capital of this state. 151,805. **3.** Gulf of, the SW part of the Gulf of Mexico.

Cam·pi·na Gran·de (kam pē/nə gran/də, -dē), a city in NE Brazil. 222,102.

Cam·pi·nas (kam pē/nəs), a city in SE Brazil, NNW of São Paulo. 566,627.

Cam·po·bel·lo (kam/pə bel/ō), an island in SE Canada, in New Brunswick province.

Cam·po Gran·de (kän/pŏŏ grän/də), the capital of Mato Grosso do Sul, in SW Brazil. 282,857.

Cam·pos (kam/pəs), a city in E Brazil, near Rio de Janeiro. 174,218.

Ca·na (kā/nə), an ancient town in N Israel, in Galilee: scene of Jesus' first miracle.

Ca·naan (kā/nən), **1.** the ancient region lying between the Jordan, the Dead Sea, and the Mediterranean: the land promised by God to Abraham. **2.** Biblical name of PALESTINE. —**Ca/naan·ite/**, *n.*, *adj.*

Can·a·da (kan/ə də), a nation in N North America: a member of the Commonwealth of Nations. 25,354,064; 3,690,410 sq. mi. (9,558,160 sq. km). *Cap.:* Ottawa. —**Ca·na·di·an** (kə nā/dē ən), *n.*, *adj.*

Cana/dian Riv/er, a river flowing E from the Rocky Mountains in NE New Mexico to the Arkansas River in E Oklahoma. 906 mi. (1460 km) long.

Canal/ Zone/, a zone in central Panama, including the Panama Canal: governed by the U.S. 1903–1979; partial control of the zone was returned to Panama, entire control to be returned by 2000; ab. 10 mi. (16 km) wide; excludes the cities of Panama and Colón. *Abbr.:* CZ, C.Z.

Canar/y Is/lands, a group of mountainous islands in the Atlantic Ocean, near the NW coast of Africa, comprising two provinces of Spain. 1,614,882; 2894 sq. mi. (7495 sq. km). Spanish, **Islas Canarias.** —**Ca·nar/i·an,** *adj., n.*

Ca·nav·er·al (kə nav/ər əl), **Cape,** a cape on the E coast of Florida: site of John F. Kennedy Space Center. Formerly (1963–73), **Cape Kennedy.**

Can·ber·ra (kan/ber ə, -bər ə), the capital of

Australia, in the SE part, in the Australian Capital Territory. 285,800 (with suburbs).

Can·cún (kan kōōn′, käng kōōn′), an island off NE Quintana Roo state, on the Yucatán Peninsula, in SE Mexico: beach resort.

Can·di·a (kan′dē ə), **1.** IRAKLION. **2.** CRETE.

Ca·ne·a (kə nē′ə), the capital of Crete, on the W part. 47,338. Greek, **Khania.**

Can·nae (kan′ē), an ancient town in SE Italy: Hannibal defeated the Romans here 216 B.C.

Cannes (kan), a city in SE France, on the Mediterranean Sea: resort; annual film festival. 72,787.

Ca·no·as (kə nō′əs), a city in SE Brazil, N of Pôrto Alegre. 214,000.

Ca·no·pus (kə nō′pəs), an ancient seacoast city in Lower Egypt, 15 mi. (24 km) E of Alexandria.

Can·so (kan′sō), **1. Cape,** a cape in SE Canada, the NE extremity of Nova Scotia. **2. Strait of.** Also called **Gut of Canso.** a channel in SE Canada that separates mainland Nova Scotia from Cape Breton Island, flowing NW from the Atlantic Ocean to Northumberland Strait. ab. 17 mi. (27 km) long and 1 mi. (1.6 km) wide.

Can·ter·bur·y (kan′tər ber′ē, -bə rē; *esp. Brit.* -brē), **1.** a city in E Kent, in SE England: early ecclesiastical center of England. 129,500. **2.** a municipality in E New South Wales, in SE Australia: suburb of Sydney. 115,100. —**Can′ter·bu′ri·an** (-byŏŏr′ē ən), *adj.*

Can·ti·gny (kän tē nyē′), a village in N France, S of Amiens: first major battle of U.S. forces in World War I, May 1918.

Can·ton (kan ton′, kan′ton *for 1;* kan′tn *for 2*), **1.** GUANGZHOU. **2.** a city in NE Ohio: location of the football Hall of Fame. 84,161.

Can′yon·lands Na′tional Park′ (kan′yən-landz′), a national park in SE Utah, at the junction of the Colorado and Green rivers: canyons, rock formations, and petroglyphs. 527 sq. mi. (1366 sq. km).

Cape′ Bret′on (brit′n, bret′n), an island forming the NE part of Nova Scotia, in SE Canada. 42,969; 3970 sq. mi. (10,280 sq. km).

Cape′ Canav′eral, CANAVERAL, Cape.

Cape′ Cod′, a sandy peninsula in SE Massachusetts between Cape Cod Bay and the Atlantic Ocean: many resort towns.

Cape′ Cod′ Bay′, a part of Massachusetts Bay, enclosed by the Cape Cod peninsula.

Cape′ Col′ony, former name of CAPE OF GOOD HOPE (def. 2).

Cape′ Cor′al, a city in SE Florida. 74,991.

Cape′ Fear′, 1. a river in SE North Carolina. 202 mi. (325 km) long. **2.** FEAR, Cape

Cape′ Horn′, a headland on a small island at the S extremity of South America: belongs to Chile.

Cape′ May′, a city in S New Jersey: seashore resort. 4853.

Cape′ of Good′ Hope′, 1. a cape in S Africa, in the SW Republic of South Africa. **2.** Also called **Cape′ Prov′ince.** Formerly, **Cape Colony.** a province in the Republic of South Africa. 7,443,500; 277,169 sq. mi. (717,868 sq. km). *Cap.:* Cape Town.

Ca·per·na·um (kə pûr′nā əm, -nē-), an ancient site in N Israel, on the Sea of Galilee: center of Jesus' ministry in Galilee.

Cape′ Town′, the legislative capital of the Republic of South Africa, in the SW part: also capital of Cape of Good Hope province. 789,580. —**Cape·to′ni·an** (-tō′nē ən), *n.*

Cape′ Verde′ (vûrd), a republic consisting of a group of islands (**Cape′ Verde′ Is′lands**) in the Atlantic, W of Senegal in W Africa: formerly an overseas territory of Portugal; gained independence 1975. 360,000; 1557 sq. mi. (4033 sq. km). *Cap.:* Praia. —**Cape′ Ver′de·an** (vûr′dē ən), *n.*

Cape′ York′ Penin′sula, a peninsula in NE Australia, in N Queensland, between the Gulf of Carpentaria and the Coral Sea.

Cap-Haï·tien (kap/hä′shən) also **Cap-Ha·i·tien** (*Fr.* ka pa syan′), a seaport in N Haiti. 64,406.

Cap·i·to·line (kap′i tl in′), one of the seven hills on which ancient Rome was built.

Cap′itol Reef′ Na′tional Park′, a national park in S central Utah: sedimentary formations and fossils. 397 sq. mi. (1028 sq. km).

Cap·o·ret·to (kä′pə ret′ō), Italian name of KOBA-RID.

Cap·pa·do·cia (kap′ə dō′shə), an ancient country in E Asia Minor: it became a Roman province in A.D. 17. —**Cap′pa·do′cian,** *adj., n.*

Ca·pri (kä′prē, kap′rē, kə prē′), an island in W Italy, in the Bay of Naples: grottoes; resort. 5½ sq. mi. (14 sq. km). —**Cap·ri·ote** (kap′rē ōt′, -ət), *n.*

Cap·u·a (kap′yoo ə), a town in S Italy, N of Naples: near site of ancient city of Capua. 18,053.

Ca·ra·cas (kə rä′kəs), the capital of Venezuela, in the N part. 1,044,851.

Car·cas·sonne (kar ka sôn′), a city in S France: medieval fortifications. 44,623.

Car·chem·ish (kär′kə mish, kär kē′-), an ancient city in what is now S Turkey, on the upper Euphrates: a chief city of the Hittite empire.

Car·diff (kär′dif), a seaport in South Glamorgan, in SE Wales. 281,500.

Car·di·gan·shire (kär′di gən shēr′, -shər), a historic county in Dyfed, in W Wales. Also called **Car′di·gan.**

Car·i·a (kâr′ē ə), an ancient district in SW Asia Minor.

Car·ib·be·an (kar′ə bē′ən, kə rib′ē-), **1.** CARIB-BEAN SEA. **2. the,** the islands and countries of the Caribbean Sea collectively.

Car′ibbe′an Sea′, a part of the Atlantic Ocean bounded by Central America, the West Indies, and South America. ab. 750,000 sq. mi. (1,943,000 sq. km); greatest known depth 22,788 ft. (6946 m). Also called **Caribbean.**

Car·i·bees (kar′ə bēz′), See under ANTILLES.

Car′i·boo Moun′tains (kar′ə boo′), a mountain range in SW Canada, in E central British Columbia, part of the Rocky Mountains: highest peak, ab. 11,750 ft. (3580 m).

Ca·rin·thi·a (kə rin′thē ə), a province in S Austria. 541,876; 3681 sq. mi. (9535 sq. km). *Cap.:* Klagenfurt. —**Ca·rin′thi·an,** *adj., n.*

Car·lisle (kär līl′, kär′līl), a city in Cumbria, in NW England. 101,500.

Car·low (kär′lō), a county in Leinster, in the SE Republic of Ireland. 40,948; 346 sq. mi. (896 sq. km). *Co. seat:* Carlow.

Carls·bad (kärlz′bad), a town in S California. 63,126.

Carls′bad Cav′erns Na′tional Park′, a national park in SE New Mexico: limestone caverns. 73 sq. mi. (189 sq. km).

Car·mar·then (kär mär′ǁ̵hən), **1.** a seaport in Dyfed, S Wales. 53,600. **2.** CARMARTHENSHIRE.

Car·mar′then·shire′ (-shēr′, -shər), a historic county in Dyfed, S Wales.

Car·mel (kär′məl, kär mel′ for 1; kär mel′ for 2), **1. Mount,** a mountain ridge in NW Israel, near the Mediterranean coast. Highest point, 1818 ft. (554 m). **2.** Also called **Car·mel′-by-the-Sea′.** a town in W California, on the Pacific Ocean: artists' colony and resort. 4707.

Car·nat·ic (kär nat′ik), a region on the SE coast of India: now in Tamil Nadu state.

Car′nic Alps′ (kär′nik), a mountain range in S Austria and N Italy, part of the E Alps. Highest peak, 9217 ft. (2809 m).

Car·ni·o·la (kär′nē ō′lə, kärn yō′-), a former duchy and crown land of Austria: now in Slovenia. —**Car′ni·o′lan,** adj.

Car·o·li·na (kar′ə lī′nə; for 1 also Sp. kä′RÔ-lē′nä), **1.** a city in NE Puerto Rico, SE of San Juan. 162,888. **2. the Carolinas,** North Carolina and South Carolina. —**Car′o·lin′i·an** (-lin′ē ən), adj., n.

Car′o·line Is′lands (kar′ə lin′, -lin), a group of islands in the W Pacific, E of the Philippines: comprises the Federated States of Micronesia and the Republic of Palau.

Car·pa′thi·an Moun′tains (kär pā′thē ən), a mountain range in central Europe extending from N Czechoslovakia to central Romania. Highest peak, Gerlachovka, 8737 ft. (2663 m). Also called **Car·pa′thi·ans.**

Car·pen·tar·i·a (kär′pən târ′ē ə), **Gulf of,** a gulf on the coast of N Australia. ab. 480 mi. (775 km) long; ab. 300 mi. (485 km) wide.

Car·ra·ra (kə rär′ə), a city in NW Tuscany, in NW Italy. 68,460. —**Car·ra′ran,** n., adj.

Car·roll·ton (kar′əl tən), a town in N Texas. 82,169.

Car·son (kär′sən), a city in SW California. 83,995.

Car′son Cit′y, a town in and the capital of Nevada, in the W part. 36,650.

Car·stensz (kär′stənz), **Mount,** former name of PUNCAK JAYA.

Car·ta·ge·na (kär/tə jē/nə, -gā/nə, -hä/-), **1.** a seaport in SE Spain. 168,809. **2.** a seaport in N Colombia. 531,426.

Car·thage (kär/thij), an ancient city-state in N Africa near modern Tunis: founded by the Phoenicians in the middle of the 9th century B.C.; destroyed in 146 B.C. in the last of the Punic Wars. —**Car/tha·gin/i·an** (-thə jin/ē ən), adj., n.

Cas·a·blan·ca (kas/ə blang/kə, kä/sə bläng/kə), a seaport in NW Morocco. 2,139,204.

Ca·sa Gran·de (kä/sə grän/dä, -dē), a national monument in S Arizona, near the Gila River: ruins of a prehistoric culture.

Cas·bah (kaz/bä, käz/-), KASBAH.

Cascade/ Range/, a mountain range extending from N California to W Canada. Highest peak, Mt. Rainier, 14,408 ft. (4392 m).

Cas/co Bay/ (kas/kō), a bay in SW Maine.

Cash·mere (kash/mēr, kazh/-, kash mēr/, kazh-), KASHMIR.

Cas·per (kas/pər), a city in central Wyoming. 46,742.

Cas/pi·an Sea/ (kas/pē ən), a salt lake between SE Europe and Asia: the largest inland body of water in the world. ab. 169,000 sq. mi. (438,000 sq. km); 85 ft. (26 m) below sea level.

Cas·si·no (kə sē/nō), a town in central Italy, NNW of Naples. 26,300.

Cas·tel Gan·dol·fo (kä stel/ gän dôl/fō), a village in central Italy, 15 mi. (24 km) SE of Rome: summer palace of the pope.

Cas·tile (ka stēl/), a former kingdom comprising most of Spain. Spanish, **Cas·ti·lla** (käs tē/lyä, -yä). —**Cas·til/ian** (-stil/yən), n., adj.

Cas·ti/lla la Nue/va (lä nwe/vä), Spanish name of NEW CASTILE.

Cas·ti/lla la Vie/ja (vye/hä), Spanish name of OLD CASTILE.

Cas·tries (kas/trēz, -trēs, kä strē/), the capital of St. Lucia, on the NW coast. 52,868.

Ca·strop-Rau·xel or **Ka·strop-Rau·xel** (kä/-strəp rouk/səl, kas/trəp-), a city in central North Rhine-Westphalia, in W Germany. 76,430.

Cat/a·li/na Is/land (kat/l ē/nə, kat/-), SANTA CATALINA. Also called **Catalina.**

Cat·a·lo·ni·a (kat/l ō/nē ə, -ōn/yə), a region in NE Spain, bordering on France and the Mediterranean: formerly a province. Spanish, **Ca·ta·lu·ña** (kä′tä lōō′nyä). —**Cat′a·lo′ni·an,** *adj., n.*

Ca·ta·nia (kə tän′yə), a seaport in E Sicily. 372,212.

Ca·tan·za·ro (kä′tän dzär′ō), a city in S Italy. 103,004.

Ca·taw·ba (kə tô′bə), a river flowing from W North Carolina into South Carolina, where it becomes the Wateree River.

Ca·thay (ka thā′), *Archaic.* China.

Cats′kill Moun′tains (kat′skil), a range of mountains in E New York: resort area. Highest peak, 4204 ft. (1281 m). Also called **Cats′kills.**

Cau·ca (kou′kä), a river in W Colombia: tributary of the Magdalena. 600 mi. (965 km) long.

Cau·ca·sus (kô′kə səs), **the, 1.** Also called **Cau′casus Moun′tains.** a mountain range in Caucasia, between the Black and Caspian seas, along the border between the Russian Federation, Georgia, and Azerbaijan. Highest peak, Mt. El-brus, 18,465 ft. (5628 m). **2.** Also, **Cau·ca·sia** (kô kā′zhə, -shə). a region between the Black and Caspian seas: divided by the Caucasus Mountains into Ciscaucasia in Europe and Transcaucasia in Asia. —**Cau·ca′sian** *adj., n.*

Cau·ver·y (kô′və rē) also **Kaveri,** a river in S India, flowing SE from the Western Ghats in Karnataka state through Tamil Nadu state to the Bay of Bengal: sacred to the Hindus. 475 mi. (765 km) long.

Cav·an (kav′ən), a county in Ulster, in the N Republic of Ireland. 53,763; 730 sq. mi. (1890 sq. km).

Ca·xi·as (kä shē′əs), a city in NE Brazil. 125,771.

Caxi′as do Sul′ (dŏŏ sōōl′), a city in S Brazil. 220,725.

Cay·enne (ki en′, kā-), the capital of French Guiana. 38,135.

Cay′man Is′lands (kā′mən′, -mən), three islands in the West Indies, NW of Jamaica: a British crown colony. 23,700; 104 sq. mi. (269 sq. km).

Ca·yu′ga Lake′ (kā yōō′gə, ki-), a lake in central New York: one of the Finger Lakes. 40 mi. (64 km) long.

Ce·a·rá (sā′ə rä′), a state on the NE coast of Brazil. 6,122,500; 57,149 sq. mi. (148,016 sq. km). *Cap.:* Fortaleza.

Ce·bú (sə bōō′, sā-), **1.** an island in the S central Philippines. 1703 sq. mi. (4411 sq. km). **2.** a seaport on this island. 610,000.

Ce·chy (che′ʜi), Czech name of BOHEMIA.

Ce′dar Rap′ids, a city in E Iowa. 108,751.

Cel·e·bes (sel′ə bēz′, sə lē′bēz), former name of SULAWESI.

Cel′ebes Sea′, an arm of the Pacific Ocean, N of Sulawesi and S of the Philippines.

Ce·nis (sə nē′), **Mont** (môn), a mountain pass between SE France and Italy, in the Alps. 6834 ft. (2083 m) high.

Cen·tral (sen′trəl), a region in central Scotland. 272,077. 1016 sq. mi. (2631 sq. km).

Cen′tral Af′rican Repub′lic, a republic in central Africa: a member of the French Community. 2,759,000; 238,000 sq. mi. (616,420 sq. km). *Cap.:* Bangui. Formerly, **Cen′tral Af′rican Em′pire, Ubangi-Shari.**

Cen′tral Amer′ica, continental North America S of Mexico, usu. considered as comprising Guatemala, Belize, El Salvador, Honduras, Nicaragua, Costa Rica, and Panama. 29,000,000; 227,933 sq. mi. (590,346 sq. km). —**Cen′tral Amer′ican,** *n., adj.*

Cen′tral Park′, a public park in central Manhattan, New York City. 840 acres (340 hectares).

Cen′tral Val′ley, the agricultural lowland of central California, comprising the Sacramento and San Joaquin river valleys.

Cen·tre (sän′tRə), a metropolitan region in central France, SW of Paris. 2,324,000; 15,390 sq. mi. (39,062 sq. km).

Ceph·a·lo·ni·a (sef′ə lō′nē ə), the largest of the Ionian Islands, off the W coast of Greece. 287 sq. mi. (743 sq. km). Greek, **Kefallinia.**

Ce·ram or **Se·ram** (si ram′, sā′räm), an island of the Moluccas in Indonesia, W of New Guinea. 7191 sq. mi. (18,625 sq. km).

Cer·nă·u·ţi (cheR′nə ŏŏts′), Romanian name of CHERNOVTSY.

Cer·ri·tos (sə rē′təs), a city in SW California. 53,240.

Cer′ro Gor·do (ser′ō gôr′dō), a mountain pass in E Mexico between Veracruz and Jalapa.

Ce·se·na (che zā′nä), a city in E central Italy. 89,640.

Čes·ké Bu·dě·jo·vi·ce (ches′ke bŏŏ′dye yô vi·tse), a city in W Czechoslovakia, on the Vltava River. 97,000. German, **Budweis.**

Ceu·ta (sā′ŏŏ tə, -tä), a seaport and enclave of Spain in N Morocco, on the Strait of Gibraltar. 71,403.

Cé·vennes (sā ven′), a mountain range in S France. Highest peak, 5753 ft. (1754 m).

Cey·lon (si lon′, sā-), former name of Sri Lanka. —**Cey·lon·ese** (sē′lə nēz′, -nēs′, sā′-), adj., n., pl. -ese.

Cha·co (chä′kō), 1. a part of the Gran Chaco region in central South America: in Bolivia, Paraguay, and Argentina. ab. 100,000 sq. mi. (259,000 sq. km). 2. Gran Chaco.

Chad (chad), 1. Lake, a lake in Africa at the junction of Cameroon, Chad, Niger, and Nigeria. 5000 to 10,000 sq. mi. (13,000 to 26,000 sq. km) (seasonal variation). 2. Republic of, a republic in N central Africa, E of Lake Chad: a member of the French Community; formerly part of French Equatorial Africa. 5,400,000; 501,000 sq. mi. (1,297,590 sq. km). Cap.: N'Djamena. French, **Tchad.** —**Chad′i·an,** n., adj.

Chaer·o·ne·a (ker′ə nē′ə), an ancient city in E Greece, in Boeotia: victory of Philip of Macedon over the Athenians, Thebans, and their allies, 338 B.C.

Cha′gos Archipel′ago (chä′gōs, -gəs), a group of islands in the British Indian Ocean Territory. ab. 75 sq. mi. (195 sq. km).

Cha·gres (chä′grās), a river in Panama, flowing through Gatun Lake into the Caribbean Sea.

Chal·ce·don (kal′si don′, kal sēd′n), an ancient city in NW Asia Minor, on the Bosporus: ecumenical council A.D. 451. —**Chal′ce·do′ni·an** (-dō′nē·ən), adj., n.

Chal·cid·i·ce (kal sid′ə sē), a peninsula in NE Greece. Greek, **Khalkidiki.**

Chal·cis (kal′sis, -kis), a city on Euboea, in SE Greece. 44,867. Greek, **Khalkis.**

Chal·de·a or **Chal·dae·a** (kal dē′ə), 1. an ancient region in the lower Tigris and Euphrates valley, in S Babylonia. 2. Babylonia.

Cha·leur′ Bay′ (shə lŏŏr′, -lûr′), an inlet of the Gulf of St. Lawrence between NE New Brunswick and SE Quebec, in SE Canada. ab. 85 mi. (135 km) long; 15–25 mi. (24–40 km) wide.

Cha·lon (sha lôn′), a city in E France, on the Saône River. 56,194. Also called **Cha·lon-sur-Saône′** (-syr sôn′).

Châ·lons (sha lôn′), a city in NE France: defeat of Attila A.D. 451. 51,137. Also called **Châ·lons-sur-Marne** (sha lôn syr marn′).

Cham·bé·ry (shän bā rē′), a city in SE France. 54,896.

Cha·mo·nix (sham′ə nē′), a mountain valley in E France, N of Mont Blanc.

Cham·pagne (sham pān′; *Fr.* shän pan/y°), a region and former province in NE France.

Cham·pagne-Ar·dennes (shän pan/y° ar den′), a metropolitan region in NE France. 1,352,500; 9887 sq. mi. (25,606 sq. km).

Cham·paign (sham pān′), a city in E Illinois, adjoining Urbana. 63,502.

Cham·plain (sham plān′), **Lake,** a lake between New York and Vermont. 125 mi. (200 km) long; ab. 600 sq. mi. (1550 sq. km).

Champs É·ly·sées (shän zā lē zā′), a boulevard in Paris, France, noted for its cafés, shops, and theaters.

Chan·cel·lors·ville (chan/sə lərz vil′, -slərz-, chän′-), a village in NE Virginia: site of a Confederate victory 1863.

Chan·der·na·gor (chun/dər nə gôr′, -gôr′) also **Chan·dar·na·gar** (-nug/ər), a port in S West Bengal, in E India, on the Hooghly River: a former French dependency. 421,256.

Chan·di·garh (chun/di gur′), a city and a union territory in N India: the joint capital of Punjab and Haryana states. 450,061; 44 sq. mi. (114 sq. km).

Chan·dler (chand/lər, chänd′-), a town in central Arizona. 90,533.

Chang·an (*Chin.* chäng′än′), former name of XIAN.

Chang·chia·k'ou (*Chin.* chäng′jyä′kō′), ZHANGJIAKOU.

Chang·chou or **Chang·chow** (*Chin.* chäng′jō′), ZHANGZHOU.

Chang·chun (chäng/chŏōn′), the capital of Jilin province, in NE China. 1,860,000.

Chang·de or **Chang·teh** (chäng/du′), a city in N Hunan province, in E China. 225,000.

Chang Jiang (chäng′ jyäng′), a river in E Asia, flowing S and then E from the Tibetan plateau to the East China Sea. ab. 3200 mi. (5150 km) long. Also called **Yangtze.**

Chang·sha (chäng′shä′), the capital of Hunan province, in SE China. 1,120,000.

Chang·zhou or **Ch'ang·chou** (chäng′jō′), a city in S Jiangsu province, in E China. 300,000.

Chan′nel Is′lands, a British island group in the English Channel, near the coast of France, consisting of Alderney, Guernsey, Jersey, and smaller islands. 126,156; 75 sq. mi. (194 sq. km).

Chan·til·ly (shan til′ē; *Fr.* shäN tē yē′), a town in N France, N of Paris: lace manufacture. 10,684.

Chao′·an or **Chao·an** (*Chin.* chou′än′), former name of CHAOZHOU.

Chao Phra·ya (chou′ prä yä′), a river in N Thailand, flowing S to the Gulf of Thailand. 150 mi. (240 km) long. Formerly, **Menam.**

Chao·zhou or **Chao·chow** (chou′zhō′), a city in E Guangdong province, in SE China. 101,000. Formerly, **Chao'an.**

Cha·pa·la (chə pä′lə), **Lake,** the largest lake in Mexico, located in Jalisco state. 651 sq. mi. (1686 sq. km).

Chap′el Hill′, a city in central North Carolina. 32,421.

Cha·pul·te·pec (chə pul′tə pek′, -pŏŏl′-), a castle-fortress and military school on the outskirts of Mexico City: captured by U.S. forces (1847) in the Mexican War; now a park.

Char·dzhou (chär jō′), a city in E Turkmenistan, on the Amu Darya. 162,000.

Cha·ri (shär′ē), SHARI.

Char·le·roi (shaR lə RwA′), a city in S Belgium. 209,000.

Charles (chärlz), **1. Cape,** a cape in E Virginia, N of the entrance to Chesapeake Bay. **2.** a river in E Massachusetts, flowing between Boston and Cambridge into the Atlantic. 47 mi. (75 km) long.

Charles·bourg (shärl bŏŏr′, chärlz′bûrg′), a city in S Quebec, in E Canada, near the city of Quebec. 68,996.

Charles·ton (chärlz′tən, chärl′stən), **1.** a seaport in SE South Carolina. 80,414. **2.** the capital of West Virginia, in the W part. 57,287.

Charles′town′, a former city in E Massachusetts: since 1874 a part of Boston; site of Battle of Bunker Hill June 17, 1775.

Char·lotte (shär′lət), a city in S North Carolina. 395,934.

Char′lotte A·ma′li·e (ə mä′lē ə), the capital of the Virgin Islands of the U.S., on St. Thomas. 12,372. Formerly, **St. Thomas.**

Char·lottes·ville (shär′ləts vil′), a city in central Virginia. 39,916.

Char′lotte·town′, the capital of Prince Edward Island, in SE Canada. 15,776.

Char·tres (shär′trə, shärt; *Fr.* shᴀʀ′tʀ°), a city in N France, SW of Paris: cathedral. 41,251.

Cha·ryb·dis (kə rib′dis), a whirlpool in the Strait of Messina off the NE coast of Sicily.

Châ·teau-Thier·ry (sha tō′tē′ə rē′, -tye rē′, shä-), a town in N France, on the Marne River: World War I battles. 13,856.

Chat·ham (chat′əm), a city in N Kent, in SE England. 56,921.

Chat′ham Is′lands, a group of islands in the S Pacific, E of and belonging to New Zealand. 372 sq. mi. (963 sq. km).

Chat·ta·hoo·chee (chat′ə hōō′chē), a river flowing S from N Georgia along part of the boundary between Alabama and Georgia into the Apalachicola River. ab. 418 mi. (675 km) long.

Chat·ta·noo·ga (chat′ə nōō′gə), a city in SE Tennessee, on the Tennessee River. 152,466. —**Chat′ta·noo′gan, Chat′ta·noo′gi·an** (-jē ən), *adj., n.*

Chau·tau·qua (shə tô′kwə, chə-), **Lake,** a lake in SW New York. 18 mi. (29 km) long.

Cheap·side (chēp′sīd′), a district and thoroughfare in London, England.

Che·bo·ksa·ry (cheb′ək sär′ē), the capital of the Chuvash Autonomous Republic, in the Russian Federation in Europe, on the Volga. 420,000.

Che·chen′-In·gush′ (or **Che·chen′o-In·gush′**) **Auton′omous Repub′lic** (chə chen′in-gōōsh′ or chə chen′ō-), an autonomous republic of the Russian Federation, in Caucasia. 1,277,000; 7,350 sq. mi. (19,300 sq. km). *Cap.:* Grozny.

Che·ju (che′jōō′), an island S of and belonging to South Korea. 718 sq. mi. (1860 sq. km). Formerly, **Quelpart.**

Che·kiang (*Chin.* ju′gyäng′), ZHEJIANG.

Chel·sea (chel′sē), a former borough in Greater London, England: now part of Kensington and Chelsea; many homes of artists and writers.

Chel·ten·ham (chelt′nəm), a city in N Gloucestershire, in W England: resort. 86,500.

Chel·ya·binsk (chel yä′binsk), a city in the S Russian Federation in Asia, E of the Ural Mountains. 1,143,000.

Chel·yus·kin (chel yŏŏs′kin), **Cape,** a cape in the N Russian Federation in Asia, on the Taimyr Peninsula: the northernmost point of the Asia mainland.

Chem·nitz (kem′nits), a city in E Germany. 314,437. Formerly (1953–90), **Karl-Marx-Stadt.**

Che·mul·po (*Korean.* che′mŏōl pô′), former name of INCHON.

Che·nab (chi näb′), a river in S Asia, flowing SW from N India to the Sutlej River in E Pakistan. ab. 675 mi. (1085 km) long.

Chen·chiang (*Chin.* jun′jyäng′), ZHENJIANG.

Cheng·chow (*Chin.* jung′jō′), ZHENGZHOU.

Cheng·de or **Cheng·teh** (chœng′dœ′), a city in NE Hebei province, in NE China: summer residence of the Manchu emperors. 316,397. Formerly, **Jehol.**

Cheng·du or **Cheng·tu** (chœng′dy′), capital of Sichuan province, in central China. 2,580,000.

Cher (shär; *Fr.* sher), a river in central France flowing NW to the Loire River. 220 mi. (355 km) long.

Cher·bourg (shär′bŏōrg; *Fr.* sher bŏōr′), a seaport in NW France. 30,112.

Che·rem·kho·vo (chə rem′kə vō′), a city in the SE Russian Federation in Asia, NW of Irkutsk. 110,000.

Che·re·po·vets (cher′ə pə vets′), a city in the NW Russian Federation, N of Rybinsk Reservoir. 315,000.

Cher·kas·sy (chûr kä′sē, -kas′ē, cher-), a city in central Ukraine, on the Dnieper River, SE of Kiev. 287,000.

Cher·kessk (chûr kesk′, cher-), the capital of the Karachai-Cherkess Autonomous Region, in the Russian Federation. 113,000.

Cher·ni·gov (chûr nē′gôf, -gof, cher-), a city in N Ukraine, on the Desna River, NE of Kiev. 291,000.

Cher·no·byl (chŭr nō′bəl, cher-), a city in N Ukraine 80 mi. NW of Kiev: nuclear-plant accident 1986.

Cher·nov·tsy (chŭr′nôf tsē′, -nof-, cher′-), a city in SW Ukraine, on the Prut river: formerly in Romania. 254,000. Romanian, **Cernăuţi.**

Ches·a·peake (ches′ə pēk′), a city in SE Virginia. 151,976.

Ches′apeake Bay′, an inlet of the Atlantic, in Maryland and Virginia. 200 mi. (320 km) long; 4–40 mi. (6–64 km) wide.

Chesh·ire (chesh′ər, -ēr), a county in NW England. 951,900; 899 sq. mi. (2328 sq. km). Formerly, **Chester.**

Ches·ter (ches′tər), **1.** a city in Cheshire, in NW England: only English city with intact Roman walls. 117,200. **2.** former name of CHESHIRE.

Che·tu·mal (che′tōō mäl′), the capital of Quintana Roo, in SE Mexico. 23,685.

Chev·i·ot Hills′ (chev′ē ət, chē′vē-), a range of hills on the boundary between England and Scotland: highest point, 2676 ft. (816 m).

Chey·enne (shī en′, -an′), the capital of Wyoming, in the S part. 50,008.

Chey′enne Riv′er, a river flowing NE from E Wyoming to the Missouri River in South Dakota. ab. 500 mi. (800 km) long.

Chia·i (jyä′ē′), a city on W Taiwan. 250,000.

Chia·mu·ssu (*Chin.* jyä′mōō′sōō′), JIAMUSI.

Chiang Mai (chyäng′ mī′), a city in NW Thailand. 101,595.

Chi·a·pas (chē ä′päs), a state in S Mexico. 2,084,717; 28,732 sq. mi. (74,415 sq. km). *Cap.:* Tuxtla Gutiérrez.

Chi·ba (chē′bä′), a city on SE Honshu in central Japan, near Tokyo. 793,000.

Chi·ca·go (shi kä′gō, -kô′-), a city in NE Illinois, on Lake Michigan: third largest city in the U.S. 2,783,726. —**Chi·ca′go·an,** *n.*

Chi·chén It·zá (chē chen′ ēt sä′, ēt′sə), the ruins of an ancient Mayan city in central Yucatán state, Mexico.

Chi·chi·haerh (chē′chē′här′), QIQIHAR.

Chi·cla·yo (chi klä′yō), a city in NW Peru. 394,800.

Chic·o·pee (chik′ə pē′), a city in S Massachusetts, on the Connecticut River. 56,632.

Chi·cou·ti·mi (shi kōō/tə mē), a city in S Quebec, in E Canada. 61,083.

Chi·hua·hua (chi wä/wä, -wə), **1.** a state in N Mexico. 2,000,000; 94,831 sq. mi. (245,610 sq. km). **2.** the capital of this state. 406,830.

Chil·e (chil/ē, chē/lā), a republic in SW South America, on the Pacific Coast. 12,680,000; 286,396 sq. mi. (741,765 sq. km). *Cap.:* Santiago. —**Chil/e·an,** *adj., n.*

Chil/koot Pass/ (chil/kōot), a mountain pass on the boundary between SE Alaska and British Columbia, Canada, in the Coast Mountains. ab. 3500 ft. (1065 m) high.

Chi·llán (chē yän/), a city in central Chile. 148,805.

Chi·lo·é/ Is/land (chil/ō ā/), an island off the SW coast of Chile. 4700 sq. mi. (12,175 sq. km).

Chil·pan·cin·go (chēl/pän sēng/gō), the capital of Guerrero in SW Mexico. 56,904.

Chi·lung (chē/lŏong/) also **Jilong, Keelung,** a seaport on the N coast of Taiwan. 543,000.

Chim·bo·ra·zo (chim/bə rä/zō), a volcano in central Ecuador, in the Andes. 20,702 ft. (6310 m).

Chim·kent (chim kent/), a city in S Kazakhstan, N of Tashkent. 321,000.

Chi·na (chi/nə), **1. People's Republic of,** a country in E Asia. 1,133,682,501; 3,691,502 sq. mi. (9,560,990 sq. km). *Cap.:* Beijing. **2. Republic of,** a republic consisting mainly of the island of Taiwan off the SE coast of mainland China: under Nationalist control since 1948 but claimed by the People's Republic of China. 19,700,000; 13,885 sq. mi. (35,960 sq. km). *Cap.:* Taipei. —**Chi·nese/,** *n., pl.* **-nese,** *adj.*

Chi/na Sea/, the East China Sea and the South China Sea, taken together.

Chin·chow (*Chin.* jin/jō/), JINZHOU.

Chin·co·teague (shing/kə tēg/, ching/-), a town on a small island in a lagoon (**Chin/coteague Bay/**) in E Virginia: annual wild pony roundup. 1607.

Chin·dwin (chin/dwin/), a river in N Burma, flowing S to the Irrawaddy River. 550 mi. (885 km) long.

Chi/nese Tur/kestan. See under TURKESTAN.

Chi/nese Wall/, GREAT WALL OF CHINA.

Ch'ing·hai (ching′hī′), QINGHAI.

Chin·go·la (ching gō′lə), a town in N central Zambia. 214,000.

Chin·huang·tao (chin′hwäng′dou′), QINHUANG-DAO.

Chin·kiang (chin′kyäng′), ZHENJIANG.

Chi·no (chē′nō), a city in SE California. 59,682.

Chin·wang·tao (chin′wäng′dou′), QINHUANGDAO.

Chi·os (kī′os, -ōs, kē′-), a Greek island in the Aegean, near the W coast of Turkey. 53,942; 322 sq. mi. (834 sq. km). Greek, **Khios.**

Chis′holm Trail′ (chiz′əm), a cattle trail leading N from San Antonio, Tex., to Abilene, Kan.: used for about 20 years after the Civil War.

Chi·și·nă·u (kē′shē nu′ōō), Romanian name of KISHINEV.

Chi·ta (chi tä′), a city in the SE Russian Federation in Asia. 349,000.

Chit·ta·gong (chit′ə gong′), a port in SE Bangladesh near the Bay of Bengal. 1,391,000.

Chiu·si (kyōō′sē), a town in central Italy, in Tuscany: Etruscan tombs. 8756. Ancient, **Clusium.**

Chka·lov (*Russ.* chkä′ləf), former name of OREN-BURG.

Choi·seul (shwä zōōl′), an island in the W central Pacific Ocean: part of the independent Solomon Islands. 8021; 1500 sq. mi. (3885 sq. km).

Cho·lu·la (chō lōō′lä), a town in S Mexico, SE of Mexico City: ancient Aztec ruins. 20,913.

Chong·jin (chung′jin′), a seaport in W North Korea. 754,128.

Chong·ju (chung′jōō′), a city in central South Korea. 252,985.

Chong·qing (chông′ching′) also **Chungking,** a city in SE Sichuan province, in S central China, on the Chang Jiang. 2,780,000.

Chon·ju (chœn′jōō′), a city in SW South Korea. 426,490.

Cho·rzów (hô′zhōōf), a city in S Poland. 156,000.

Cho·sen (chō′sen′), Japanese name of KOREA.

Christ·church (krist′chûrch′), a city on E South Island, in New Zealand. 325,710.

Chris·ti·an·i·a (kris′chē an′ē ə, -ä′nē ə, kris′tē-), former name of OSLO.

Christ′mas Is′land, 1. an Australian island in

the Indian Ocean. ab. 190 mi. (300 km) S of Java. 3300; 52 sq. mi. (135 sq. km). **2.** former name of KIRITIMATI.

Chuan·chow (*Chin.* chwän′jō′), QUANZHOU.

Chu·chow (*Chin.* jōō′jō′), ZHUZHOU.

Chud·sko·ye O·ze·ro (chyōōt skô′yə ô′zyi rə), Russian name of PEIPUS.

Chuk′chi Penin′sula (chŏŏk′chē), a peninsula in the NE Russian Federation across the Bering Strait from Alaska.

Chuk′chi Sea′, a part of the Arctic Ocean, N of the Bering Strait.

Chu Kiang (*Chin.* jōō′ gyäng′), ZHU JIANG.

Chu·la Vis·ta (chōō′lə vis′tə), a city in SW California near San Diego. 135,163.

Chun·chon (chōōn′chun′), a city in N South Korea. 155,214.

Chung·king (chŏŏng′king′), CHONGQING.

Chur (kŏŏr), the capital of Grisons, in E Switzerland. 32,600.

Church·ill (chûr′chil, -chəl), **1.** a river in Canada flowing NE from E Saskatchewan through Manitoba to Hudson Bay. ab. 1000 mi. (1600 km) long. **2.** Formerly, **Hamilton.** a river in S central Labrador, Newfoundland, in E Canada, flowing E to Lake Melville. 208 mi. (335 km) long. **3.** a seaport and railway terminus in NE Manitoba, on Hudson Bay at the mouth of the Churchill River. 1435.

Church′ill Falls′, waterfalls near the head of the Churchill River in SW Labrador, Newfoundland, in E Canada. ab. 200 ft. (60 m) wide; 316 ft. (96 m) high. Formerly, **Grand Falls.**

Chu·vash′ Auton′omous Repub′lic (chōō-väsh′), an autonomous republic in the Russian Federation in Europe. 1,336,000; 7064 sq. mi. (18,300 sq. km). *Cap.:* Cheboksary.

Cic·e·ro (sis′ə rō′), a city in NE Illinois, near Chicago. 67,436.

Cien·fue·gos (syen fwä′gōs), a seaport in S Cuba. 109,300.

Ci·li·cia (si lish′ə), an ancient country in SE Asia Minor: at one time a Roman province. —**Ci·li′cian,** *adj., n.*

Cili′cian Gates′, a mountain pass in SE Asia Minor connecting Cappadocia and Cilicia.

Cim·ar·ron (sim′ə ron′, -rōn′, -ər ən), a river

flowing E from NE New Mexico to the Arkansas River in Oklahoma. 600 mi. (965 km) long.

Cin·cin·nat·i (sin/sə nat/ē), a city in SW Ohio, on the Ohio River. 364,040.

Cir·cas·sia (sər kash/ə, -ē ə), a region in the S Russian Federation in Europe bordering on the NE coast of the Black Sea. —**Cir·cas/sian,** n., adj.

Ci·re·bon (chir/ə bôn/), a seaport on N Java, in S central Indonesia. 178,529.

Cir·e·na·i·ca (sir/ə nā/i kə, si/rə-), CYRENAICA.

C.I.S. Commonwealth of Independent States.

Cis·al·pine Gaul/ (sis al/pin, -pin), See under GAUL.

Cis·cau·ca·sia (sis/kô kā/zhə, -shə), the part of Caucasia north of the Caucasus Mountains.

Cis·kei (sis/kī), a self-governing black homeland in SE South Africa, on the Indian Ocean: granted independence in 1981. 2,000,000; 3205 sq. mi. (8300 sq. km). *Cap.:* Bisho. —**Cis·kei/an,** adj., n.

Ci·tlal·té·petl (sē/tläl tā/pet/l), ORIZABA (def. 1).

Cit·tà del Va·ti·ca·no (chē tä/ del vä/tē kä/nō), Italian name of VATICAN CITY.

Ciu·dad Bo·lí·var (syoo ŧħäŧħ/ bō lē/vär), a port in E Venezuela, on the Orinoco River. 182,941.

Ciudad/ Gua·ya·na (gwä yä/nä), a city in NE Venezuela, on the Orinoco River. 314,041.

Ciudad/ Juá/rez (hwär/es, -ez), a city in N Mexico across the Rio Grande from El Paso, Texas. 570,000.

Ciudad/ Vic·to·ria (bēk tôr/yä, vik tôr/ē ə), the capital of Tamaulipas state in NE Mexico. 153,206.

Clack·man·nan (klak man/ən), a historic county in central Scotland. Also called **Clack·man·nan·shire** (klak man/ən shēr/, -shər).

Clare (klâr), a county in W Republic of Ireland. 87,489; 1231 sq. mi. (3190 sq. km).

Clarks·ville (klärks/vil), a city in N Tennessee. 75,494.

Clear·wa·ter (klēr/wô/tər, -wot/ər), a city in W Florida. 98,784.

Clear/water Moun/tains, a group of mountains in N Idaho.

Cler·mont-Fer·rand (klɛʀ môn fe ʀäN/), a city in central France. 161,203.

Cleve·land (klēv/lənd), **1.** a port in NE Ohio, on

Lake Erie. 505,616. **2.** a county in N England. 400,000; 225 sq. mi. (583 sq. km).

Cleve·land Heights′, a city in NE Ohio, near Cleveland. 54,052.

Cli·chy (klē shē′), an industrial suburb of Paris, France, on the Seine. 47,956.

Clif·ton (klif′tən), a city in NE New Jersey. 71,742.

Cling′mans Dome′ (kling′mənz), a mountain on the border between North Carolina and Tennessee: the highest peak in the Great Smoky Mountains. 6642 ft. (2024 m).

Clo·vis (klō′vis), a city in central California. 50,323.

Cluj-Na·po·ca (klŏŏzh′nä pô′kä), a city in NW Romania. 309,843. Hungarian, **Kolozsvár.** Formerly, **Cluj** (klŏŏzh).

Clu·ny (klōō′nē; *Fr.* klv nē′), a town in E France, N of Lyons: ruins of a Benedictine abbey. 4335.

Clu·si·um (klōō′sē əm), ancient name of Chiusi.

Clu·tha (klōō′thə), a river in S New Zealand, on SE South Island, flowing SE to the Pacific Ocean. ab. 200 mi. (320 km) long.

Clw·yd (klōō′id), a county in N Wales. 402,800; 937 sq. mi. (2426 sq. km).

Clyde (klīd), **1.** a river in S Scotland, flowing NW into the Firth of Clyde. 106 mi. (170 km) long. **2. Firth of,** an inlet of the Atlantic, in SW Scotland. 64 mi. (103 km) long.

Cni·dus (nī′dəs), an ancient city in SW Asia Minor, in Caria: the Athenians defeated the Spartans in a naval battle near here 394 B.C.

Cnos·sus (nos′əs), Knossos.

CO, Colorado.

Co·a·hui·la (kō′ä wē′lä), a state in N Mexico. 1,557,265; 58,067 sq. mi. (150,395 sq. km). *Cap.:* Saltillo.

Coast′ Moun′tains, a mountain range in W British Columbia, Canada: N continuation of the Cascade Range.

Coast′ Rang′es, a series of mountain ranges along the Pacific coast of North America, extending from S California to SE Alaska.

Cóbh (kōv), a seaport in S Republic of Ireland: port for Cork. 6586. Formerly, **Queenstown.**

Co·blenz or **Ko·blenz** (kō′blents), a city in W

Germany, at the junction of the Rhine and Moselle rivers. 110,300.

Co·cha·bam·ba (kō′chä bäm′bä), a city in central Bolivia. 317,251; 8394 ft. (2558 m) above sea level.

Co·chin (kō′chin), a seaport in W Kerala, in SW India: first European fort in India, built by Portuguese 1503. 686,000.

Co′chin-Chi′na (kō′chin-, koch′in-), a former state in S French Indochina: now part of Vietnam. French, **Co·chin·chine** (kô shan shēn′).

Co·co (kō′kō), a river rising in N Nicaragua and flowing NE along the Nicaragua-Honduras border to the Caribbean Sea. ab. 300 mi. (485 km) long. Also called **Segovia**.

Co′cos Is′lands (kō′kōs), a group of 27 coral islands in the Indian Ocean, SW of Java, administered by Australia. 569; 5.5 sq. mi. (14 sq. km). Also called **Keeling Islands**.

Cod (kod), **Cape**, CAPE COD.

Co·im·ba·tore (kō im′bä tôr′, -tōr′), a city in W Tamil Nadu state, in SW India. 917,000.

Col., Colorado.

Col·ches·ter (kōl′ches·tər, -chə stər), a city in NE Essex, in E England. 147,200.

Col·chis (kol′kis), an ancient country in Asia S of the Caucasus and bordering on the Black Sea: the land of the Golden Fleece and of Medea in Greek legend.

Co·li·ma (kō lē′mä), **1.** a state in SW Mexico, on the Pacific Coast. 346,293; 2010 sq. mi. (5205 sq. km). **2.** the capital of this state, in the E part. 73,000. **3.** a volcano NW of this city, in Jalisco state. 12,631 ft. (3850 m).

Col′lege Sta′tion, a city in E central Texas. 52,456.

Colo., Colorado.

Co·logne (kə lōn′), a city in W Germany. 914,300. German, **Köln.**

Co·lom·bi·a (kə lum′bē ə), a republic in NW South America. 27,900,000; 439,828 sq. mi. (1,139,155 sq. km). *Cap.:* Bogotá. —**Co·lom′·bi·an,** *adj., n.*

Co·lom·bo (kə lum′bō), the capital of Sri Lanka, on the W coast. 587,647.

Co·lón (kə lōn′), a seaport in Panama at the Atlantic end of the Panama Canal. 85,600.

Colón′ Archipel′ago, GALÁPAGOS ISLANDS.

Col·o·phon (kol′ə fon′), an ancient city in Asia Minor: one of the 12 Ionian cities. —**Col′o·pho′ni·an** (-fō′nē ən), *n., adj.*

Col·o·rad·o (kol′ə rad′ō, -rä′dō), **1.** a state in the W United States. 3,294,394; 104,247 sq. mi. (270,000 sq. km). *Cap.:* Denver. *Abbr.:* CO, Col., Colo. **2.** a river flowing SW from N Colorado through Utah and Arizona into the Gulf of California. 1450 mi. (2335 km) long. **3.** a river flowing SE from W Texas to the Gulf of Mexico. 840 mi. (1350 km) long. —**Col′o·rad′an, Col′o·rad′o·an,** *adj., n.*

Col′orad′o Des′ert, an arid region in SE California, W of the Colorado River. ab. 2500 sq. mi. (6475 sq. km).

Colorad′o Plateau′, a plateau in the SW United States, in N Arizona, NW New Mexico, S Utah, and W Colorado.

Col′orad′o Springs′, a city in central Colorado: U.S. Air Force Academy. 281,140.

Co·lum·bi·a (kə lum′bē ə), **1.** a river in SW Canada and the NW United States, flowing S and W from SE British Columbia through Washington along the boundary between Washington and Oregon and into the Pacific. 1214 mi. (1955 km) long. **2.** the capital of South Carolina, in the central part. 98,052. **3.** a city in central Maryland. 75,883. **4.** a city in central Missouri. 69,101. **5.** the United States of America. —**Co·lum′bi·an,** *adj.*

Co·lum·bus (kə lum′bəs), **1.** the capital of Ohio, in the central part. 632,910. **2.** a city in W Georgia. 178,681.

Com′monwealth of In′dependent States′, an alliance of former Soviet republics formed in December 1991, including: Armenia, Azerbaijan, Belarus, Kazakhstan, Kyrgyzstan, Moldova, Russian Federation, Tajikistan, Turkmenistan, Ukraine, and Uzbekistan. *Abbr:* C.I.S.

Com′monwealth of Na′tions, a voluntary association of independent nations and their dependencies linked by historical ties as parts of the former British Empire and cooperating on matters of mutual concern. Formerly, **British Commonwealth of Nations.**

Com′munism Peak′, a peak in the Pamirs in NE Tajikistan. 24,590 ft. (7495 m).

Co·mo (kō′mō), **1. Lake,** a lake in N Italy, in Lombardy. 35 mi. (56 km) long; 56 sq. mi. (145 sq. km). **2.** a city at the SW end of this lake. 97,169.

Com·o·rin (kom′ər in), **Cape,** a cape on the S tip of India, extending into the Indian Ocean.

Com′o·ro Is′lands (kom′ə rō′), a group of islands in the Indian Ocean between N Madagascar and E Africa: formerly an overseas territory of France; now divided between the Comoros and France. 511,466; 863 sq. mi. (2235 sq. km).

Com·o·ros (kom′ə rōz′), **Federal Islamic Republic of the,** a republic comprising three of the Comoro Islands: a former overseas territory of France; declared independence 1975. 434,166; 719 sq. mi. (1862 sq. km). *Cap.:* Moroni.

Com·piè·gne (kôN pyen′yᵉ), a city in N France, on the Oise River: nearby were signed the armistices between the Allies and Germany 1918, and between Germany and France 1940. 40,720.

Comp·ton (komp′tən), a city in SW California. 90,454.

Com′stock Lode′ (kum′stok, kom′-), a rich deposit of silver and gold ore: discovered in 1859 by Henry T. P. Comstock near Virginia City, Nev.

Co·na·kry (kon′ə krē), the capital of Guinea, in NW Africa. 705,280.

Con·cep·ción (kən sep′sē ōn′), a city in central Chile, near the mouth of the Bío-Bío River. 294,375.

Con·chos (kon′chōs, -chəs), a river in NE Mexico flowing E and N to the Rio Grande. ab. 350 mi. (565 km) long.

Con·cord (kong′kərd *for 1, 3;* kon′kôrd, kong′- *for 2),* **1.** a city in W California, near San Francisco. 111,348. **2.** the capital of New Hampshire, in the S part. 30,400. **3.** a town in E Massachusetts, NW of Boston: second battle of the Revolution fought here April 19, 1775. 16,293.

Co′ney Is′land (kō′nē), an area in S Brooklyn in New York City: amusement park and beach.

Con·ga·ree (kong′gə rē′), a river flowing E in central South Carolina, joining with the Wateree River to form the Santee. ab. 60 mi. (97 km) long.

Con·go (kong′gō), **1. People's Republic of the.** Formerly, **French Congo, Middle Congo.** a re-

public in central Africa, W of Zaire: a former
French territory; gained independence 1960.
2,270,000; 132,046 sq. mi. (341,999 sq. km).
Cap.: Brazzaville. **2. Democratic Republic of
the,** a former name of Zaire (def. 1). **3.** Also
called Zaire. a river in central Africa, flowing in a
great loop from SE Zaire to the Atlantic. ab. 3000
mi. (4800 km) long. —**Con'go·lese'**, *n., pl.* **-lese,**
adj.

Con·go Free' State', a former name of Zaire
(def. 1).

Conn., Connecticut.

Con·naught (kon'ôt), a province in the NW Republic of Ireland. 430,726; 6610 sq. mi. (17,120
sq. km). Irish, **Con'nacht** (-əкнt, -ət).

Con·nect·i·cut (kə net'i kət), **1.** a state in the
NE United States. 3,287,116; 5009 sq. mi.
(12,975 sq. km). *Cap.:* Hartford. *Abbr.:* Conn., Ct.,
CT **2.** a river flowing S from N New Hampshire
along the boundary between New Hampshire and
Vermont and then through Massachusetts and
Connecticut into Long Island Sound. 407 mi. (655
km) long.

Con·stance (kon'stəns), **1. Lake.** German, **Bo-
densee.** a lake in W Europe, bounded by Germany, Austria, and Switzerland. 46 mi. (74 km)
long; 207 sq. mi. (536 sq. km). **2.** German, **Kon-
stanz.** a city in S Germany, on this lake: church
council 1414–18. 68,305.

Con·stan·tine (kon'stən tēn'), a city in NE Algeria. 448,578.

Con·stan·ti·no·ple (kon'stan tn ō'pəl), former
name of Istanbul.

Con·ti·nent (kon'tn ənt), **the,** the mainland of
Europe, as distinguished from the British Isles.

Conti·nen'tal Divide', the watershed in North
America formed by the Rocky Mountains, separat-
ing streams flowing west from those flowing east.
Also called **Great Divide.**

Cooch Be·har (kōōch' bə här'), a former state
in NE India: now part of West Bengal.

Cook (kōōk), **Mount,** a mountain in New Zealand,
on South Island. 12,349 ft. (3764 m). Also called
Aorangi.

Cook' In'let, an inlet of the Gulf of Alaska. 150
mi. (240 km) long.

Cook' Is'lands, a group of islands in the S Pacific belonging to New Zealand. 21,317; 99 sq. mi. (256 sq. km).

Cook/ Strait/, a strait in New Zealand between North and South Islands.

Coon Rap·ids (kōōn/ rap/idz), a city in E Minnesota. 52,978.

Coorg (kŏŏrg), a former province in SW India; now part of Karnataka state. 1593 sq. mi. (4126 sq. km).

Co·pen·ha·gen (kō/pən hā/gən, -hä/-, kō/pənhā/-, -hä/-), the capital of Denmark, on the E coast of Zealand. 802,391; with suburbs, 1,380,204. Danish, **København.**

Cop·per·mine (kop/ər min/), a river in N Canada, central Northwest Territories, flowing N to the Arctic Ocean. 525 mi. (845 km) long.

Cop/per Riv/er, a river in S Alaska flowing through the SE part. 300 mi. (483 km) long.

Co·quil·hat·ville (kô kē yA vēl/), former name of MBANDAKA.

Cor·al Ga·bles (kôr/əl gā/bəlz, kor/-), a city in SE Florida near Miami. 43,241.

Cor/al Sea/, a part of the S Pacific bounded by NE Australia, New Guinea, the Solomon Islands, and Vanuatu.

Cor/al Springs/, a town in SE Florida. 79,443.

Cor·co·va·do (kôr/kô vä/dŏŏ), a mountain in SE Brazil, S of Rio de Janeiro: statue of Christ on peak. 2310 ft. (704 m).

Cor·cy·ra (kôr si/rə), ancient name of CORFU.

Cor·di·lle·ra Cen·tral (kôr/dl yär/ə sen träl/, -ᵗhē yär/ə), **1.** a mountain range in Colombia: part of the Andes. Highest peak, Huila, 18,700 ft. (5700 m). **2.** a mountain range in the Dominican Republic. Highest peak, 10,414 ft. (3174 m). **3.** a mountain range in N Peru, E of the Marañón River: part of the Andes. **4.** a mountain range in Luzon, Philippines. Highest peak, 9606 ft. (2928 m). **5.** a mountain range in central Puerto Rico. Highest peak, 4389 ft. (1338 m).

Cor·di·lle/ra Oc·ci·den·tal/ (ŏk/sē ᵗhen täl/), the W coastal ranges of the Andes, in Peru and Colombia.

Cor·di·lle/ra O·rien·tal/ (ôr/ē en täl/), the E ranges of the Andes, in Bolivia, Colombia, and Peru.

Cor·di·lle/ra Re·al/ (rā äl/), **1.** a range of the

Andes, in Bolivia. Highest peak, Illimani, 21,188 ft. (6458 m). **2.** a range of the Andes, in Ecuador. Highest peak, Chimborazo, 20,702 ft. (6310 m).

Cor·dil·le·ras (kôr'dl yâr'əz, -thē yâr'äs, kôr-dil'ər əz), the entire chain of mountain ranges parallel to the Pacific coast, extending from Cape Horn to Alaska. —**Cor'dil·le'ran,** *adj.*

Cór·do·ba (kôr'də bə, -və), **1.** Also, **Cor'do·ba, Cor'do·va** (-və). a city in S Spain on the Guadalquivir River: the capital of Spain under Moorish rule. 304,826. **2.** a city in central Argentina. 982,018.

Cor·fu (kôr'fōō, -fyōō, kôr fōō'), **1.** Ancient, **Cor·cyra.** one of the Ionian Islands, off the NW coast of Greece. 89,664; 229 sq. mi. (593 sq. km). **2.** a seaport on this island. 33,561. Greek, **Kerkyra.**

Cor·inth (kôr'inth, kor'-), **1.** an ancient city in Greece, on the Isthmus of Corinth. **2.** a port in the NE Peloponnesus, in S Greece: NE of the site of ancient Corinth. **3. Gulf of.** Also called **Gulf of Lepanto.** an arm of the Ionian Sea, N of the Peloponnesus. **4. Isthmus of,** an isthmus at the head of the Gulf of Corinth, connecting the Peloponnesus with central Greece. —**Co·rin·thi·an** (kə rin'thē ən), *adj., n.*

Cork (kôrk), **1.** a county in Munster province in S Republic of Ireland. 279,427; 2881 sq. mi. (7460 sq. km). **2.** a seaport in and the county seat of Cork, in the S part. 133,196.

Corn·wall (kôrn'wôl; *esp. Brit.* -wəl), **1.** a county in SW England. 453,100; 1369 sq. mi. (3545 sq. km). **2.** a city in SE Ontario, in S Canada, SW of Ottawa, on the St. Lawrence. 51,000. —**Cor·nish** (kôr'nish), *adj.*

Cor'o·man'del Coast' (kôr'ə man'dl, kor'-, kôr'-, kor'-), a coastal region in SE India S of the Krishna River.

Co·ro·na (kə rō'nə), a city in SE California. 76,095.

Cor·pus Chris·ti (kôr'pəs kris'tē), a seaport in S Texas. 257,453.

Cor'pus Chris'ti Bay', a bay in S Texas at the mouth of the Nueces River.

Cor·reg·i·dor (kə reg'i dôr', -dōr'), an island in Manila Bay, in the Philippines: U.S. forces defeated by the Japanese in May, 1942. 2 sq. mi. (5 sq. km).

Cor·ri·en·tes (kôr′ē en′tās), a port in NE Argentina, on the Paraná River. 180,000.

Cor·si·ca (kôr′si kə), a French island in the Mediterranean, N of Sardinia: constitutes a metropolitan region of France. 248,700; 3367 sq. mi. (8720 sq. km). *Cap.:* Ajaccio. French, **Corse** (kôrs). —**Cor′si·can,** *adj., n.*

Co·ru·ña (kə rōōn′yə) also **Co·run′na** (-run′ə), LA CORUÑA.

Cor·val·lis (kôr val′is), a city in W Oregon. 41,800.

Cos (kos, kôs), Kos.

Co·sen·za (kō zen′tsä), a city in S Italy. 105,913.

Cos′ta Bra′va (kos′tə brä′və, kô′stə, kō′-), a coastal region in NE Spain, on the Mediterranean, extending NE from Barcelona: resorts.

Cos′ta del Sol′ (del sōl′), a coastal region in S Spain, on the Mediterranean, extending E from Gibraltar: resorts.

Cos′ta Me′sa (mā′sə), a city in SW California near Los Angeles. 96,357.

Cos′ta Ri′ca (rē′kə), a republic in Central America, between Panama and Nicaragua. 2,810,000; 19,238 sq. mi. (49,825 sq. km). *Cap.:* San José. —**Cos′ta Ri′can,** *adj., n.*

Cos·ter·mans·ville (kos′tər mənz vil′, kô′stər-), former name of BUKAVU.

Côte d′A·zur (kōt dA zYR′), the part of the French Riviera E of Cannes.

Côte d′I·voire (kōt dē vwAR′), French name of IVORY COAST.

Côte d′Or (kōt dôr′), a range of hills in E France SW of Dijon: vineyards.

Co·to·nou (kō′tə nōō′), a seaport in SE Benin. 487,020.

Co·to·pax·i (kō′tə pak′sē, -pä′hē), a volcano in central Ecuador, in the Andes: the highest active volcano in the world. 19,498 ft. (5943 m).

Cots·wolds (kots′wōldz, -wəldz), a range of hills in SW England, in Gloucestershire. Also called **Cots′wold Hills′.**

Cott·bus (kot′bəs, -bŏŏs), a city in E Germany, on the Spree River. 125,784.

Cot′ti·an Alps′ (kot′ē ən), a mountain range in SW Europe, in France and Italy: a part of the Alps. Highest peak, 12,602 ft. (3841 m).

Cot′ton Belt′, (*sometimes l.c.*) the part of the

southern U.S. where cotton is grown, orig. Alabama, Georgia, and Mississippi, but now often extended to include parts of Texas and California.

Coun/cil Bluffs/, a city in SW Iowa, across the Missouri River from Omaha, Neb. 54,315.

Cour·an·tyne (kôr/ən tin/, kôr/-), a river in N South America, flowing N along the Guyana-Suriname border to the Atlantic Ocean. ab. 450 mi. (725 km) long.

Cour·land or **Kur·land** (kōōr/lənd), a former duchy on the Baltic: later, a province of Russia and, in 1918, incorporated into Latvia.

Cour·trai (*Fr.* kōōr tre/), a city in W Belgium, on the Lys River: important medieval city. 43,364. Flemish, **Kortrijk.**

Cov/ent Gar/den (kuv/ənt, kov/-), a district in central London, England, formerly a vegetable and flower market: a historic theater.

Cov·en·try (kuv/ən trē, kov/-), a city in West Midlands, in central England: cathedral. 337,000.

Cov·ing·ton (kuv/ing tən), a city in N Kentucky, on the Ohio River. 49,585.

Cowes (kouz), a seaport on the Isle of Wight, in S England: resort. 19,663.

Co·zu·mel (kō/zōō mel/), an island off NE Quintana Roo state on the Yucatán Peninsula in SE Mexico: resort.

Cra·cow (krak/ou, krä/kou), **KRAKÓW.**

Cra·io·va (krä yô/vä), a city in SW Romania. 275,098.

Cran·ston (kran/stən), a city in E Rhode Island, near Providence. 76,060.

Cra/ter Lake/, a lake in the crater of an extinct volcano in SW Oregon. 1932 ft. (589 m) deep.

Cra/ter Lake/ Na/tional Park/, a national park in SW Oregon, in the Cascade Range: Crater Lake. 286 sq. mi. (741 sq. km).

Cra/ter Mound/, a bowl-shaped depression in the earth in central Arizona: believed to have been made by the impact of a meteoroid. 4000 ft. (1220 m) wide; 600 ft. (183 m) deep.

Cra/ters of the Moon/, a national monument in S Idaho: site of scenic lava-flow formations.

Cré·cy or **Cres·sy** (kres/ē), a village in N France, NNW of Reims: English victory over the French 1346.

Cre·mo·na (kri mōʹnə), a city in N Italy, on the Po River. 82,411.

Crete (krēt), a Greek island in the Mediterranean, SE of mainland Greece. 502,165; 3235 sq. mi. (8380 sq. km). *Cap.:* Canea. Also called **Candia.** —**Creʹtan,** *adj., n.*

Cri·me·a (krī mēʹə, krī-), **the,** a peninsula in SE Ukraine, between the Black Sea and the Sea of Azov. Russian, **Krim, Krym.** —**Cri·meʹan,** *adj.*

Cripʹple Creekʹ, a town in central Colorado: gold rush 1891. 655; 9600 ft. (2925 m) above sea level.

Cris·to·bal (kri stōʹbəl), a seaport in Panama at the Atlantic end of the Panama Canal, adjacent to Colón. 11,600. Spanish, **Cris·tó·bal** (krēs tôʹbäl).

Cro·a·tia (krō āʹshə, -shē ə), a republic in SE Europe: includes the historical regions of Dalmatia, Istria, and Slavonia; formerly part of Yugoslavia. 4,660,000; 21,835 sq. mi. (56,555 sq. km). *Cap.:* Zagreb. —**Croʹat** (krōʹat, -ät), *n.* —**Cro·aʹtian,** *adj., n.*

Crocʹodile Rivʹer, Limpopo.

Crownʹ Pointʹ, a village in NE New York, on Lake Champlain: the site of a strategic fort in the French and Indian and the Revolutionary wars. 1837.

Croy·don (kroidʹn), a borough of Greater London, England. 324,900.

CT or **Ct.,** Connecticut.

Ctes·i·phon (tesʹə fon′), a ruined city in Iraq, on the Tigris, near Baghdad: an ancient capital of Parthia.

Cu·ba (kyooʹbə), **1.** an island in the West Indies, S of Florida. **2.** a republic in the Caribbean, including this island and several nearby islands. 10,240,000; 44,206 sq. mi. (114,524 sq. km). *Cap.:* Havana. —**Cuʹban,** *adj., n.*

Cu·ban·go (koo bängʹgoo), Portuguese name of Okavango.

Cú·cu·ta (kooʹkoo tä′), a city in E Colombia. 379,478.

Cuen·ca (kwengʹkä), a city in SW Ecuador. 272,397.

Cuer·na·va·ca (kwerʹnə väʹkə), the capital of Morelos, in central Mexico. 357,600.

Cu·ia·bá (kooʹyə bäʹ), the capital of Mato Grosso, in W Brazil. 167,894.

Cu·lia·cán (kōō/lyä kän/), the capital of Sinaloa state, in NW Mexico. 358,800.

Cu·mae (kyōō/mē), an ancient city in SW Italy, on the coast of Campania: believed to be the earliest Greek colony in Italy or Sicily. —**Cu·mae/an**, *adj.*

Cu·ma·ná (kōō/mä nä/), a seaport in N Venezuela. 179,814.

Cum·ber·land (kum/bər lənd), **1.** a former county in NW England, now part of Cumbria. **2.** a city in NW Maryland, on the Potomac River. 23,230. **3.** a river flowing W from SE Kentucky through N Tennessee into the Ohio River. 687 mi. (1106 km) long.

Cum/berland Gap/, a pass in the Cumberland Mountains at the junction of the Virginia, Kentucky, and Tennessee boundaries. 1315 ft. (401 m) high.

Cum/berland Moun/tains, a plateau largely in Kentucky and Tennessee, a part of the Appalachian Mountains: highest point, ab. 4000 ft. (1220 m). Also called **Cum/berland Plateau/.**

Cum·bri·a (kum/brē ə), a county in NW England. 486,900; 2659 sq. mi. (6886 sq. km).

Cu·nax·a (kyōō nak/sə), an ancient town in Babylonia, near the Euphrates: site of defeat of Cyrus the Younger by Artaxerxes II in 401 B.C.

Cu·ra·çao (kŏŏr/ə sou/, -sō/, kyŏŏr/-; kŏŏr/ə-sou/, -sō/, kyŏŏr/-), **1.** the main island of the Netherlands Antilles, off the NW coast of Venezuela. 159,072; 173 sq. mi. (448 sq. km). *Cap.:* Willemstad. **2.** former name of NETHERLANDS ANTILLES.

Cu·ri·ti·ba (kŏŏr/i tē/bə), the capital of Paraná, in SE Brazil. 1,052,147.

Cus·co (kōōs/kō), Cuzco.

Cush or **Kush** (kŏŏsh, kush), an ancient kingdom in North Africa, in the region of Nubia.

Cutch (kuch), KUTCH.

Cut·tack (kut/ək), a city in E Orissa, in NE India. 326,000.

Cux·ha·ven (kŏŏks/hä/fən), a seaport in NW Germany, at the mouth of the Elbe River. 60,200.

Cuz·co or **Cus·co** (kōōs/kō), a city in S Peru: Inca ruins. 255,300.

Cyc·la·des (sik/lə dēz/), a group of Greek islands in the S Aegean. 88,458; 1023 sq. mi. (2650 sq. km). —**Cy·clad·ic** (si klad/ik, si-), *adj.*

Cy·prus (sī′prəs), an island republic in the Mediterranean, S of Turkey: formerly a British colony; independent since 1960. 680,400; 3572 sq. mi. (9250 sq. km). *Cap.:* Nicosia. —**Cyp·ri·ot** (sip′rē-ət), **Cyp′ri·ote′** (-ōt′, -ət), *n., adj.*

Cyr·e·na·i·ca or **Cir·e·na·i·ca** (sir′ə nā′i kə, sī′rə-), **1.** an ancient district in N Africa. **2.** the E part of Libya.

Cy·re·ne (sī rē′nē), an ancient Greek city in N Africa, in Cyrenaica.

Cyth·er·a (si thēr′ə), KITHIRA.

Cyz·i·cus (siz′i kəs), an ancient city in NW Asia Minor, in Mysia, on a peninsula in the Sea of Marmara.

CZ or **C.Z.,** Canal Zone.

Czech·o·slo·va·ki·a (chek′ə slə vä′kē ə, -vak′-ē ə), a republic in central Europe: formed in 1918; comprises Bohemia, Moravia, Slovakia, and part of Silesia; a federal republic since 1968. 15,520,839; 49,383 sq. mi. (127,903 sq. km). *Cap.:* Prague. Official name, **Czech′ and Slo′vak Fed′erative Repub′lic.** Formerly (1948–89), **Czech′o·slo′vak So′cialist Repub′lic** (chek′ə-slō′väk, -vak). —**Czech′o·slo·va′ki·an,** *adj., n.*

Czech′ Repub′lic, a constituent republic of Czechoslovakia, in the W part: includes the regions of Bohemia, Moravia, and part of Silesia. 10,343,398; 30,449 sq. mi. (78,864 sq. km). *Cap.:* Prague. —**Czech,** *n., adj.*

D

Dac·ca (dak′ə, dä′kə), DHAKA.

Da·chau (dä′кнои), a city in S Germany, near Munich: site of Nazi concentration camp. 33,950.

Da·ci·a (dä′shē ə, -shə), an ancient kingdom and later a Roman province in S Europe between the Carpathian Mountains and the Danube, corresponding generally to modern Romania and adjacent regions. —**Da′ci·an,** *adj., n.*

Da·dra and Na·gar Ha·ve·li (də drä′ ən nug′ər hə vā′lē), a union territory in W India, between Gujarat and Maharashtra. 103,700; 189 sq. mi. (491 sq. km).

Dag·en·ham (dag′ə nəm), a former borough in Greater London, now a part of Barking and Redbridge.

Da·ge·stan (dä′gə stän′, dag′ə stan′), an autonomous republic in the SW Russian Federation on the W shore of the Caspian Sea. 1,800,000; 19,421 sq. mi. (50,300 sq. km). *Cap.:* Makhachkala. Formerly, **Dagestan′ Auton′omous So′viet So′cialist Repub′lic.**

Da·ho·mey (də hō′mē), former name of BENIN (def. 1). —**Da·ho′me·an, Da·ho′man,** *adj., n.*

Dai·ren (di′ren′), former Japanese name of DA-LIAN.

Da·kar (dä kär′), a seaport in and the capital of Senegal. 1,380,000; 68 sq. mi. (176 sq. km).

Dakh·la (däкн′lä), **1.** an oasis in S Egypt: source of ocher. **2.** Formerly, **Villa Cisneros.** a seaport in Western Sahara: former capital of Río de Oro in the former Spanish Sahara. 7000.

Da·ko·ta (də kō′tə), **1.** a former territory in the U.S.: divided into the states of North Dakota and South Dakota 1889. **2. the Dakotas,** North Dakota and South Dakota. —**Da·ko′tan,** *adj., n.*

Da·lian (dä′lyän′), a seaport in S Liaoning province, in NE China. 1,630,000. Formerly, *Japanese,* **Dairen;** *Russian,* **Dalny.** Compare LÜDA.

Dal·las (dal′əs), a city in NE Texas. 1,006,877.

Dal·ma·tia (dal mā′shə), a region along the Adriatic coast of Croatia: a former Austrian crownland.

Dal·ny (däl′nē), former Russian name of DALIAN.

Da·ly Cit·y (dā/lē), a city in central California, S of San Francisco. 92,311.

Dam·an (də män/), **1.** a district on the coast of Gujarat state in W India: part of the union territory of Daman and Diu. **2.** the capital of Daman and Diu. 21,000.

Daman/ and Di/u, a union territory in W India: formerly part of Portuguese India; annexed by India in 1961 and formed part of union territory of Goa, Daman, and Diu until 1987. 78,981; 42 sq. mi. (110 sq. km). *Cap.:* Daman.

Da·man·hur (dä/män hŏŏr/), a city in N Egypt, near Alexandria. 225,900.

Da·mas·cus (də mas/kəs), the capital of Syria, in the SW part: reputed to be the oldest continuously existing city in the world. 1,251,000.

Dam·i·et·ta (dam/ē et/ə), a city in NE Egypt, in the Nile delta. 121,200. Arabic, **Dumyat.**

Dan (dan), the northernmost city of ancient Palestine.

Da·nang or **Da Nang** (də näng/, -nang/, dä-), a seaport in central Vietnam. 500,000. Formerly, **Tourane.**

Dan·bur·y (dan/ber/ē, -bə rē), a city in SW Connecticut. 65,585.

Dan·dong (dän/dông/), a seaport in SE Liaoning province, in NE China, at the mouth of the Yalu River. 537,745. Formerly, **Antung.**

Dan/ger Cave/, a deep, stratified site in the E Great Basin, in Utah, occupied by Amerindian cultures from at least 7000 B.C. to historic times.

Dan/ish West/ In/dies, former name of the VIRGIN ISLANDS OF THE UNITED STATES.

Dan·ube (dan/yōōb), a river in central and SE Europe, flowing E from S Germany to the Black Sea. 1725 mi. (2775 km) long. German, **Donau.** Hungarian, **Duna.** Czech and Slovak, **Dunaj.** Romanian, **Dunărea.** —**Dan·u/bi·an,** *adj.*

Dan·ville (dan/vil), a city in S Virginia. 53,056.

Dan·zig (dan/sig, dän/-), **1.** German name of GDAŃSK. **2. Gulf of,** an inlet of the Baltic Sea, in N Poland. ab. 60 mi. (95 km) wide.

Dap·sang (dəp sung/), See K2.

Dar·da·nelles (där/dn elz/), the strait between European and Asian Turkey, connecting the Aegean Sea with the Sea of Marmara. 40 mi. (64 km) long; 1–5 mi. (1.6–8 km) wide. Ancient, **Hellespont.**

Dar es Sa·laam or **Dar-es-Sa·laam** (där′ es sə läm′), a seaport in Tanzania, on the Indian Ocean. 757,346.

Dar·fur (där foõr′), a province in the W Sudan. 3,093,699; 191,650 sq. mi. (496,374 sq. km).

Dar·ien or **Dar·ién** (där yen′), **Gulf of,** an arm of the Caribbean between NE Panama and NW Colombia.

Dar·jee·ling (där jē′ling), a town in West Bengal, in NE India: mountain resort. 42,700.

Darling (där′ling), a river in SE Australia, flowing SW into the Murray River. 1160 mi. (1870 km) long.

Dar′ling Range′, a range of low mountains along the SE coast of Australia. Highest peak, 1910 ft. (580 m).

Dar·ling·ton (där′ling tən), a city in S Durham, in NE England. 97,800.

Darm·stadt (därm′stat, -shtät′), a city in SW central Germany, S of Frankfurt. 133,600.

Dart·moor (därt′mŏŏr, -môr, -mōr), a rocky plateau in SW England, in Devonshire. ab. 20 mi. (30 km) long.

Dart·mouth (därt′məth), a coastal city in S Nova Scotia, in SE Canada, on Halifax harbor across from Halifax. 65,243.

Dar·win (där′win), a seaport in and the capital of Northern Territory, in N Australia. 73,300.

Dasht-i-Ka·vir (däsht′ē kə vēr′), a salt desert in N central Iran. ab. 18,000 sq. mi. (46,620 sq. km). Also called **Kavir Desert, Great Salt Desert.**

Dasht-i-Lut (däsht′ē lōōt′), a desert in E central Iran. ab. 20,000 sq. mi. (52,000 sq. km).

Da·tong (dä′tông′) also **Tatung,** a city in N Shanxi province, in NE China. 967,608.

Dau·ga·va (dou′gä vä′), Latvian name of DVINA.

Dau·gav·pils (dou′gäf pēls′), a city in SE Latvia, on the Dvina. 128,200. Russian, **Dvinsk.**

Dau·phi·né (dō fē nā′), a historical region and former province of SE France.

Da·vao (dä vou′, dä′vou), a seaport on SE Mindanao, in the S Philippines. 850,000.

Davao′ Gulf′, a gulf of the Pacific Ocean on the SE coast of Mindanao, Philippines.

Dav·en·port (dav′ən pôrt′, -pōrt′), a city in E Iowa, on the Mississippi River. 95,333.

Da′vis Strait′ (dā′vis), a strait between Canada and Greenland, connecting Baffin Bay and the Atlantic. 200–500 mi. (320–800 km) wide.

Daw·son (dô′sən), a town in NW Canada: former capital of Yukon Territory. 1650.

Daw′son Creek′, a city in NE British Columbia, Canada: SE terminus of the Alaska Highway. 10,544.

Day·ton (dāt′n), a city in SW Ohio. 182,044.

Day·to′na Beach′ (dā tō′nə), a city in NE Florida: seashore resort. 61,921.

DC or **D.C.,** District of Columbia.

DE, Delaware.

Dead′ Sea′, a salt lake between Israel and Jordan: the lowest lake in the world. ab. 390 sq. mi. (1010 sq. km); 1293 ft. (394 m) below sea level.

Dear·born (dēr′bərn, -bôrn), a city in SE Michigan, near Detroit. 89,286.

Dear′born Heights′, a city in SE Michigan, near Detroit. 60,838.

Death′ Val′ley, an arid basin in E California and S Nevada: lowest land in North America. 280 ft. (85 m) below sea level.

Death′ Val′ley Na′tional Mon′ument, a national monument in E California, including most of Death Valley. 3231 sq. mi. (8368 sq. km).

Deau·ville (dō′vil, dō vēl′), a coastal resort in NW France, S of Le Havre. 5655.

De·bre·cen (deb′rət sen′), a city in E Hungary. 217,000.

De·cap·o·lis (di kap′ə lis), a region in the NE part of ancient Palestine: confederacy of ten cities in the 1st century B.C.

De·ca·tur (di kā′tər), a city in central Illinois. 83,885.

Dec·can (dek′ən), **1.** the peninsula of India S of the Narbada River. **2.** a plateau region in S India between the Narbada and Krishna rivers.

Dee (dē), **1.** a river in NE Scotland, flowing E into the North Sea at Aberdeen. 90 mi. (145 km) long. **2.** a river in N Wales and W England, flowing E and N into the Irish Sea. ab. 70 mi. (110 km) long.

Deep′ South′, the southeastern section of the

U.S., usu. including South Carolina, Georgia, Alabama, Mississippi, and Louisiana.

De·hi·wa·la-Mount La·vin·i·a (de/hi wä/lə-mount/ lə vin/ē ə), a city in SW Sri Lanka, on the Indian Ocean. 173,529.

Deh·ra Dun (dā/rə dōōn/), a city in NW Uttar Pradesh, in N India. 294,000.

Del., Delaware.

Del·a·go·a Bay/ (del/ə gō/ə, del/-), an inlet of the Indian Ocean, in S Mozambique. 55 mi. (89 km) long.

Del·a·ware (del/ə wâr/), **1.** a state in the E United States, on the Atlantic coast. 666,168. 2057 sq. mi. (5330 sq. km). *Cap.:* Dover. *Abbr.:* DE, Del. **2.** a river flowing S from SE New York, along the boundary between Pennsylvania and New Jersey into Delaware Bay. 296 mi. (475 km) long.

Del/aware Bay/, an inlet of the Atlantic between E Delaware and S New Jersey. ab. 70 mi. (115 km) long.

Delft (delft), a city in W Netherlands. 88,074.

Del·ga·do (del gä/dō), **Cape,** a cape at the NE extremity of Mozambique.

Del·hi (del/ē), **1.** a union territory in N India. 6,220,400; 574 sq. mi. (1487 sq. km). **2.** the capital of this territory: former capital of the old Mogul Empire; administrative headquarters of British India 1912–29. 5,714,000. Compare NEW DELHI.

Del·mar/va Penin/sula (del mär/və), a peninsula between Chesapeake and Delaware bays including most of Delaware and those parts of Maryland and Virginia E of Chesapeake Bay. Compare EASTERN SHORE.

De·los (dē/los, del/ōs), a Greek island in the Cyclades, in the SW Aegean: legendary birthplace of Apollo and Artemis. —**De·li·an** (dē/lē ən, dēl/-yən), *adj., n.*

Del·phi (del/fī), an ancient city in central Greece: site of an oracle of Apollo. —**Del/phi·an** (-fē ən), *n., adj.* —**Del/phic** (-fik), *adj.*

Dem·a·vend (dem/ə vend/), a mountain in N Iran: highest peak of the Elburz Mountains. 18,606 ft. (5671 m).

Dem·e·ra·ra (dem/ə rär/ə, -râr/ə), a river in E Guyana flowing N into the Atlantic Ocean at Georgetown. 215 mi. (346 km) long.

De·na·li Na·tion·al Park (də nä′lē), a national park in S central Alaska, including Mount McKinley. 7370 sq. mi. (19,088 sq. km). Formerly, **Mount McKinley National Park.**

Den·bigh·shire (den′bē shēr′, -shər), a historic county in Clwyd in N Wales. Also called **Den′bigh.**

Den Haag (den häкн′), a Dutch name of The HAGUE.

Den·mark (den′märk), a kingdom in N Europe, on the Jutland peninsula and adjacent islands. 5,130,000; 16,576 sq. mi. (42,930 sq. km). *Cap.:* Copenhagen. —**Dane** (dān), *n.* —**Dan′ish**, *adj.*

Den′mark Strait′, a strait between Iceland and Greenland. 130 mi. (210 km) wide.

Den·pa·sar or **Den Pa·sar** (den pä′sär), a city on S Bali, in S Indonesia. 261,263.

Den·ton (den′tn), a city in N Texas. 66,270.

D'En′tre·cas′teaux Is′lands (dän′trə kas′tō, dän′-), a group of islands in Papua New Guinea, off the E tip of New Guinea.

Den·ver (den′vər), the capital of Colorado, in the central part. 467,610.

Der·bent (dər bent′), a seaport in SE Dagestan in the SW Russian Federation, on the Caspian Sea. 69,000.

Der·by (dûr′bē; *Brit.* där′-), **1.** a city in Derbyshire, in central England. 215,200. **2.** DERBYSHIRE.

Der′by·shire′ (-shēr′, -shər), a county in central England. 918,700; 1060 sq. mi. (2630 sq. km). Also called **Derby.**

Der·ry (der′ē), **1.** a county in NW Northern Ireland. 186,800; 798 sq. mi. (2067 sq. km). **2.** its county seat, a seaport. 62,700. Also called **Londonderry.**

Der·went (dûr′wənt), a river in S Australia, in S Tasmania, flowing SE to the Tasman Sea. 107 mi. (170 km) long.

Des·chutes (dā shōōt′), a river flowing N from the Cascade Range in central Oregon to the Columbia River. 250 mi. (400 km) long.

Des Moines (də moin′), **1.** the capital of Iowa, in the central part, on the Des Moines River. 193,187. **2.** a river flowing SE from SW Minnesota through Iowa to the Mississippi River. ab. 530 mi. (850 km) long. —**Des Moines′i·an**, *n.*

De·sna (də snä′), a river in the W Russian Federation flowing S to join the Dnieper River near Kiev in Ukraine. ab. 500 mi. (800 km) long.

Des Plaines (des plānz′), a city in NE Illinois, near Chicago. 53,223.

Des·sau (des′ou), a city in NE central Germany, SW of Berlin. 102,000.

De·troit (di troit′), **1.** a city in SE Michigan, on the Detroit River. 1,027,974. **2.** a river in SE Michigan, flowing S from Lake St. Clair to Lake Erie, forming part of the boundary between the U.S. and Canada. ab. 32 mi. (52 km) long.

Deutsch·land (doich′länt′), German name of GERMANY.

De·ven·ter (dā′vən tər), a city in E Netherlands. 66,062.

Dev′il's Is′land, a small island off the coast of French Guiana: former French penal colony. French, **Île du Diable.**

Dev·on·shire (dev′ən shēr′, -shər), a county in SW England. 1,010,000; 2591 sq. mi. (6710 sq. km). Also called **Dev′on.**

De·zhnev (dezh′nef, -nē ôf′, -of′), **Cape,** a cape in the NE Russian Federation in Asia, on the Bering Strait: the northeasternmost point of Asia.

D.F., Distrito Federal.

Dhah·ran (dä rän′), a city in E Saudi Arabia: oil center. 12,500.

Dha·ka or **Dac·ca** (dak′ə, dä′kə), the capital of Bangladesh, in the central part. 3,440,147.

Dhau·la·gi·ri (dou′lə gēr′ē), a mountain in W central Nepal: a peak of the Himalayas. 26,826 ft. (8180 m).

Dia′mond Bar′ (dī′mənd, dī′ə-), a city in SW California. 53,672.

Dia′mond Head′, a promontory on SE Oahu Island in central Hawaii. 761 ft. (232 m) high.

Dié·go-Sua·rez (dyā′gō swär′es), a seaport on N Madagascar. 46,000.

Dien Bien Phu (dyen′ byen′ foo′), a town in NW Vietnam: site of defeat of French forces by Vietminh 1954, bringing to an end the French rule of Indochina.

Di·eppe (dē ep′), a seaport in N France, on the English Channel. 26,111.

Di·jon (dē zhôn′), a city in E central France. 145,569.

Di·li (dil′ē), a city on NE Timor, in S Indonesia. 60,150.

Di·nar′ic Alps′ (di nar′ik), a mountain range extending S along the Adriatic coast of Croatia to Albania.

Di′nosaur Na′tional Mon′ument, a national monument in NE Utah and NW Colorado: site of prehistoric animal fossils. 322 sq. mi. (834 sq. km).

Di′o·mede Is′lands (dī′ə mēd′), two islands in the Bering Strait, one belonging to the Russian Federation (**Big Diomede**), ab. 15 sq. mi. (39 sq. km), and one belonging to the U.S. (**Little Diomede**), ab. 4 sq. mi. (10 sq. km): separated by the International Date Line.

Di·re·da·wa or **Di·re Da·wa** (dē′rä də wä′), a city in E Ethiopia. 98,104.

Dis′mal Swamp′, a swamp in SE Virginia and NE North Carolina. ab. 30 mi. (48 km) long; ab. 600 sq. mi. (1500 sq. km).

Dis′trict of Colum′bia, a federal area in the E United States, on the Potomac, coextensive with the federal capital, Washington. 606,900; 69 sq. mi. (179 sq. km). *Abbr.:* DC, D.C.

Dis·tri·to Fe·de·ral (dēs trē′tô fe′the räl′), *Spanish.* Federal District. *Abbr.:* D.F.

Di·u (dē′oō), a small island off the coast of Gujarat state in W India: part of the union territory of Daman and Diu.

Dix·ie (dik′sē), the southern states of the United States, esp. those that were part of the Confederacy. Also called **Dix′ie·land′, Dix′ie Land′.**

Di·yar·ba·kir (di yär′buk ər, -yär′bä kēr′), a city in SE Turkey in Asia, on the Tigris River. 305,259.

Dja′ja Peak′ (jä′yä), Puncak Jaya.

Dja·ja·pu·ra (jä′yə pŏŏr′ə), Jayapura.

Dja·kar·ta (jə kär′tə), Jakarta.

Djam·bi (jäm′bē), Jambi.

Djeb·el Druze (jeb′əl drŏŏz′), Jebel ed Druz.

Djer·ba or **Jer·ba** (jer′bə), an island off the SE coast of Tunisia: Roman ruins. 92,269; 197 sq. mi. (510 sq. km).

Dji·bou·ti (ji bōō′tē), **1.** Formerly, **French Somaliland, French Territory of the Afars and Issas.** a republic in E Africa, on the Gulf of Aden: a former overseas territory of France; gained independence 1977. 484,000; 8960 sq. mi.

(23,200 sq. km). **2.** the capital of this republic, in the SE part. 290,000. —**Dji·bou′ti·an,** *adj., n.*

Djok·ja·kar·ta (jōk′yä kär′tä), JOGJAKARTA.

Dnepr (*Russ.* dnyepr), DNIEPER.

Dne·pro·dzer·zhinsk (nep′rō dər· zhinsk′), a city in E central Ukraine, on the Dnieper River, W of Dnepropetrovsk. 279,000.

Dne·pro·pe·trovsk (nep′rō pi trôfsk′), a city in E central Ukraine, on the Dnieper River. 1,179,000.

Dnes·tr (*Russ.* dnyestr), DNIESTER.

Dnie·per or **Dne·pr** (nē′pər; *Russ.* dnyepr), a river rising in the W Russian Federation, flowing S through Belarus and Ukraine to the Black Sea. 1400 mi. (2250 km) long.

Dnies·ter or **Dnes·tr** (nē′stər; *Russ.* dnyestr), a river rising in SW Ukraine, flowing SE from the Carpathian Mountains through Ukraine and Moldova to the Black Sea. ab. 875 mi. (1410 km) long.

Do·bru·ja (dō′brŏŏ jə), a region in SE Romania and NE Bulgaria, between the Danube River and the Black Sea. 2970 sq. mi. (7690 sq. km). Romanian, **Do·bro·gea** (dô′bro jä′).

Do·dec·a·nese (dō dek′ə nēs′, -nēz′, dō′-dek ə-), a group of 12 Greek islands in the Aegean, off the SW coast of Turkey: belonged to Italy 1911–45. 1035 sq. mi. (2680 sq. km).

Dodge′ Cit′y, a city in SW Kansas, on the Arkansas River: important frontier town and railhead on the old Santa Fe route. 18,001.

Do·do·ma (dō′dō mä, dō′də-), the capital of Tanzania, in the NE central part. 45,703.

Do·do·na (də dō′nə), an ancient town in NW Greece, in Epirus: the site of an oracle of Zeus.

Dog′ger Bank′ (dô′gər, dog′ər), a shoal in the North Sea, between N England and Denmark: fishing grounds.

Do·ha (dō′hä), the capital of Qatar, on the Persian Gulf. 217,294.

Do·lo·mites (dō′lə mits′, dol′ə-), a mountain range in N Italy: a part of the Alps. Highest peak, Marmolada, 11,020 ft. (3360 m). Also called **Do′lomite Alps′.**

Dom·i·ni·ca (dom′ə nē′kə, də min′i kə), **Commonwealth of,** an island republic, one of the Windward Islands, in the E West Indies: a former

British colony; gained independence 1978. 94,191; 290 sq. mi. (751 sq. km). *Cap.*: Roseau. —**Dom·i·ni·can**, *adj.*, *n.*

Do·min·i·can Repub·lic (də min'i kən), a republic in the West Indies, occupying the E part of the island of Hispaniola. 6,700,000; 19,129 sq. mi. (49,545 sq. km). *Cap.*: Santo Domingo. Formerly, **Santo Domingo, San Domingo.** —**Do·min·i·can**, *adj.*, *n.*

Don (don), a river flowing generally S from Tula in the Russian Federation in Europe to the Sea of Azov. ab. 1200 mi. (1930 km) long.

Do·nau (dō'nou), German name of the DANUBE.

Don·cas·ter (dong'kas tər; *Brit.* dong'kə stər), a city in South Yorkshire, in N England. 290,100.

Don·e·gal (don'i gôl', don'i gôl'), a county in the N Republic of Ireland. 129,428; 1865 sq. mi. (4830 sq. km).

Do·nets (də nets'), **1.** a river rising in the SW Russian Federation near Belgorod, flowing SE through Ukraine to the Don River. ab. 650 mi. (1045 km) long. **2.** Also called **Donets' Ba'sin.** an area S of this river, in E Ukraine: coal-mining region. 9650 sq. mi. (24,995 sq. km).

Do·netsk (də netsk'), a city in E Ukraine, in the Donets Basin. 1,110,000. Formerly, **Stalin, Stalino, Yuzovka.**

Dong·ting (dông'ting'), also **Tungting,** a lake in SE China, in Hunan province. 1450 sq. mi. (3755 sq. km).

Don'ner Pass' (don'ər), a mountain pass in the Sierra Nevada, in E California. 7135 ft. (2175 m) high.

Dor·ches·ter (dôr'ches/tər, -chə stər), a town in S Dorsetshire, in S England: named *Casterbridge* in Thomas Hardy's novels. 14,049.

Dor·dogne (dôr dôn'yə), a river in SW France, flowing W to the Gironde estuary. 300 mi. (485 km) long.

Dor·drecht (dôr'drɛкнt), a city in SW Netherlands, on the Waal River. 108,041. Also called **Dort.**

Do·ris (dôr'is, dōr'-, dor'-), **1.** an ancient region in central Greece: traditionally the earliest home of the Dorians. **2.** a region in SW Asia Minor, on the coast of Caria: Dorian settlements. —**Do·ri·an** (dôr'ē ən, dōr'-), *adj.*, *n.*

Dor·set·shire (dôr′sit shēr′, -shər), a county in S England. 648,600; 1024 sq. mi. (2650 sq. km). Also called **Dor′set.**

Dort (dôrt), DORDRECHT.

Dort·mund (dôrt′mənd), a city in W Germany. 583,600.

Do·than (dō′thən), a city in SE Alabama. 53,589.

Dou·ai or **Dou·ay** (dōō ā′), a city in N France, SE of Calais. 44,515.

Dou·a·la (dōō ä′lä), a seaport in W Cameroon. 637,000.

Doubs (dōō), a river in E France, flowing into the Saône River. ab. 260 mi. (420 km) long.

Doug·las (dug′ləs), the capital of the Isle of Man: resort. 20,368.

Dou·ro (Port. dō′rōō), a river in SW Europe, flowing W from N Spain through N Portugal to the Atlantic. ab. 475 mi. (765 km) long. Spanish, **Duero.**

Do·ver (dō′vər), **1.** a seaport in E Kent, in SE England: point nearest the coast of France. 104,300. **2. Strait of.** French, **Pas de Calais.** a strait between England and France, connecting the English Channel and the North Sea: narrowest point 20 mi. (32 km). **3.** the capital of Delaware, in the central part. 23,512.

Down (doun), a county in SE Northern Ireland. 339,200; 945 sq. mi. (2448 sq. km).

Down′ East′, 1. NEW ENGLAND. **2.** the state of Maine. —**Down′-East′er,** n.

Dow·ney (dou′nē), a city in SW California, near Los Angeles. 91,444.

Downs (dounz), **the, 1.** a range of low ridges in S and SW England. **2.** a roadstead in the Strait of Dover, between SE England and Goodwin Sands.

Dra·kens·berg (drä′kənz bûrg′), a mountain range in the E Republic of South Africa: highest peak, 10,988 ft. (3350 m). Also called **Quathlamba.**

Drake′ Pas′sage (drāk), a strait between S South America and the South Shetland Islands, connecting the Atlantic and Pacific oceans.

Dra·va (drä′və), a river in S central Europe, flowing SE from the Alps in S Austria, along the border between Croatia and Hungary, into the Danube, in Yugoslavia. 450 mi. (725 km) long. German, **Drau** (drou).

Dren·the (dren/tə), a province in E Netherlands. 436,586; 1011 sq. mi. (2620 sq. km).

Dres·den (drez/dən), the capital of Saxony in E Germany, on the Elbe River. 518,057.

Drin (drēn), a river in S Europe, flowing generally NW from SW Macedonia through N Albania into the Adriatic. 180 mi. (290 km) long.

Dri·na (drē/nə, -nä), a river in SE Europe, flowing N along the border between Serbia and Bosnia and Herzegovina to the Sava River. 160 mi. (258 km) long.

Drog·he·da (drô/i də), a seaport in the E Republic of Ireland, on the Boyne River: captured by Cromwell in 1649. 23,173.

Dru/ry Lane/ (drŏŏr/ē), a street in London, England, formerly notable for its theaters.

Dry/ Tor·tu/gas (tôr tōō/gəz), a group of ten small islands at the entrance to the Gulf of Mexico W of Key West: a part of Florida; the site of Fort Jefferson.

Du·bai (dōō bī/), **1.** an emirate in the NE United Arab Emirates, on the Persian Gulf. 419,104. **2.** the capital of the emirate of Dubai. 265,702.

Dub·lin (dub/lin), **1.** the capital of the Republic of Ireland, in the E part, on the Irish Sea. 422,220. **2.** a county in E Republic of Ireland. 1,001,985; 356 sq. mi. (922 sq. km). *Co. seat:* Dublin. Irish, **Baile Átha Cliath. —Dub/lin·er,** *n.*

Du·brov·nik (dōō/brôv nik), a seaport in S Croatia, on the Adriatic: resort. 58,920. Italian, **Ra·gusa.**

Du·buque (də byōōk/), a city in E Iowa, on the Mississippi River. 57,546.

Dud·ley (dud/lē), a borough in West Midlands, central England, near Birmingham. 302,600.

Due·ro (dwe/rô), Spanish name of **Douro.**

Duis·burg (dys/bŏŏrk), a city in W Germany, at the junction of the Rhine and Ruhr rivers: the largest river port in Europe. 525,200.

Du·luth (də lōōth/), a port in E Minnesota, on Lake Superior. 85,493.

Dum·bar·ton (dum bär/tn), **1.** Also, **Dunbarton.** Also called **Dum·bar·ton·shire** (dum bär/tn-shēr/, -shər) a historic county in W Scotland. **2.** a city in W Scotland, near the Clyde River. 80,105.

Dum·fries (dum frēs/), **1.** Also called **Dum·fries/shire/** (-shēr/, -shər). a historic county in S

Scotland. **2.** a burgh of Dumfries and Galloway in S Scotland: burial place of Robert Burns. 57,149.

Dumfries′ and Gal′loway, a region in S Scotland. 147,036; 2460 sq. mi. (6371 sq. km).

Dum·yat (dōōm yät′), Arabic name of DAMIETTA.

Du·na (dōō′no), Hungarian name of the DANUBE.

Du·naj (dōō′nī), Czech and Slovak name of the DANUBE.

Du·nă·rea (dōō′no Ryä), Romanian name of the DANUBE.

Dun·bar (dun bär′), a town in the Lothian region, in SE Scotland, at the mouth of the Firth of Forth: site of Cromwell's defeat of the Scots 1650. 4586.

Dun·bar·ton (dun bär′tn), DUMBARTON (def. 1).

Dun·dalk (dun′dôk), a town in Maryland, near Baltimore. 65,800.

Dun·dee (dun dē′, dun′dē), a seaport in E Scotland, on the Firth of Tay. 175,748.

Dun·e·din (dun ē′din), a seaport on SE South Island in New Zealand. 137,393.

Dun·kirk (dun′kûrk), a seaport in N France: site of the evacuation of Allied forces under German fire 1940. 73,618. French, **Dun·kerque** (dœn-kerk′).

Dun·si·nane (dun′so nän′, dun′so nän′), a hill NE of Perth, in central Scotland. 1012 ft. (308 m).

Du·que de Ca·xi·as (dōō′ki di kä shē′äs), a city in SE Brazil: a suburb of Rio de Janeiro. 537,308.

Du·ran·go (də rang′gō, -räng′-). **1.** a state in N Mexico. 1,182,320; 47,691 sq. mi. (123,520 sq. km). **2.** the capital of this state, in the S part. 321,148.

Du·raz·zo (dōō Rät′tsô), Italian name of DURRËS.

Dur·ban (dûr′bən), a seaport in SE Natal, in the E Republic of South Africa. 982,075.

Dur·ham (dûr′əm, dur′-). **1.** a county in NE England. 598,700; 940 sq. mi. (2435 sq. km). **2.** a city in N North Carolina. 136,611.

Dur·rës (dōōr′əs), a seaport in W Albania, on the Adriatic: important ancient city. 217,000. Italian, **Durazzo.**

Du·shan·be (dōō shän′bə, -shäm′-, dyōō-), the capital of Tajikistan, in the E part. 595,000. Formerly, **Dyushambe** (before 1929), **Stalinabad** (1929–61).

Düs·sel·dorf (dōōs′əl dôrf′), the capital of North

Rhine–Westphalia, in W Germany, on the Rhine. 563,400.

Dust/ Bowl/, the region in the S central U.S. that suffered from dust storms in the 1930s.

Dutch/ Bor/neo, the former name of the southern and larger part of the island of Borneo: now part of Indonesia.

Dutch/ East/ In/dies, a former name of the Republic of INDONESIA.

Dutch/ Guian/a, a former name of SURINAME.

Dutch/ New/ Guin/ea, a former name of IRIAN JAYA.

Dutch/ West/ In/dies, a former name of the NETHERLANDS ANTILLES.

Dvi·na (dvē/nə), **1.** Also called **Western Dvina.** Latvian, **Daugava.** a river rising in the Valdai Hills in the W Russian Federation, flowing W through Belarus and Latvia to the Baltic Sea at Riga. ab. 640 mi. (1030 km) long. **2.** Also called **Northern Dvina.** a river in the N Russian Federation in Europe, flowing NW into the White Sea. ab. 470 mi. (750 km) long.

Dvi/na Bay/, an arm of the White Sea, in the NW Russian Federation in Europe. Formerly, **Gulf of Archangel.**

Dvinsk (dvyēnsk), Russian name of DAUGAVPILS.

Dy·fed (duv/id), a county in Wales. 343,200; 2227 sq. mi. (5767 sq. km).

Dyu·sham·be (dyōō shäm/bə), a former name of DUSHANBE.

Dzer·zhinsk (dər zhinsk/), a city in the central Russian Federation in Europe, W of Nizhni Novgorod. 281,000.

Dzham·bul (jäm bōōl/, jum-), a city in S Kazakhstan, NE of Chimkent. 315,000.

Dzi·bil·chal·tun (dzē bēl/chäl tōōn/), a large ancient Mayan ceremonial and commercial center near Mérida, Mexico, founded perhaps as early as 3000 B.C. and in continuous use until the 16th century.

Dzun·ga·ri·a (dzōōng gâr/ē ə, zōōng-), a region in N Xinjiang Uygur, China: a Mongol kingdom during the 11th to 14th centuries.

E

Ea·ling (ē/ling), a borough of Greater London, England. 297,600.

East/ An/glia, 1. a region in E England, consisting chiefly of Norfolk and Suffolk. **2.** a kingdom of the Anglo-Saxon heptarchy in E Britain. —**East/ An/glian,** *adj., n.*

East/ Bengal/, a part of the former Indian province of Bengal: now coextensive with Bangladesh. Compare BENGAL (def. 1).

East/ Berlin/. See under BERLIN.

East·bourne (ēst/bôrn, -bōrn, -bərn), a seaport in East Sussex, in SE England: resort. 72,700.

East/ Chi/na Sea/, a part of the N Pacific, bounded by China, Japan, the Ryukyu Islands, and Taiwan. 480,000 sq. mi. (1,243,200 sq. km).

East/ Coast/, the region of the U.S. bordering on the Atlantic Ocean.

Eas/ter Is/land, an island in the S Pacific, ab. 2000 mi. (3180 km) W of and belonging to Chile: gigantic statues. 1867; ab. 45 sq. mi. (117 sq. km). Also called **Rapa Nui.** Spanish, **Isla de Pascua.**

East/ern Ghats/, a low mountain range in S India along the E margin of the Deccan plateau and parallel to the coast of the Bay of Bengal.

East/ern Hem/isphere, the part of the globe east of the Atlantic, including Asia, Africa, Australia, and Europe, their islands, and surrounding waters.

East/ern Ro/man Em/pire, the eastern part of the Roman Empire, esp. after the division in A.D. 395, having its capital at Constantinople: survived the fall of the Western Roman Empire in A.D. 476. Compare BYZANTINE EMPIRE.

East/ern shore/, the eastern shore of Chesapeake Bay, including parts of Maryland, Delaware, and Virginia.

East/ern Thrace/, See under THRACE (def. 2).

East/ Flan/ders, a province in W Belgium. 1,325,419; 1150 sq. mi. (2980 sq. km). *Cap.:* Ghent.

East/ Fri/sians, See under FRISIAN ISLANDS.

East/ Ger/many, a former country in central Europe, created in 1949 from the Soviet zone of occupied Germany established in 1945: reunited with West Germany in 1990. 16,340,000; 41,827 sq. mi. (108,333 sq. km). *Cap.:* East Berlin. Official name, **German Democratic Republic.** Compare GERMANY. —**East′ Ger′man,** *adj., n.*

East′ Ham′, a former borough, now part of Newham, in SE England, near London.

East′ Hart′ford, a town in central Connecticut. 52,563.

East′ In′dies, (esp. formerly) **1.** the Malay Archipelago. **2.** SE Asia, including India, Indonesia, and the Malay Archipelago. Also called **East′ In′dia, the Indies.** —**East′ In′dian,** *adj., n.*

East′ Lan′sing, a city in S Michigan. 50,677.

East′ Lon′don, a seaport in the SE Cape of Good Hope province, in the S Republic of South Africa. 130,000.

East′ Los′ An′geles, an urban community in SW California, near Los Angeles. 126,379.

East′ Lo′thi·an (lō′thē ən), a historic county in SE Scotland.

East′ Or′ange, a city in NE New Jersey, near Newark. 73,552.

East′ Pak′istan, a former province of Pakistan: since 1971 constitutes the country of Bangladesh.

East′ Prov′idence, a town in NE Rhode Island, near Providence. 50,380.

East′ Prus′sia, a former province in NE Germany: an enclave separated from Germany by the Polish Corridor; now divided between Poland and the Russian Federation. *Cap.:* Königsberg. —**East′ Prus′sian,** *adj., n.*

East′ Punjab′, the E part of the former province of Punjab, in British India: now part of Punjab state, India. —**East′ Punja′bi,** *adj., n.*

East′ Ri′ding (rī′ding), a former administrative division of Yorkshire, in NE England.

East′ Riv′er, a strait in SE New York separating Manhattan Island from Long Island and connecting New York Bay and Long Island Sound.

East′ Sus′sex, a county in SE England. 698,000; 693 sq. mi. (1795 sq. km).

Eau Claire (ō′ klâr′), a city in W Wisconsin. 56,856.

Eb·la (eb′lə, ē′blə), an ancient city whose remains are located near Aleppo in present-day Syria, the site of the discovery in 1974–75 of cuneiform tablets documenting a culture of the 3rd millennium B.C.

Eb·o·ra·cum (eb′ə rā′kəm), ancient name of YORK, England.

E·bro (ē′brō, ā′brō), a river flowing SE from N Spain to the Mediterranean. ab. 470 mi. (755 km) long.

Ec·bat·a·na (ek bat′n ə), the ancient capital of Media: at the site of modern Hamadan in W Iran.

Ec·ua·dor (ek′wə dôr′), a republic in NW South America. 9,640,000; 109,483 sq. mi. (283,561 sq. km). *Cap.:* Quito. —**Ec′ua·do′ran, Ec′ua·do′re·an, Ec′ua·do′ri·an,** *adj., n.*

Ed′dy·stone Rocks′ (ed′ə stən), a group of rocks near the W end of the English Channel, SW of Plymouth, England.

E·de (ā dā′, ā′dā *for 1;* ā′də *for 2),* **1.** a city in SW Nigeria. 221,900. **2.** a city in the central Netherlands. 91,246.

E·der (ā′dər), a river in central Germany, mainly in Hesse and flowing E to Kassel. 110 mi. (177 km) long.

E·des·sa (i des′ə), an ancient city in NW Mesopotamia, on the site of modern Urfa, in Turkey: an early center of Christianity.

E·di·na (i dī′nə), a city in SE Minnesota, near Minneapolis. 46,073.

Ed·in·burgh (ed′n bûr′ə, -bur′ə; *esp. Brit.* -brə), the capital of Scotland, in the SE part, in the Lothian region. 470,085.

E·dir·ne (ā dēr′nə), a city in NW Turkey, in the European part. 71,927. Also called **Adrianople.**

Ed·i·son (ed′ə sən), a township in central New Jersey. 70,193.

Ed·mond (ed′mənd), a city in central Oklahoma. 52,315.

Ed·mon·ton (ed′mən tən), the capital of Alberta, in the central part, in SW Canada. 573,982.

E·do (ed′ō, ā′dō), a former name of TOKYO.

E·dom (ē′dəm), an ancient country between the Dead Sea and the Gulf of Aqaba, bordering ancient Palestine. —**E′dom·ite′,** *n., adj.*

Ed′sel Ford′ Range′ (ed′səl), a mountain range in Antarctica, E of the Ross Sea.

Ed·ward (ed′wərd), **Lake,** a lake in central Africa, between Uganda and Zaire: a source of the Nile. 830 sq. mi. (2150 sq. km).

Ed·wards Plateau, a highland area in SW Texas. 2000–5000 ft. (600–1500 m) high.

EEC, European Economic Community.

E·ga·di (eg′ə dē), a group of islands in the Mediterranean Sea off the coast of W Sicily. 15 sq. mi. (39 sq. km). Also called **Aegadian Islands.** Ancient, **Aegates.**

E·ger (ā′gər), German name of OHŘE.

E·gypt (ē′jipt), a country in NE Africa on the Mediterranean and Red seas. 49,280,000; 386,198 sq. mi. (1,000,252 sq. km). *Cap.:* Cairo. Arabic, **Misr.** Official name, **Arab Republic of Egypt.** —E·gyp′tian, *n., adj.*

Ei·lat (ā lät′), ELAT.

Eind·ho·ven (int′hō′vən), a city in S Netherlands. 195,669.

Eir·e (âr′ə, i′rə, âr′ē, i′rē), **1.** the Irish name of IRELAND. **2.** a former name of the Republic of IRELAND.

E·ka·te·rin·burg (i kat′ər in bûrg′), a city in the Russian Federation in Asia, in the Ural Mountains. 1,367,000. Formerly (1924–91), **Sverdlovsk.**

El Aa·iún (el′ ä yōōn′), the capital of Western Sahara. 96,784

El A·la·mein (el ä′lä mān′, -ä′lə-), a town in the N coast of Egypt, ab. 70 mi. (113 km) W of Alexandria: decisive British victory in World War II, 1942. Also called **Alamein.**

E·lam (ē′lam), an ancient kingdom E of Babylonia and N of the Persian Gulf. *Cap.:* Susa. —E′lam·ite′, *n., adj.*

E·lat or **Ei·lat** or **E·lath** (ā lät′), a seaport at the N tip of the Gulf of Aqaba, in S Israel: resort. 19,600.

E·lâ·zığ (ā′lə zē′), a city in central Turkey. 483,715.

El·ba (el′bə), an Italian island in the Mediterranean, between Corsica and Italy: site of Napoleon's first exile 1814–15. 26,830; 94 sq. mi. (243 sq. km).

El·be (el′bə), elb), a river in central Europe, flowing from W Czechoslovakia NW through Germany to the North Sea. 725 mi. (1165 km) long. Czech, **Labe.**

El·bert (el′bərt), **Mount,** a mountain in central Colorado, in the Sawatch range: highest peak of the Rocky Mountains. 14,431 ft. (4399 m).

El·blag (el′blôngk), a seaport in N Poland. 118,500.

El·brus (el broos′), a mountain in the S Russian Federation in Europe, in the Caucasus: highest peak in Europe, 18,465 ft. (5628 m).

El·burz′ Moun′tains (el boorz′), a mountain range in N Iran, along the S coast of the Caspian Sea. Highest peak, Mt. Demavend, 18,606 ft. (5671 m).

El Ca·jon (el′ kə hōn′), a city in SW California. 88,693.

El Cap·i·tan (el kap′i tan′), a mountain in E California, in the Sierra Nevada Mountains: precipice rises over 3300 ft. (1000 m).

El·che (el′che), a city in E Spain. 173,392.

E·le·a (ē′lē ə), an ancient Greek city in SW Italy, on the coast of Lucania.

E·lek·tro·stal (i lek′trə stäl′), a city in the Russian Federation, in Europe, E of Moscow. 147,000.

El′ephant Butte′, a dam and irrigation reservoir in SW New Mexico, on the Rio Grande. Dam, 309 ft. (94 m) high.

E·leu·sis (i loo′sis), an ancient city in Greece, in Attica.

El Fai·yum (el′ fī yoom′, fā-), FAIYUM (def. 2).

El Fer·rol (el fə rōl′), a seaport in NW Spain. 88,101.

El·gin (el′jin), a city in NE Illinois. 77,010.

El Gi·za or **El Gi·zeh** (el gē′zə), GIZA.

El·gon (el′gon), an extinct volcano in E Africa, on the boundary between Uganda and Kenya. 14,176 ft. (4321 m).

El Ha·sa (el hä′sə), HASA.

E·lis (ē′lis), an ancient country in W Greece, in the Peloponnesus: site of the ancient Olympic Games.

E·liz·a·beth (i liz′ə bəth), a city in NE New Jersey. 110,002.

El Ja·di·da (el′ zhə dē′də), a city on the W central coast of Morocco. 102,000.

El Kha·lil (el′ kä lēl′), Arabic name of HEBRON.

El·las (el′las), Modern Greek name of GREECE.

Elles′mere Is′land (elz′mēr), an island in the Arctic Ocean, NW of Greenland: a part of Canada. 76,600 sq. mi. (198,400 sq. km).

El′lice Is′lands (el′is), former name of TUVALU.

El·lis Is·land (el'is), an island in upper New York Bay: a national monument and museum; former U.S. immigrant examination station.

El·lo·ra (e lôr'ə, e lōr'ə) also **Elura,** a village in S central India: important Hindu archaeological site.

El Man·su·ra (el' man sŏŏr'ə), a city in NE Egypt, in the Nile delta: scene of the defeat of the Crusaders 1250 and the capture of Louis IX by the Mamelukes. 215,000. Also called **Mansura.**

El Mis·ti (el mēs'tē), a volcano in S Peru, in the Andes. 19,200 ft. (5880 m). Also called **Misti.**

El Mon·te (el mon'tē), a city in SW California, near Los Angeles. 106,209.

El O·beid (el' ō bād') also **Al-Ubayyid,** a city in the central Sudan. 140,024.

El Pas·o (el pas'ō), a city in W Texas, on the Rio Grande. 515,342.

El Sal·va·dor (el sal'və dôr'), a republic in NW Central America. 5,480,000; 13,176 sq. mi. (34,125 sq. km). *Cap.:* San Salvador. Also called **Salvador.**

El·se·ne (el'sə nə), Flemish name of IXELLES.

El·si·nore (el'sə nôr', -nōr'), HELSINGØR.

E·lu·ra (e lŏŏr'ə), ELLORA.

E·ly (ē'lē), **Isle of,** a former county in E England: now part of Cambridgeshire.

E·lyr·i·a (i lēr'ē ə), a city in N Ohio. 56,746.

Em·bar·ca·de·ro (em bär'kə dâr'ō), a waterfront section in San Francisco.

E·mi·lia-Ro·ma·gna (ā mēl'yə rō män'yə), a region in N Italy. 3,924,199; 8547 sq. mi. (22,135 sq. km).

Emp·ty Quar'ter, RUB' AL KHALI.

En·ci·ni·tas (en'sə nē'təs), a city in SW California. 55,386.

En·der·by Land' (en'dər bē), a part of the coast of Antarctica, E of Queen Maud Land: discovered 1831.

En·field (en'fēld'), a borough of Greater London, England. 261,900.

En·ga·dine (eng'gə dēn', eng'gə dēn'), the valley of the Inn River in E Switzerland: resorts. 60 mi. (97 km) long.

En·gels (eng'gəlz), a city in the Russian Federation in Europe, on the Volga River opposite Saratov. 182,000.

Eng·land (ing′glənd *or, often,* -lənd), the largest division of the United Kingdom, constituting, with Scotland and Wales, the island of Great Britain. 55,780,000; 50,360 sq. mi. (130,439 sq. km). *Cap.:* London. —**Eng′lish,** *adj., n.*

Eng′lish Chan′nel, an arm of the Atlantic between S England and N France, connected with the North Sea by the Strait of Dover. 350 mi. (565 km) long; 20–100 mi. (32–160 km) wide.

E·nid (ē′nid), a city in N Oklahoma. 45,309.

En·i·we·tok (en′ə wē′tok), an atoll in the NW Marshall Islands: site of atomic and hydrogen bomb tests 1947–52.

En·na (en′ə), a city in central Sicily, in SW Italy. 27,705.

En·sche·de (en′sкнə dā′), a city in E Netherlands. 144,346.

En·se·na·da (en′sə nä′də), a seaport in N Baja California, in NW Mexico. 175,400.

En·teb·be (en teb′ə, -teb′ē), a town in S Uganda, on Lake Victoria: former capital. 21,096.

E·nu·gu (ā nōō′gōō), a city in SE Nigeria. 228,400.

Eph·e·sus (ef′ə səs), an ancient city in W Asia Minor, S of Smyrna (Izmir): famous temple of Artemis, or Diana; early Christian community. —**E·phe·sian** (i fē′zhən), *adj., n.*

Ep·i·dau·rus (ep′i dôr′əs), an ancient town in S Greece, in Argolis: sanctuary of Asclepius; outdoor theater.

É·pi·nal (ā pē nAl′), a city in NE France. 42,810.

E·pi·rus (i pī′rəs), **1.** an ancient district in what is now NW Greece and S Albania. **2.** a modern region in NW Greece. 324,541; 3573 sq. mi. (9255 sq. km). —**E·pi·rote** (-rōt), *n.*

Ep′ping For′est (ep′ing), a park in E England, NE of London: formerly a royal forest.

Ep·som (ep′səm), a town in Surrey, SE England, S of London: site of a racetrack (**Ep′som Downs′**) where the annual Derby is held. 71,100. Official name, **Ep′som and Ew′ell** (yōō′əl).

E′quato′rial Guin′ea, a republic in W equatorial Africa: formerly a Spanish colony; gained independence 1968. 400,000; 10,824 sq. mi. (28,034 sq. km). *Cap.:* Malabo. Formerly, **Spanish Guinea.**

Er·bil (ēr′bil, âr′-) also **Arbil,** a town in N Iraq: built on the site of ancient Arbela. 333,903.

Er·e·bus (er′ə bəs), **Mount,** a volcano in Antarctica, on Ross Island. 13,202 ft. (4024 m).

E·rech (ē′rek, er′ek), Biblical name of Uruk.

Er·furt (ɛR′fŏŏRt), the capital of Thuringia in central Germany. 220,016.

E·rie (ēr′ē), **1. Lake,** a lake between the NE central United States and SE central Canada: the southernmost lake of the Great Lakes. 239 mi. (385 km) long; 9940 sq. mi. (25,745 sq. km). **2.** a port in NW Pennsylvania, on Lake Erie. 108,718.

E·rie Canal′, a canal in New York between Albany and Buffalo, connecting the Hudson River with Lake Erie: completed in 1825; now constitutes the major part of the New York State Barge Canal. 363 mi. (584 km) long.

Er·i·man·thos (er′ə man′thəs; *Gk.* e Rē′mänthôs), ERYMANTHUS.

Er·in (er′in), *Literary.* Ireland.

Er·i·tre·a (er′i trē′ə), a province of Ethiopia, on the Red Sea: formerly an Italian colony. 2,614,700; 47,076 sq. mi. (121,927 sq. km). *Cap.:* Asmara. —**Er′i·tre′an,** *adj., n.*

Er·na·ku·lam (er nä′kə ləm), a city in S Kerala, in SW India, on the Malabar Coast. 213,811.

Er Rif (er rif), RIF.

Er·y·man·thus (er′ə man′thəs), a mountain in S Greece, in the NW Peloponnesus. 7295 ft. (2225 m). Greek, **Erimanthos.**

Erz·ge·bir·ge (ârts′gə bēr′gə), a mountain range in central Europe, on the boundary between Germany and Czechoslovakia. Highest peak, 4080 ft. (1245 m).

Er·zu·rum or **Er·ze·rum** (er′zə rŏŏm′), a city in NE Turkey in Asia. 252,648.

Es·bjerg (es′byɛʀ), a seaport in SW Denmark. 81,385.

Es·bo (es′bŏō), Swedish name of ESPOO.

Es·caut (es kō′), French name of SCHELDT.

Es·con·di·do (es′kən dē′dō), a city in SW California. 108,635.

Es·dra·e·lon (es′drā ē′lon, -drə-, ez′-), a plain in N Israel, extending from the Mediterranean near Mt. Carmel to the Jordan River: scene of ancient battles. Also called **Plain of Jezreel.**

E·skil·stu·na (es′kil stoo̅′nə, -styoo̅′-), a city in SE Sweden, W of Stockholm. 88,508.

Es·ki·şe·hir or **Es·ki·she·hir** (es kē′shə hēr′), a city in W Turkey in Asia. 367,328.

Es·me·ral·das (es′me räl′däs), a seaport in NW Ecuador. 141,030.

Es·pa·ña (es pä′nyä), Spanish name of SPAIN.

Es·pí·ri·to San·to (e spir′i too̅ san′too̅), a state in E Brazil. 2,382,000; 15,196 sq. mi. (39,360 sq. km). *Cap.:* Vitória.

Es·poo (es′pō), a city in S Finland, W of Helsinki. 164,569. Swedish, **Esbo.**

Es·qui·line (es′kwə lin′), one of the seven hills on which ancient Rome was built.

Es·sen (es′ən), a city in W Germany in the Ruhr River valley. 623,000.

Es·se·qui·bo (es′i kwē′bō), a river flowing from S Guyana N to the Atlantic. ab. 550 mi. (885 km) long.

Es·sex (es′iks), **1.** a county in SE England. 1,521,800; 1418 sq. mi. (3670 sq. km). **2.** a kingdom of the Anglo-Saxon heptarchy in SE England.

Es′tes Park′ (es′tēz), a summer resort in N Colorado. 2703.

Es·to·ni·a (e stō′nē ə, e stōn′yə), a republic in N Europe, on the Baltic, S of the Gulf of Finland: an independent republic 1918–40; annexed by the Soviet Union 1940; regained independence 1991. 1,573,000; 17,413 sq. mi. (45,100 sq. km). *Cap.:* Tallinn. —**Es·to′ni·an,** *n., adj.*

Es·to·ril (esh′tə rēl′), a town in W Portugal, W of Lisbon: seaside resort. 24,300.

Es·tre·ma·du·ra (es′trə mə door′ə) also **Extremadura,** a region in W Spain, formerly a province.

E·thi·o·pi·a (ē′thē ō′pē ə), **1.** Formerly, **Abyssinia.** a republic in E Africa: formerly a monarchy. 46,000,000; 471,800 sq. mi. (1,221,900 sq. km). *Cap.:* Addis Ababa. **2.** Also called **Abyssinia.** an ancient kingdom in NE Africa, bordering on Egypt and the Red Sea. —**E′thi·o′pi·an,** *adj., n.*

Et·na (et′nə), **Mount,** an active volcano in E Sicily. 10,758 (3280 m).

E·ton (ēt′n), a town in Berkshire, in S England, on the Thames River, W of London: site of Eton College, a boys' preparatory school. 3954.

E·tru·ri·a (i troor′ē ə), an ancient country located

between the Arno and Tiber rivers, roughly corresponding to modern Tuscany in W Italy. —**E·tru′-ri·an,** *adj.,* *n.*

Eu·boe·a (yōō bē′ə), a Greek island in the W Aegean Sea. 188,410; 1586 sq. mi. (4110 sq. km). *Cap.:* Chalcis. Modern Greek, **Evvoia.** —**Eu·boe′an,** *adj.,* *n.* —**Eu·bo′ic** (-bō′ik), *adj.*

Eu·clid (yōō′klid), a city in NE Ohio, near Cleveland. 54,875.

Eu·gene (yōō jēn′), a city in W Oregon. 112,669.

Eu·phra·tes (yōō frā′tēz), a river in SW Asia, flowing from E Turkey through Syria and Iraq, joining the Tigris to form the Shatt-al-Arab near the Persian Gulf. 1700 mi. (2735 km) long. —**Eu·phra′te·an,** *adj.*

Eur·a·sia (yōō rā′zhə, -shə, yə-), Europe and Asia considered together as one continent. —**Eur·a′-sian,** *adj.,* *n.*

Eu·rope (yōōr′əp, yûr′-), a continent in the W part of the landmass lying between the Atlantic and Pacific oceans, separated from Asia by the Ural Mountains on the E and the Caucasus Mountains and the Black and Caspian seas on the SE. ab. 4,017,000 sq. mi. (10,404,000 sq. km). —**Eu′ro·pe′an,** *n.,* *adj.*

Europe′an Commu′nity, an association of W European countries that includes the European Atomic Energy Community (Euratom), the European Economic Community, the European Parliament, and allied organizations. *Abbr.:* EC

Europe′an Econom′ic Commu′nity, *n.* an association for economic cooperation established in 1957 by Belgium, France, Italy, Luxembourg, the Netherlands, and West Germany: later joined by the United Kingdom, the Republic of Ireland, Denmark, Greece, Spain, and Portugal; the Common Market. *Abbr.:* EEC

Eux′ine Sea′ (yōōk′sin, -sin), BLACK SEA.

Ev·ans·ton (ev′ən stən), a city in NE Illinois, on Lake Michigan, near Chicago. 73,233.

Ev·ans·ville (ev′ənz vil′), a city in SW Indiana, on the Ohio River. 126,272.

Ev·er·est (ev′ər ist, ev′rist), **Mount,** a mountain in S Asia, on the boundary between Nepal and Tibet, in the Himalayas: the highest mountain in the world. 29,028 ft. (8848 m).

Ev·er·ett (ev′ər it, ev′rit), a seaport in NW Washington, on Puget Sound. 69,961.

Ev·er·glades (ev′ər glādz′), a partly forested marshland in S Florida, mostly S of Lake Okeechobee. over 5000 sq. mi. (12,950 sq. km).

Ev′erglades Na′tional Park′, a national park in the Everglades region of S Florida. 2186 sq. mi. (5662 sq. km).

Eve·sham (ēv′shəm, ē′shəm, ē′səm), a town in Hereford and Worcester county, in W England: battle 1265. 15,271.

Ev·voi·a (e′vē ä), Modern Greek name of Euboea.

Ex·e·ter (ek′si tər), a city in Devonshire, in SW England. 99,700.

Ex·tre·ma·du·ra (*Sp.* es′tre mä ᵺōō′Rä), Estremadura.

Eyre (âr), **Lake,** a shallow salt lake in NE South Australia. 3430 sq. mi. (8885 sq. km).

Eyre′ Penin′sula, a peninsula in S South Australia, E of the Great Australian Bight.

F

Fa·en·za (fä en/zə, -ent/sə), a city in N Italy, SE of Bologna. 55,612.

Faer/oe (or **Far/oe**) **Is/lands** (fâr/ō), a group of 21 islands in the N Atlantic between Great Britain and Iceland, belonging to Denmark but having home rule. 46,312; 540 sq. mi. (1400 sq. km). *Cap.:* Thorshavn. Also called **Faer/oes, Faroes.** Danish, **Faer·ö·er·ne** (fɛr œ/ɛr nə).

Fair·banks (fâr/bangks/), a city in central Alaska, on the Tanana River. 30,843.

Fair·field (fâr/fēld/), **1.** a city in central California. 77,211. **2.** a town in SW Connecticut. 54,849.

Fair/ Oaks/, a locality in E Virginia, near Richmond: battle 1862.

Fair·weath·er (fâr/weṯh/ər), **Mount,** a mountain in SE Alaska. 15,292 ft. (4660 m).

Fai·sa·la·bad (fī sä/lə bäd/, -sal/ə bad/), a city in NE Pakistan. 1,092,000. Formerly, **Lyallpur.**

Fai·yum (fī yōōm/, fä-), **1.** a province in N central Egypt: many archaeological remains. 691 sq. mi. (1790 sq. km). **2.** Also called **El Faiyum.** the capital of this province, SW of Cairo. 227,300.

Fal·kirk (fôl/kûrk), a city in S central Scotland, W of Edinburgh: Scots under Wallace defeated by the English 1298. 37,489.

Falk/land Is/lands (fôk/lənd), a group of islands in the SW Atlantic, E of Argentina, constituting a self-governing British colony. 2000; 4618 sq. mi. (11,961 sq. km). *Cap.:* Stanley. Spanish, **Islas Malvinas.**

Fall/ Riv/er, a seaport in SE Massachusetts, on an arm of Narragansett Bay. 92,703.

Fal·mouth (fal/məth), **1.** a seaport in S Cornwall, in SW England: resort. 17,883. **2.** a town in SE Massachusetts. 23,640.

Fal·ster (fäl/stər), an island in SE Denmark. 198 sq. mi. (513 sq. km).

Far/ East/, the countries of E Asia, including China, Japan, Korea, and sometimes adjacent areas. —**Far/ East/ern,** *adj.*

Fare·well (fâr/wel/), **Cape,** a cape in S Greenland: the most southerly point of Greenland.

Far·go (fär/gō), a city in SE North Dakota. 74,111.

Farm·ing·ton (fär′ming tən), a city in NW New Mexico. 38,470.

Farm′ington Hills′, a city in SE Michigan. 74,652.

Far′oe Is′lands (fâr′ō), Faeroe Islands. Also called **Far′oes. —Far′o·ese′,** *n., pl.* **-ese,** *adj.*

Fars (färs), a province in SW Iran. 3,193,769; 51,466 sq. mi. (133,297 sq. km).

Far′ West′, the area of the U.S. west of the Great Plains. **—Far′ West′ern,** *adj.*

Fa·sho·da (fə shō′də), a village in the SE Sudan, on the White Nile: conflict of British and French colonial interests 1898 **(Fasho′da In′cident).** Modern name, **Kodok.**

Fá·ti·ma (fä′ti mə), a village in central Portugal, N of Lisbon: Roman Catholic shrine.

Fat·shan (fät′shän′), Foshan.

Fay·ette·ville (fā′it vil′), **1.** a city in S North Carolina. 75,695. **2.** a city in NW Arkansas. 36,608.

Fa·yum (fī yōōm′, fā-), Faiyum.

Fear (fēr), **Cape,** a cape in SE North Carolina at the mouth of Cape Fear River.

Fed′eral Cap′ital Ter′ritory, former name of Australian Capital Territory.

Fed′eral Dis′trict, a district in which the national government of a country is located, esp. one in Latin America.

Fed′eral Repub′lic of Ger′many, 1. official name of Germany. **2.** official name of West Germany.

Fed′erated Ma′lay States′, a former federation of four states in British Malaya: Negri Sembilan, Pahang, Perak, and Selangor.

Feng·jie or **Feng·chieh** (fung′jyu′), also **Feng·kieh** (-gyu′, -jyu′), a city in E Sichuan province, in S central China, on the Chang Jiang. 250,000. Formerly, **Guizhou.**

Feng·tien (fung′tyen′), **1.** a former name of Shenyang. **2.** former name of Liaoning.

Fens (fenz), **the,** a marshy region W and S of the Wash, in E England.

Fer·ga·na (fer gä′nə, fər-), a city in E Uzbekistan, SE of Tashkent. 203,000.

Fer·man·agh (fər man′ə), a county in SW Northern Ireland. 51,008; 647 sq. mi. (1676 sq. km).

Fer·nan·do de No·ro·nha (fər nan′dō də nô·rōn′yə, fər nän′-), a Brazilian island in the S Atlantic, NE of Natal: with nearby islands it constitutes a territory of Brazil. 1342; 10 sq. mi. (26 sq. km).

Fer·nan·do Po (fər nan′dō pō′) also **Fernan′do Po′o** (pō′ō), a former name of Bioko.

Fer·ra·ra (fə rär′ə), a city in N Italy, near the Po River. 143,046.

Fer·rol (fe rōl′), El Ferrol.

Fer′tile Cres′cent, a crescent-shaped agricultural region of the ancient Near East beginning at the Mediterranean Sea and extending between the Tigris and Euphrates rivers to the Persian Gulf.

Fez (fez), a city in N Morocco. 448,823.

Fez·zan (fez zän′), a former province in SW Libya: a part of the Sahara with many oases. 220,000 sq. mi. (570,000 sq. km).

Fia·na·ran·tso·a (fyə när′ənt sō′ə, -sōō′ə), a city in E central Madagascar. 111,000.

Fie·so·le (fyä′zə lē), a town in central Italy, near Florence: Etruscan and ancient Roman ruins. 14,138.

Fife (fīf), a region in E Scotland: formerly a county. 336,339; 504 sq. mi. (1305 sq. km). Also called **Fife·shire** (fīf′shēr, -shər).

Fi·ji (fē′jē), a republic consisting of an archipelago of some 332 islands in the S Pacific, N of New Zealand, composed of the Fiji Islands and a smaller group to the NW: formerly a British colony. 715,375; 7078 sq. mi. (18,333 sq. km). *Cap.:* Suva. —**Fi′ji·an,** *n., adj.*

Fi′ji Is′lands, a group of islands in the S Pacific constituting most of the country of Fiji.

Filch′ner Ice′ Shelf′ (filk′nər, filkh′-), an ice barrier in Antarctica, in the SE Weddell Sea.

Fin′gal's Cave′ (fing′gəlz), a cave on the island of Staffa, in the Hebrides, Scotland. 227 ft. (69 m) long; 42 ft. (13 m) wide.

Fin′ger Lakes′, a group of elongated glacial lakes in central and W New York: resort region.

Fin·land (fin′lənd), **1.** Finnish, **Suomi.** a republic in N Europe, on the Baltic. 4,940,000; 130,119 sq. mi. (337,010 sq. km). *Cap.:* Helsinki. **2. Gulf of,** an arm of the Baltic, S of Finland. —**Finn,** *n.* —**Finn′ish,** *adj.*

Fin·ster·aar·horn (fin′stər är′hôrn), a mountain

in S central Switzerland: highest peak of the Bernese Alps, 14,026 ft. (4275 m).

Fire′ Is′land, a narrow barrier beach off S Long Island, New York: summer resort. 30 mi. (48 km) long.

Fi·ren·ze (fē ren′dze), Italian name of FLORENCE.

Fiu·me (fyōō′me), Italian name of RIJEKA.

FL or **Fla.,** Florida.

Flag·staff (flag′staf′, -stäf′), a city in central Arizona. 34,743. ab. 6900 ft. (2100 m) high.

Fla·min′i·an Way′ (flə min′ē ən), an ancient Roman road extending N from Rome to what is now Rimini. 215 mi. (345 km) long.

Flan·ders (flan′dərz), a medieval country in W Europe, extending along the North Sea from the Strait of Dover to the Scheldt River: the corresponding modern regions include the provinces of East Flanders and West Flanders in W Belgium and the adjacent parts of N France and SW Netherlands.

Fleet′ Street′, a street in central London, England: location of many newspaper offices.

Fle·vo·land (flē′vō land′), a province in the central Netherlands. 193,739.

Flin′ders Range′ (flin′dərz), a mountain range in S Australia. Highest peak, 3900 ft. (1190 m).

Flint (flint), **1.** a city in SE Michigan. 140,761. **2.** FLINTSHIRE.

Flint·shire (flint′shēr, -shər), a historic county in Clwyd, in NE Wales.

Flod·den (flod′n), a hill in NE England, in Northumberland county: the invading Scots were defeated here by the English, 1513.

Flor·ence (flôr′əns, flor′-), a city in Tuscany, in central Italy, on the Arno River. 421,299. Italian, **Firenze.** —**Flor′en·tine′,** *adj., n.*

Flo·res (flôr′is, -ēz, flōr′- for 1; Port. flô′rish for 2), **1.** one of the Lesser Sunda Islands in Indonesia, separated from Sulawesi by the Flores Sea. ab. 200,000 with adjacent islands; 7753 sq. mi. (20,080 sq. km). **2.** the westernmost island of the Azores, in the N Atlantic. 55 sq. mi. (142 sq. km).

Flo′res Sea′ (flôr′is, -ēz, flōr′-), a sea between Sulawesi and the Lesser Sunda Islands in Indonesia. ab. 180 mi. (290 km) wide.

Flo·ri·a·nóp·o·lis (flôr′ē ə nop′ə lis, flōr′-), the

capital of Santa Catarina state, on an island off the S coast of Brazil. 196,055.

Flor·i·da (flôr′i də, flor′-), a state in the SE United States between the Atlantic and the Gulf of Mexico. 12,937,926; 58,560 sq. mi. (151,670 sq. km). *Cap.:* Tallahassee. *Abbr.:* FL, Fla. —**Flo·rid·i·an** (flə rid′ē ən), **Flor′i·dan,** *adj., n.*

Flor′ida Keys′, a chain of small islands and reefs off the coast of S Florida. ab. 225 mi. (362 km) long.

Flor′ida Strait′, a strait between Florida, Cuba, and the Bahamas, connecting the Gulf of Mexico and the Atlantic.

Flor·is·sant (flôr′ə sənt), a city in E Missouri, near St. Louis. 51,206.

Flush·ing (flush′ing), a seaport in the SW Netherlands. 46,055. Dutch, **Vlissingen.**

Fly (flī), a river in New Guinea, flowing SE from the central part to the Gulf of Papua. ab. 800 mi. (1290 km) long.

Fog·gia (fôd′jä), a city in SE Italy. 159,192.

Fog′gy Bot′tom, a low-lying area bordering the Potomac River in Washington, D.C.: U.S. Department of State office building.

Fon·se·ca (fon sā′kə), **Gulf of,** a bay of the Pacific Ocean in W Central America, bordered by El Salvador on the W, Honduras on the NE, and Nicaragua on the S. ab. 700 sq. mi. (1800 sq. km).

Fon·taine·bleau (fon′tin blō′), a town in N France, SE of Paris: residence of French kings. 19,595.

Fon·tan·a (fon tan′ə), a city in S California. 87,535.

Foo·chow (fōō′jō′, -chou′), Fuzhou.

For·a·ker (fôr′ə kər, for′-), **Mount,** a mountain in central Alaska, in the Alaska Range, near Mt. McKinley. 17,280 ft. (5267 m).

Forbid′den Cit′y, a walled section of Beijing, built in the 15th century, containing the imperial palace and other buildings of the imperial government of China.

For′est Hills′, a residential area in New York City, New York.

For′est of Dean′, a royal forest in Gloucestershire, in W England. ab. 180 sq. mi. (475 sq. km).

For·far (fôr′fər, -fär), **1.** a town in the Tayside region, in E Scotland. 12,742. **2.** former name of Angus.

For·lì (fôr lē′), a city in N Italy, SE of Bologna. 110,334.

For·mo·sa (fôr mō′sə), Taiwan.

Formo′sa Strait′, former name of Taiwan Strait.

For·ta·le·za (fôr′tl ā′zə), the capital of Ceará, in NE Brazil. 648,851.

Fort′ Col′lins (kol′inz), a city in N Colorado. 87,758.

Fort-de-France (fôr də fräns′), the capital of Martinique, in the French West Indies. 97,000.

Forth (fôrth, fōrth), **1. Firth of,** an arm of the North Sea, in SE Scotland: estuary of Forth River. 48 mi. (77 km) long. **2.** a river in S central Scotland, flowing E into the Firth of Forth. 116 mi. (187 km) long.

Fort′ Jef′ferson, a national monument in Dry Tortugas, Fla.: a federal prison 1863–73; now a marine museum.

Fort′ Knox′ (noks), a military reservation in N Kentucky, SSW of Louisville: federal gold depository.

Fort-La·my (Fr. fôr lA mē′), former name of N′Djamena.

Fort′ Lau′der·dale (lô′dər dāl′), a city in SE Florida: seashore resort. 149,377.

Fort′ Mc·Hen′ry (mək hen′rē), a fort in N Maryland, at the entrance to Baltimore harbor: Francis Scott Key wrote *The Star-Spangled Banner* during British bombardment in 1814.

Fort′ Peck′ (pek), a dam on the Missouri River in NE Montana.

Fort′ Pu·las′ki (pə las′kē), a fort in E Georgia, at the mouth of the Savannah River: captured by Union forces in 1862; now a national monument.

Fort′ Smith′ (smith), a city in W Arkansas, on the Arkansas River. 72,798.

Fort′ Sum′ter (sum′tər, sump′-), a fort in SE South Carolina, in the harbor of Charleston: its bombardment by the Confederates opened the Civil War on April 12, 1861.

Fort′ Wayne′ (wān), a city in NE Indiana. 173,072.

Fort′ Wil′liam (wil′yəm), See under Thunder Bay.

Fort′ Worth′ (wûrth), a city in N Texas. 447,619.

Fosh·an (fush′än′) also **Fatshan,** a city in S central Guangdong province, in SE China, near Guangzhou. 285,540.

Foth·er·ing·ghay (foth′ə ring gā′), a village in NE Northamptonshire, England: Mary Queen of Scots imprisoned here and executed 1587.

Foun′tain Val′ley,. a city in SW California. 53,691.

Fram·ing·ham (frā′ming ham′), a town in E Massachusetts. 65,113.

France (frans, fräns), a republic in W Europe. 56,560,000; 212,736 sq. mi. (550,985 sq. km). *Cap.*: Paris. —**French,** *adj., n.*

Franche-Com·té (fränsh kôn tā′), **1.** a historic region and former province in E France: once a part of Burgundy. **2.** a metropolitan region in E France; l,085,900; 6256 sq. mi. (16,202 sq. km).

Fran·co·ni·a (frang kō′nē ə, -kôn′yə, fran-), a medieval duchy in Germany, largely in the valley of the Main River.

Frank·fort (frangk′fərt), the capital of Kentucky, in the N part. 25,973.

Frank·furt (frangk′fûrt, frängk′foŏrt), **1.** Also called **Frank·furt am Main** (fRängk′foŏrt äm min′). a city in W Germany, on the Main River. 618,500. **2.** Also called **Frank·furt an der O·der** (fRängk′foŏrt än dəR ō′dəR). a city in NE Germany, on the Oder River. 85,158.

Frank·lin (frangk′lin), a district in extreme N Canada, in the Northwest Territories, including the Boothia and Melville peninsulas, Baffin Island, and other Arctic islands. 549,253 sq. mi. (1,422,565 sq. km).

Franz Jo·sef Land (frants′ jō′zəf land′, jō′səf, fränts′), an archipelago in the Arctic Ocean, E of Spitsbergen and N of Novaya Zemlya: belongs to the Russian Federation. Also called **Fridtjof Nansen Land.**

Fra·ser (frā′zər), a river in SW Canada, flowing S through British Columbia to the Pacific. 695 mi. (1119 km) long.

Frau·en·feld (frou′ən felt′), the capital of Thurgau, in N Switzerland. 18,400.

Fred·er·icks·burg (fred′riks bûrg′, fred′ər iks-), a city in NE Virginia: scene of a Confederate victory 1862. 15,322.

Fred·er·ic·ton (fred′rik tən, fred′ər ik-), the capital of New Brunswick, in SE Canada, on the St.
John River. 44,352.

Fre·de·riks·berg (fred′riks bûrg′, fred′ər iks-), a
city in E Denmark: a part of Copenhagen. 93,692.

Free·town (frē′toun′), the capital of Sierra Leone, in W Africa. 469,776.

Frei·burg (frī′bŏŏrk′), **1.** a city in SW Baden-
Württemberg, in SW Germany. 178,700. **2.** German name of FRIBOURG.

Fre·man·tle (frē′man′tl), a seaport in SW Australia, near Perth. 25,990.

Fre·mont (frē′mont), **1.** a city in W California
near San Francisco Bay. 173,339. **2.** a city in E
Nebraska, on the Platte River, near Omaha.
23,780.

French′ Cameroons′, CAMEROUN (def. 2).

French′ Commu′nity, an association of France
and its former colonies, territories, and overseas
departments, formed in 1958.

French′ Con′go, a former name of the People's
Republic of the CONGO.

French′ Equato′rial Af′rica, a former federation of French territories in central Africa, including Chad, Gabon, Middle Congo (now People's
Republic of the Congo), and Ubangi-Shari (now
Central African Republic): each became independent in 1960.

French′ Gui·an·a (gē an′ə, -ä′nə), an overseas
department of France, on the NE coast of South
America: formerly a French colony. 73,012;
35,135 sq. mi. (91,000 sq. km). *Cap.:* Cayenne.
—**French′ Guianese′, French′ Guian′an,**
adj., n.

French′ Guin′ea, former name of GUINEA (def.
2).

French′ In′dia, a former French territory in India.

French′ Indochi′na, an area in SE Asia, formerly a French colonial federation: now comprising the three independent states of Vietnam,
Cambodia, and Laos.

French′ Moroc′co, See under MOROCCO.

French′ Ocean′ia, former name of FRENCH
POLYNESIA.

French′ Polyne′sia, a French overseas territory
in the S Pacific, including the Society Islands,

Marquesas Islands, and other scattered island groups. 191,400; 1544 sq. mi. (4000 sq. km). *Cap.:* Papeete.

French/ Soma/liland, a former name of Dji-bouti (def. 1).

French/ Sudan/, former name of Mali.

French/ West/ Af/rica, a former French federation in W Africa, including Dahomey (now Benin), French Guinea (now Guinea), French Sudan (now Mali), Ivory Coast, Mauritania, Niger, Senegal, and Upper Volta (now Burkina Faso).

French/ West/ In/dies, the French islands in the Lesser Antilles of the West Indies, including Martinique and Guadeloupe and the five dependencies of Guadeloupe: administered as two overseas departments.

Fres·no (frez/nō), a city in central California. 354,202.

Fri·bourg (*Fr.* frē bōōr/), **1.** a canton in W Switzerland. 197,200; 644 sq. mi. (1668 sq. km). **2.** the capital of this canton. 40,500. German, **Freiburg.**

Fridt/jof Nan/sen Land/ (frit/yôf nän/sən, nan/-), Franz Josef Land.

Friend/ly Is/lands, Tonga.

Fries·land (frēz/lənd, -land/, frēs/-), a province in the N Netherlands. 599,104; 1431 sq. mi. (3705 sq. km). *Cap.:* Leeuwarden.

Fris·co (fris/kō), *Informal.* San Francisco.

Fri/sian Is/lands (frizh/ən, frē/zhən), a chain of islands in the North Sea, extending along the coasts of the Netherlands, Germany, and Denmark: includes groups belonging to the Netherlands (**West Frisians**) and to Germany (**East Frisians**) and a group divided between Germany and Denmark (**North Frisians**).

Fri·u·li-Ve·ne·zi·a Giu·lia (frē ōō/lē və nāt/sē ə jōōl/yə), a region in NE Italy: formerly part of Venezia Giulia, most of which was ceded to Yugoslavia. 1,242,987; 2947 sq. mi. (7630 sq. km).

Fro/bish·er Bay/ (frō/bi shər, frob/i-), an arm of the Atlantic Ocean extending NW into SE Baffin Island, Northwest Territories, Canada.

Front/ Range/, a mountain range extending from central Colorado to S Wyoming: part of the Rocky Mountains. Highest peak, Grays Peak, 14,274 ft. (4350 m).

Frun·ze (froon′zə), a former name (1926–91) of
BISHKEK.

Fu·ji (foo′jē), a dormant volcano in central Japan,
on Honshu island: highest mountain in Japan.
12,395 ft. (3778 m). Also called **Fu·ji·ya·ma**
(foo′jē yä′mə), **Fu·ji·san** (foo′jē sän′).

Fu·jian (fy′jyän′) also **Fu·kien** (foo′kyen′), a
province in SE China opposite Taiwan.
27,490,000; 47,529 sq. mi. (123,000 sq. km).
Cap.: Fuzhou.

Fu·ji·sa·wa (foo′jē sä′wə), a city on E Honshu in
Japan, S of Tokyo. 333,000.

Fu·ku·o·ka (foo′koo ō′kə), a city on N Kyushu, in
SW Japan. 1,142,000.

Ful·ler·ton (fool′ər tən), a city in SW California,
SE of Los Angeles. 114,144.

Fu·na·fu·ti (foo′nə foo′tē), the capital of Tuvalu.
1328.

Fun·chal (*Port.* foon shäl′), the capital of the Madeira islands, on SE Madeira: resort. 48,638.

Fun·dy (fun′dē), **Bay of,** an inlet of the Atlantic
in SE Canada, between New Brunswick and Nova
Scotia, having swift tidal currents.

Fü·nen (fy′nən), German name of FYN.

Fürth (fyrt), a city in S Germany, near Nurem-
berg. 97,331.

Fu·shun (fy′shyn′), a city in E Liaoning province,
in NE China. 1,700,000.

Fu·xin (fy′shin′) also **Fu·sin** (foo′sin′), a city in
central Liaoning province, in NE China. 644,200.

Fu·zhou (fy′jō′) also **Foochow,** the capital of
Fujian province, in SE China, opposite Taiwan.
1,190,000.

Fyn (fyn), an island of Denmark, between Jutland
and Zealand. 446,233; 1149 sq. mi. (2975 sq.
km). German, **Fünen.**

G

GA or **Ga.,** Georgia.

Ga·be·ro·nes (gä'bə rō'nes, gab'ə-), former name of GABORONE.

Ga·bès (gä'bes), **Gulf of,** a gulf of the Mediterranean on the E coast of Tunisia.

Ga·bon (gä bôn') also **Gabun, 1.** Official name, **Gab·o·nese Repub'lic,** a republic in W equatorial Africa: formerly a part of French Equatorial Africa; member of the French Community. 1,220,000; 102,290 sq. mi. (264,931 sq. km). *Cap.:* Libreville. **2.** an estuary in W Gabon. ab. 40 mi. (65 km) long. —**Gab·o·nese** (gab'ə nēz', -nēs', gä'-bə-), *adj., n., pl.* **-nese.**

Ga·bo·ro·ne (gä'bə rō'nē, gab'ə-), the capital of Botswana, in the SE part. 110,973. Formerly, **Gaberones.**

Ga·bun (gə bōōn'), GABON.

Gail·lard' Cut' (gil yärd', gā'lärd), the SE section of the Panama Canal, formed by an artificial channel cut through the mountains NW of Panama City. 8 mi. (13 km) long.

Gaines·ville (gānz'vil), a city in N Florida. 84,770.

Ga·lá'pa·gos Is'lands (gə lä'pə gōs', -gəs, -lap'ə-), an archipelago on the equator in the Pacific, ab. 600 mi. (965 km) W of and belonging to Ecuador: many unique species of animal life. 4058; 3029 sq. mi. (7845 sq. km). Also called **Colón Archipelago.**

Ga·la·ta (gä'lä tä), the chief commercial section of Istanbul, Turkey.

Ga·la·ţi (gä läts', -lä'tsē) also **Ga·latz** (-läts'), a port in E Romania, on the Danube River. 292,805.

Ga·la·tia (gə lā'shə, -shē ə), an ancient country in central Asia Minor: later a Roman province; site of an early Christian community. —**Ga·la'tian,** *adj., n.*

Ga·li·ci·a (gə lish'ē ə, -lish'ə; *for 2 also Sp.* gä-lē'thyä, -syä), **1.** a region in E central Europe: a former crown land of Austria, included in S Poland after World War I, and now partly in Ukraine. ab. 30,500 sq. mi. (79,000 sq. km). **2.** a maritime region in NW Spain: a former kingdom, and later a province. 11,256 sq. mi. (29,153 sq. km). —**Ga·li′ci·an**, *adj., n.*

Gal·i·lee (gal′ə lē′), **1.** an ancient Roman province in what is now N Israel. **2. Sea of.** Also called **Lake Tiberias.** a lake in NE Israel through which the Jordan River flows. 14 mi. (23 km) long; 682 ft. (208 m) below sea level. —**Gal′i·le′an**, *adj., n.*

Gal·li·a (gäl′lē ä), Latin name of GAUL.

Ga·lli·nas (gä yē′näs), **Pun·ta** (pōōn′tä), a cape in NE Colombia: northernmost point of South America.

Gal·lip′o·li Penjn′sula (gə lip′ə lē), a peninsula in European Turkey, between the Dardanelles and the Aegean Sea. 60 mi. (97 km) long.

Gal·lo·way (gal′ə wā′), a historic region in SW Scotland.

Gal·ves·ton (gal′və stən), a seaport in SE Texas, on an island at the mouth of Galveston Bay. 59,070.

Gal′veston Bay′, an inlet of the Gulf of Mexico.

Gal·way (gôl′wā), **1.** a county in S Connaught, in W Republic of Ireland. 171,836; 2293 sq. mi. (5940 sq. km). **2.** its county seat: a seaport in the W part. 47,104.

Gam·bi·a (gam′bē ə), **1.** a river in W Africa, flowing W to the Atlantic. 500 mi. (800 km) long. **2. the,** a republic in W Africa: formerly a British crown colony and protectorate; gained independence 1965; member of the Commonwealth of Nations. 788,163; 4003 sq. mi. (10,368 sq. km). *Cap.:* Banjul. —**Gam′bi·an**, *adj., n.*

Gam′bier Is′lands (gam′bēr), a group of islands in French Polynesia, belonging to the Tuamotu Archipelago. 8226; 12 sq. mi. (31 sq. km).

Gand (gän), French name of GHENT.

Gan·der (gan′dər), a town in E Newfoundland, in Canada: airport on the great circle route between New York and N Europe. 10,207.

Gan·dhi·na·gar (gun′di nug′ər), the capital of Gujarat, in W India. 62,443.

Gan·ges (gan′jēz), a river flowing SE from the Himalayas in N India into the Bay of Bengal: sacred to Hindus. 1550 mi. (2495 km) long. —**Gan·get′ic** (-jet′ik), *adj.*

Gang·tok (gung′tok′), the capital of Sikkim, in NE India. 36,768.

Gan·su (gän/sy/) also **Kansu,** a province in N central China. 20,710,000; 141,500 sq. mi. (366,500 sq. km). *Cap.:* Lanzhou.

Gao·xiong (*Chin.* qou/shyông/), KAOHSIUNG.

Gar·da (gär/də), **Lake,** a lake in N Italy: the largest lake in Italy. 35 mi. (56 km) long; 143 sq. mi. (370 sq. km).

Gar·de·na (gär dē/nə), a city in SW California, near Los Angeles. 49,847.

Gar/den Grove/, a city in SW California. 143,050.

Gar·land (gär/lənd), a city in NE Texas, near Dallas. 180,650.

Ga·ronne (gA Rôn/), a river in SW France, flowing NW from the Pyrenees to the Gironde River. 350 mi. (565 km) long.

Gar·y (gâr/ē, gar/ē), a port in NW Indiana, on Lake Michigan. 116,646.

Gas·co·ny (gas/kə nē), a former province in SW France. French, **Gas·cogne** (gA skôn/y°). —**Gas/con,** *adj., n.*

Gas·pé/ Penin/sula (ga spā/), a peninsula in SE Canada, in Quebec province, between New Brunswick and the St. Lawrence River.

Gas·to·ni·a (ga stō/nē ə), a city in S North Carolina, W of Charlotte. 54,732.

Gates·head (gāts/hed/), a seaport in NE England, on the Tyne River opposite Newcastle. 222,000.

Gat·i·neau (gat/n ō/, gat/n ō/), a city in S Quebec, in E Canada, near Hull. 81,244.

Ga·tun/ Lake/ (gä tōōn/), an artificial lake in central Panama, forming part of the Panama Canal: created by a dam (**Gatun/ Dam/**) across the Chagres River. 164 sq. mi. (425 sq. km).

Gau·ga·me·la (gô/gə mē/lə), an ancient village in Assyria, E of Nineveh: site of defeat of the Persians by Alexander the Great 331 B.C., often called "battle of Arbela."

Gaul (gôl), an ancient region in W Europe, including the modern areas of N Italy, France, Belgium, and the S Netherlands: consisted of two main divisions, one part S of the Alps (**Cisalpine Gaul**) and another part N of the Alps (**Transalpine Gaul**). Latin, **Gallia.**

Gäv·le (yāv/le), a seaport in E Sweden. 87,378.

Ga·ya (gä′yə, gi′ə, gə yä′), a city in central Bihar, in NE India: Hindu center of pilgrimage. 246,778.

Ga·za (gä′zə, gaz′ə, gä′zə), a seaport on the Mediterranean Sea, in the Gaza Strip, adjacent to SW Israel: ancient trade-route center. 118,300.

Ga′za Strip′, a coastal area on the E Mediterranean: formerly in the Palestine mandate, occupied by Israel 1967.

Ga·zi·an·tep (gä′zē än tep′), a city in S Turkey in Asia. 466,302. Formerly, **Aintab.**

Gdańsk (gə dänsk′, -dansk′), a seaport in N Poland, on the Baltic Sea. 467,000. German, **Danzig.**

Gdy·nia (gə din′ē ə, -yə), a seaport in N Poland, on the Gulf of Danzig. 243,000.

Gee·long (ji lông′), a seaport in SE Australia, SW of Melbourne. 148,300.

Geel′vink Bay′ (*Du.* ᴋнäl′vingk), former (Dutch) name of Sᴀʀᴇʀᴀ Bᴀʏ.

Gel·der·land (gel′dər land′), a province in E Netherlands. 1,783,610; 1965 sq. mi. (5090 sq. km). *Cap.:* Arnhem. Also called **Guelders.**

Gel·sen·kir·chen (gel′zən kir′ᴋнən), a city in W Germany, in the Ruhr valley. 287,600.

Ge·ne·ral San·tos (hā′nä räl′ sän′tōs), a city in the Philippines, on S Mindanao. 250,000.

Gen·e·see (jen′ə sē′), a river flowing N from N Pennsylvania into Lake Ontario. 144 mi. (230 km) long.

Ge·ne·va (jə nē′və), **1.** the capital of the canton of Geneva, in SW Switzerland, on the Lake of Geneva. 160,900. **2.** a canton in SW Switzerland. 365,200; 109 sq. mi. (282 sq. km). **3. Lake of.** Also called **Lake Leman.** a lake between SW Switzerland and France. 45 mi. (72 km) long; 225 sq. mi. (583 sq. km). French, **Ge·nève** (zhə nev′), (for defs. 1, 2). German, **Genf** (genf), (for defs. 1, 2). —**Ge·ne′van, Gen·e·vese** (jen′ə vēz′, -vēs′), *adj., n., pl.* **-vans, -vese.**

Gen·o·a (jen′ō ə), a seaport in NW Italy, S of Milan. 762,895. Italian, **Ge·no·va** (je′nô vä′). —**Gen′o·ese′,** *n., pl.* **-ese,** *adj.*

Gent (ᴋнent), Flemish name of Gнᴇɴᴛ.

George (jôrj), **Lake,** a lake in E New York. 36 mi. (58 km) long.

Geor′ges Bank′ (jôr′jiz), a bank extending generally NE from Nantucket: fishing grounds. 150 mi. (240 km) long.

George·town (jôrj′toun′), **1.** Also, **George′ Town′.** the capital of the state of Penang, in NW Malaysia. 250,578. **2.** the capital of Guyana, at the mouth of the Demerara. 182,000. **3.** a residential section in the District of Columbia. **4.** the capital of the Cayman Islands, West Indies, on Grand Cayman. 12,000.

Geor·gia (jôr′jə), **1.** a state in the SE United States. 6,478,216; 58,876 sq. mi. (152,489 sq. km). *Cap.:* Atlanta. *Abbr.:* GA, Ga. **2.** Also called **Geor′gian Repub′lic.** a republic in Transcaucasia, bordering on the Black Sea, N of Turkey and Armenia. 5,449,000; 26,872 sq. mi. (69,700 sq. km). *Cap.:* Tbilisi. **3. Strait of,** an inlet of the Pacific in SW Canada between Vancouver Island and the mainland. 150 mi. (240 km) long. —**Geor′gian,** *adj., n.*

Ge·ra (gâr′ə), a city in E central Germany. 134,834.

Ger·la·chov·ka (gɛʀ′lä ꭓôf′kä), a mountain in N Slovakia, in E Czechoslovakia: highest peak of the Carpathian Mountains. 8737 ft. (2663 m).

Ger′man Af′rica (jûr′mən), the former German colonies in Africa, comprising German East Africa, German Southwest Africa, Cameroons, and Togoland.

Ger′man Democrat′ic Repub′lic, official name of EAST GERMANY.

Ger′man East′ Af′rica, a former German territory in E Africa, now comprising continental Tanzania, Rwanda, and Burundi.

Ger′man O′cean, former name of the NORTH SEA.

Ger′man South′west Af′rica, a former name (1884–1919) of NAMIBIA.

Ger·man·town (jûr′mən toun′), a NW section of Philadelphia, Pa.: American defeat by British 1777.

Ger·ma·ny (jûr′mə nē), a republic in central Europe: after World War II divided into four zones, British, French, U.S., and Soviet, and in 1949 into East Germany and West Germany; East and West Germany were reunited in 1990. 78,420,000; 137,852 sq. mi. (357,039 sq. km). *Cap.:* Berlin. Official name, **Federal Republic of Germany.** German, **Deutschland.** —**Ger′man,** *n., adj.*

Ger·mis·ton (jûr′mə stən), a city in S Transvaal, in the NE Republic of South Africa. 221,972.

Get·tys·burg (get′iz bûrg′), a borough in S Pennsylvania: Confederate forces defeated in a Civil War battle fought near here on July 1–3, 1863; national cemetery and military park. 7194.

Ge·zer (gē′zər), an ancient Canaanite town, NW of Jerusalem.

Ge·zi·ra (jə zēr′ə), a region in central Sudan, S of Khartoum, between the Blue Nile and the White Nile: a former province.

Gha·na (gä′nə, gan′ə), **1.** a republic in W Africa comprising the former colonies of the Gold Coast and Ashanti, the protectorate of the Northern Territories, and the U.N. trusteeship of British Togoland: member of the Commonwealth of Nations since 1957. 13,800,000; 91,843 sq. mi. (237,873 sq. km). *Cap.:* Accra. **2. Kingdom of,** a medieval W African empire extending from near the Atlantic coast almost to Timbuktu: flourished about 9th–12th centuries. —**Gha′na·ian, Gha′ni·an,** *n., adj.*

Ghats (gôts, gots), **1.** Eastern Ghats. **2.** Western Ghats.

Ghent (gent), a port in NW Belgium, at the confluence of the Scheldt and Lys rivers: treaty 1814. 232,620. French, **Gand.** Flemish, **Gent.**

Gi·ants′ Cause′way, a large body of basalt, unusual in displaying perfect columnar jointing, exposed on a promontory on the N coast of Northern Ireland.

Gib·e·on (gib′ē ən), a town in ancient Palestine, NW of Jerusalem.

Gi·bral·tar (ji brôl′tər), **1.** a British crown colony comprising a fortress and seaport located on a narrow promontory near the S tip of Spain. 30,689; 2½ sq. mi. (6½ sq. km). **2. Rock of.** Ancient, **Calpe.** a long, precipitous mountain nearly coextensive with this colony: one of the Pillars of Hercules. 1396 ft. (426 m) high. **3. Strait of,** a strait between Europe and Africa connecting the Atlantic to the Mediterranean. 8–23 mi. (14–37 km) wide. —**Gi·bral·tar′i·an** (-tär′ē ən), *adj., n.*

Gib′son Des′ert (gib′sən), a desert in W central Australia: scrub; salt marshes. ab. 85,000 sq. mi. (220,000 sq. km).

Gi·fu (gē′fōō′), a city on S Honshu, in central Japan. 410,368.

Gi·jón (hē hôn′), a seaport in NW Spain, on the Bay of Biscay. 259,226.

Gi·la (hē′lə), a river flowing W from SW New Mexico across S Arizona to the Colorado River. 630 mi. (1015 km) long.

Gil′bert and El′lice Is′lands (gil′bərt; el′is), a former British colony, comprising the Gilbert Islands (now Kiribati), the Ellice Islands (now Tuvalu), and other widely scattered islands in the central Pacific Ocean.

Gil′bert Is′lands, former name of KIRIBATI.

Gil·e·ad (gil′ē əd), **1.** a district of ancient Palestine, E of the Jordan River, in present N Jordan. **2. Mount,** a mountain in NW Jordan. 3596 ft. (1096 m). —**Gil′e·ad·ite′**, n.

Gin·za (gin′zə), **the,** a district in Tokyo, Japan, noted for its department stores, nightclubs, and bars.

Gir·gen·ti (jēr jen′tē), former name of AGRIGENTO.

Gi·ronde (jə rond′; Fr. zhē RÔND′), an estuary in SW France, formed by the junction of the Garonne and Dordogne rivers. 45 mi. (72 km) long.

Gi·za or **Gi·zeh** (gē′zə), a city in N Egypt, a suburb of Cairo across the Nile: the ancient Egyptian pyramids and the Sphinx are located nearby. 1,230,446. Also called **El Giza, El Gizeh.**

Gla′cier Bay′ Na′tional Park′, a national park in SE Alaska, made up of large tidewater glaciers. 4381 sq. mi. (11,347 sq. km).

Gla′cier Na′tional Park′, a national park in the Rocky Mountains in NW Montana, containing lakes and glaciers: part of Waterton-Glacier International Peace Park. 1584 sq. mi. (4102 sq. km).

Gla·mor·gan (glə môr′gən), a historic county in SE Wales, now part of Mid, South, and West Glamorgan. Also called **Gla·mor′gan·shire′** (-shēr′, -shər).

Gla·rus (glär′əs, -ŏŏs), **1.** a canton in E central Switzerland. 35,900; 264 sq. mi. (684 sq. km). **2.** its capital, E of Lucerne. 6100.

Glas·gow (glas′gō, -kō, glaz′gō), a seaport in SW Scotland, on the Clyde River. 880,617. —**Glas·we′gian** (glas wē′jən), adj., n.

Glas·ton·bur·y (glas′tən ber′ē, -bə rē), a borough of SW England: excavations of an important

Iron Age lake village and ancient abbey; linked in folklore with King Arthur. 6773.

Glen·dale (glen′dāl′), **1.** a city in SW California, near Los Angeles. 180,038. **2.** a city in central Arizona, near Phoenix. 148,134.

Gli·wi·ce (glē vē′tse), a city in SW Poland. 213,000. German, **Glei·witz** (glī′vits).

Glom·ma (glôm′mä), a river in E Norway, flowing S into the Skagerrak. 375 mi. (605 km) long.

Glos·sa (glô′sə), **Cape,** a promontory in SW Albania.

Glouces·ter (glos′tər, glô′stər), **1.** a seaport in W Gloucestershire in SW England, on the Severn River. 90,700. **2.** GLOUCESTERSHIRE.

Glouces·ter·shire (glos′tər shēr′, -shər, glô′stər-), a county in SW England. 522,200; 1255 sq. mi. (2640 sq. km). *Co. seat:* Gloucester. Also called **Gloucester.**

Go·a (gō′ə), a state in SW India, on the Arabian Sea: formerly a part of Portuguese India; then part of the union territory of Goa, Daman, and Diu (1961–87). 1,007,749; 1429 sq. mi. (3702 sq. km). *Cap.:* Panaji.

Go′a, Daman′, and Di′u, a former territory of India, in the W part: now divided into the state of Goa, and the territory of Daman and Diu.

Go·bi (gō′bē), a desert in E Asia, mostly in Mongolia. ab. 500,000 sq. mi. (1,295,000 sq. km). —**Go′bi·an,** *adj.*

Go·da·va·ri (gō dä′və rē), a river flowing SE from W India to the Bay of Bengal. 900 mi. (1450 km) long.

Go·des·berg (gō′dəs bûrg′, -berg′), a city in W Germany, SE of Bonn. 73,512. Official name, **Bad Godesberg.**

Godt·håb (gôt′hôp′, got′hop′), the capital of Greenland, in the SW part. 12,209. Also called **Nuuk.**

God·win Aus·ten (god′win ô′stən). See K2.

Goi·â·ni·a (goi ä′nē ə), the capital of Goiás, in central Brazil, SW of Brasília. 738,117.

Goi·ás (goi äs′), a state in central Brazil. 4,638,800; 247,826 sq. mi. (641,870 sq. km). *Cap.:* Goiânia.

Gol·con·da (gol kon′də), a ruined city in S India, near the modern city of Hyderabad.

Gold′ Coast′, a former British territory in W Africa: now a part of Ghana.

Gold′en Gate′, a strait in W California, between San Francisco Bay and the Pacific. 2 mi. (3.2 km) wide.

Gold′en Horn′, an inlet of the Bosporus, in European Turkey: forms the inner part of Istanbul.

Golfe du Li·on (gôlf dʏ lē ôn′), French name of the Gulf of Lions.

Gol·go·tha (gol′gə thə), Calvary.

Go·mel (gō′məl), a city in SE Belarus, on a tributary of the Dnieper. 500,000.

Go·mor·rah (gə môr′ə, -mor′ə), an ancient city destroyed, with Sodom, because of its wickedness. —**Go·mor′re·an,** adj.

Go·na·ïves (Fr. gô nA ēv′), a seaport in W Haiti. 144,081.

Go·nâve (gō näv′), **1.** an island in the Gulf of Gonâve, in W Haiti. 287 sq. mi. (743 sq. km). **2. Gulf of,** an inlet of the Caribbean Sea, between the two peninsulas of W Haiti.

Gon·dar (gon′dər), a city in NW Ethiopia, N of Lake Tana: a former capital. 68,958.

Good′ Hope′, Cape of, Cape of Good Hope.

Good′win Sands′ (goŏd′win), a line of shoals at the N entrance to the Strait of Dover, off the SE coast of England. 10 mi. (16 km) long.

Goose′ Bay′, an air base in S central Labrador, in Newfoundland, in E Canada: used as a fuel stop by some transatlantic airplanes.

Go·rakh·pur (gôr′ək pŏŏr′, -), a city in SE Uttar Pradesh, in N India. 306,000.

Gor·ki (gôr′kē), former name (1932–91) of Nizhni Novgorod.

Gör·litz (gœr′lits), a city in E Germany, on the Neisse River. 79,506.

Gor·lov·ka (gôr lôf′kə, -lof′-), a city in SE Ukraine, N of Donetsk. 345,000.

Gor′no-Al′tai Auton′omous Re′gion (gôr′nō-al′tī), an autonomous region in the Russian Federation, in the Altai territory bordering China and Mongolia. 192,000; 35,753 sq. mi. (92,600 sq. km).

Gor′no-Ba·dakh·shan′ Auton′omous Re′gion (gôr′nō bə däk shän′), an autonomous region in SE Tajikistan. 161,000; 24,590 sq. mi. (63,700 sq. km).

Gö·te·borg (yœ/tə bôr/y°) also **Goth·en·burg** (goth/ən bûrg/, got/n-), a seaport in SW Sweden, on the Kattegat. 431,273.

Got·land or **Gott·land** (got/lənd), an island in the Baltic, forming a province of Sweden. 56,840; 1212 sq. mi. (3140 sq. km) *Cap.:* Visby. **—Got/land·er,** *n.*

Göt·tin·gen (gœt/ing ən), a city in central Germany. 114,900.

Gou·da (gou/də, gōō/-), a city in the W Netherlands, NE of Rotterdam. 62,321.

Go·ver·na·dor Va·la·da·res (gô/vir nə dôr/ vä/lə dä/Ris), a city in E Brazil. 173,699.

Gra/ham Land/ (grā/əm, gram), a part of the British Antarctic Territory, in the N section of the Antarctic Peninsula.

Gram·pi·an (gram/pē ən), a region in E Scotland. 502,863; 3361 sq. mi. (8704 sq. km).

Gram·pi·ans (gram/pē ənz), **the,** a range of low mountains in central Scotland, separating the Highlands from the Lowlands. Highest peak, Ben Nevis, 4406 ft. (1343 m). Also called **Gram/pian Hills/.**

Gra·na·da (grə nä/də), **1.** a medieval kingdom along the Mediterranean coast of S Spain. **2.** a city in S Spain: the capital of this former kingdom and last stronghold of the Moors in Spain; site of the Alhambra. 280,592. **3.** a city in SW Nicaragua, near Lake Nicaragua. 88,636.

Gran Ca·na·ria (*Sp.* grän/ kä nä/Ryä), an island in the Atlantic belonging to Spain, one of the Canary Islands. 592 sq. mi. (1533 sq. km). *Cap.:* Las Palmas. Also called **Grand Canary.**

Gran Cha·co (grän chä/kô), an extensive subtropical region in central South America, in Argentina, Bolivia, and Paraguay. 300,000 sq. mi. (777,000 sq. km). Also called **Chaco.**

Grand/ Baha/ma, an island in the NW Bahamas. 33,102; 430 sq. mi. (1115 sq. km).

Grand/ Banks/ (or **Bank/**), an extensive shoal SE of Newfoundland: fishing grounds. 350 mi. (565 km) long; 40,000 sq. mi. (104,000 sq. km).

Grand/ Canal/, 1. a canal in E China, extending S from Tianjin to Hangzhou. 900 mi. (1450 km) long. **2.** a canal in Venice, Italy, forming the main city thoroughfare.

Grand/ Canar/y, GRAN CANARIA.

Grand′ Can′yon, a gorge of the Colorado River in N Arizona. over 200 mi. (320 km) long; 1 mi. (1.6 km) deep.

Grand′ Can′yon Na′tional Park′, a national park in N Arizona, including part of the Grand Canyon and the area around it. 1009 sq. mi. (2615 sq. km).

Grand′ Cay′man, the largest of the Cayman Islands, West Indies. 8932; 76 sq. mi. (197 sq. km).

Grand′ Cou′lee (kōō′lē), **1.** a dry canyon in central Washington: cut by the Columbia River in the glacial period. 52 mi. (84 km) long; over 400 ft. (120 m) deep. **2.** a dam on the Columbia River at the N end of this canyon. 550 ft. (168 m) high.

Gran·de (grand, gran′dē, grän′dä), **Rio,** Rio GRANDE.

Grande-Terre (*Fr.* gränd teR′), See under GUADELOUPE.

Grand′ Falls′, former name of CHURCHILL FALLS.

Grand′ Forks′, a town in E North Dakota. 43,765.

Grand′ Is′land, a city in S Nebraska. 39,100.

Grand′ Ma·nan′ (mə nan′), a Canadian island at the entrance to the Bay of Fundy: a part of New Brunswick. 57 sq. mi. (148 sq. km).

Grand′ Prai′rie, a city in NE Texas. 99,616.

Grand Pré (gran′ prā′), a village in central Nova Scotia, on Minas Basin: locale of Longfellow's *Evangeline.*

Grand′ Rap′ids, a city in SW Michigan. 189,126.

Grand′ Riv′er, 1. former name of the Colorado River above its junction with the Green River in SE Utah. **2.** a river in SW Michigan flowing W to Lake Michigan. 260 mi. (420 km) long.

Grand′ Te′ton Na′tional Park′, a national park in NW Wyoming, including a portion of the Teton Range. 472 sq. mi. (1222 sq. km).

Grand′ Turk′, 1. an island in the Turks and Caicos Islands of the West Indies. 7 mi. (11 km) long. **2.** capital of the Turks and Caicos Islands, on Grand Turk. 3098.

Gra·ni·cus (grə ni′kəs), a river in NW Turkey, flowing N to the Sea of Marmara: battle 334 B.C. 45 mi. (70 km) long.

Gran·ta (gran′tə), CAM.

Gras·mere (gras′mēr, gräs′-), **1.** a lake in the Lake District, in NW England. 1 mi. (1.6 km) long.

2. a village on this lake: Wordsworth's home 1790–1808.

Grasse (gräs), a city in S France. 37,673.

Grau·bün·den (grou′byn′dən), German name of GRISONS.

Gra·ven·ha·ge, 's (sкнrä′vən hä′кнə), a Dutch name of The HAGUE.

Graves·end (grāvz′end′), a seaport in NW Kent, in SE England, on the Thames River. 94,300.

Graz (gräts), a city in SE Austria. 243,405.

Great′ Ab′a·co (ab′ə kō′), See under ABACO.

Great′ Austral′ian Bight′ (bit), a wide bay in S Australia.

Great′ Bar′rier Reef′, a coral reef parallel to the coast of Queensland, in NE Australia. 1250 mi. (2010 km) long.

Great′ Ba′sin, a region in the western U.S. that has no drainage to the ocean: includes most of Nevada and parts of Utah, California, Oregon, Wyoming, and Idaho. 210,000 sq. mi. (544,000 sq. km).

Great′ Ba′sin Na′tional Park′, a national park in E Nevada: site of Lehman Caves. 120 sq. mi. (312 sq. km).

Great′ Bear′ Lake′, a lake in NW Canada, in the Northwest Territories. 12,275 sq. mi. (31,792 sq. km).

Great′ Brit′ain, an island in NW Europe, separated from the mainland by the English Channel and the North Sea: comprising England, Scotland, and Wales. 55,780,000; 88,790 sq. mi. (229,979 sq. km). Compare UNITED KINGDOM.

Great′ Divide′, CONTINENTAL DIVIDE.

Great′ Divid′ing Range′, a mountain range extending along the E coast of Australia: vast watershed region. 100 to 200 mi. (160–320 km) wide.

Great′er Antil′les, See under ANTILLES.

Great′er Lon′don, LONDON (def. 4).

Great′er Man′chester, a metropolitan county in central England, with the city of Manchester as its center. 2,708,900; 498 sq. mi. (1290 sq. km).

Great′er New′ York′, NEW YORK (def. 3).

Great′er Sun′da Is′lands, See under SUNDA ISLANDS.

Great′ Falls′, a city in central Montana, on the Missouri River. 55,097.

Great′ Lakes′, a series of five lakes between the U.S. and Canada, comprising Lakes Erie, Huron, Michigan, Ontario, and Superior: connected with the Atlantic by the St. Lawrence River.

Great′ Ouse′, Ouse (def. 1).

Great′ Plains′, a semiarid region E of the Rocky Mountains, in the U.S. and Canada.

Great′ Rift′ Val′ley, a series of rift valleys running from the Jordan Valley in SW Asia to Mozambique in SE Africa.

Great′ Salt′ Des′ert, Dasht-i-Kavir.

Great′ Salt′ Lake′, a shallow salt lake in NW Utah. 2300 sq. mi. (5950 sq. km); 80 mi. (130 km) long; maximum depth 60 ft. (18 m).

Great′ Salt′ Lake′ Des′ert, an arid region in NW Utah, extending W from the Great Salt Lake to the Nevada border. ab. 4000 sq. mi. (10,360 sq. km).

Great′ Sand′y Des′ert, 1. a desert in NW Australia. ab. 160,000 sq. mi. (414,400 sq. km). **2.** Rub′ al Khali.

Great′ Slave′ Lake′, a lake in NW Canada, in the Northwest Territories. 11,172 sq. mi. (28,935 sq. km).

Great′ Smok′y Moun′tains, a range of the Appalachian Mountains in North Carolina and Tennessee: most of the range is included in Great Smoky Mountains National Park. 720 sq. mi. (1865 sq. km). Highest peak, Clingmans Dome, 6642 ft. (2024 m). Also called **Smoky Mountains, Great′ Smok′ies.**

Great′ Smo′ky Moun′tains Na′tional Park′, a national park in SE Tennessee and SW North Carolina, including most of the Great Smoky Mountains: hardwood forest. 808 sq. mi. (2092 sq. km).

Great St. Bernard, St. Bernard (def. 1).

Great′ Victo′ria Des′ert, a desert in SW central Australia. 125,000 sq. mi. (324,000 sq. km).

Great′ Wall′ of Chi′na, a system of fortified walls with a roadway along the top, constructed as a defense for China against the nomads of the regions that are now Mongolia and Manchuria: completed in the 3rd century B.C., but later repeatedly modified and rebuilt. 2000 mi. (3220 km) long. Also called **Chinese Wall.**

Great/ White/ Way/, the theater district along Broadway, near Times Square in New York City.

Great/ Yar/mouth, a seaport in E Norfolk, in E England. 77,200.

Greece (grēs), a republic in S Europe at the S end of the Balkan Peninsula. 9,990,000; 50,147 sq. mi. (129,880 sq. km). *Cap.:* Athens. Ancient Greek, **Hellas.** Modern Greek, **Ellas.** —**Gre·cian** (grē/shən), *adj.* —**Greek** (grēk), *adj., n.*

Gree·ley (grē/lē), a city in N Colorado. 60,536.

Green (grēn), a river flowing S from W Wyoming to join the Colorado River in SE Utah. 730 mi. (1175 km) long.

Green/ Bay/, 1. an arm of Lake Michigan, in NE Wisconsin. 120 mi. (195 km) long. **2.** a port in E Wisconsin at the S end of this bay. 96,466.

Green·land (grēn/lənd, -land/), a self-governing island belonging to Denmark, located NE of North America: the largest island in the world. 55,558; ab. 844,000 sq. mi. (2,186,000 sq. km); over 700,000 sq. mi. (1,800,000 sq. km) icecapped. *Cap.:* Godthåb. —**Green/land·er,** *n.* —**Green·land/ic,** *adj.*

Green/land Sea/, a part of the Arctic Ocean, NE of Greenland and N of Iceland.

Green/ Moun/tains, a mountain range in Vermont: a part of the Appalachian system. Highest peak, Mt. Mansfield, 4393 ft. (1339 m).

Green·ock (grē/nək, gren/ək), a seaport in SW Scotland, on the Firth of Clyde. 69,171.

Greens·bo·ro (grēnz/bûr/ō, -bur/ō), a city in N North Carolina. 183,521.

Green·ville (grēn/vil), **1.** a city in NW South Carolina. 58,282. **2.** a city in W Mississippi, on the Mississippi River. 40,613.

Green·wich (grin/ij, -ich, gren/- *for 1;* gren/ich, grin/-, grēn/wich *for 2*), **1.** a borough in SE London, England: located on the prime meridian from which geographic longitude is measured; formerly the site of the Royal Greenwich Observatory. 216,600. **2.** a town in SW Connecticut. 59,578.

Green/wich Vil/lage (gren/ich, grin/-), a section of New York City, in lower Manhattan: inhabited and frequented by artists, writers, and students.

Gre·na·da (gri nā/də), **1.** one of the Windward

Islands, in the E West Indies. **2.** an independent country comprising this island and the S Grenadines: a former British colony; gained independence 1974. 107,779; 133 sq. mi. (344 sq. km). *Cap.:* St. George's. —**Gre·na·di·an** (gri nā/dē ən), *adj.*, *n.*

Gren·a·dines (gren/ə dēnz/, gren/ə dēnz/), a chain of about 600 islands in the E West Indies in the Windward Islands: a former British colony; now divided between Grenada and St. Vincent and the Grenadines.

Gre·no·ble (grə nō/bəl), a city in SE France, on the Isère River. 169,740.

Gresh·am (gresh/əm), a city in NW Oregon. 68,235.

Gret·na Green/ (gret/nə), a village in S Scotland, near the English border, to which many English couples formerly eloped to be married.

Grims·by (grimz/bē), a seaport in Humberside county, in E England at the mouth of the Humber estuary. 93,800.

Gri·sons (*Fr.* grē zôN/), a canton in E Switzerland. 167,100; 2747 sq. mi. (7115 sq. km). *Cap.:* Chur. German, **Graubünden.**

Gro·dno (grod/nō), a city in W Belarus, on the Neman River: formerly in Poland. 263,000.

Gro·ning·en (grō/ning ən), **1.** a province in the NE Netherlands. 556,757. **2.** the capital of this province. 167,929.

Groz·ny (grôz/nē), the capital of the Chechen-Ingush Autonomous Republic in the Russian Federation in Europe. 401,000.

Grub/ Street/, a street in London, England formerly inhabited by impoverished writers and literary hacks.

GU, Guam.

Gua·da·la·ja·ra (gwäd/l ə här/ə), the capital of Jalisco, in W Mexico. 2,244,715.

Gua·dal·ca·nal (gwäd/l kə nal/), the largest of the Solomon Islands, in the W central Pacific. 47,000; ab. 2500 sq. mi. (6475 sq. km).

Gua·dal·qui·vir (gwäd/l kē vēr/), a river in S Spain, flowing W to the Gulf of Cádiz. 374 mi. (602 km) long.

Gua·da·lupe Hi·dal·go (gwäd/l ōōp/ hi däl/gō, gwäd/l ōō/pē), a city in the Federal District of

Mexico: famous shrine; peace treaty 1848. 1,182,895. Official name, **Gustavo A. Madero.**

Gua·dalupe Moun·tains, a mountain range in S New Mexico and SW Texas, part of the Sacramento Mountains. Highest peak, Guadalupe Peak, 8751 ft. (2667 m).

Gua·dalupe Moun·tains Na·tional Park, a national park E of El Paso, Texas: limestone fossil reef. 129 sq. mi. (334 sq. km).

Gua·de·loupe (gwäd/l ōōp/), two islands (**Basse-Terre** and **Grande-Terre)** separated by a narrow channel in the Leeward Islands of the West Indies: together with five dependencies they form an overseas department of France. 334,900; 687 sq. mi. (1179 sq. km). *Cap.:* Basse-Terre.

Gua·di·a·na (gwä/dē ä/nə, gwəd yä/-), a river flowing S from central Spain through SE Portugal to the Gulf of Cádiz. 515 mi. (830 km) long.

Guai·ra (gwī/rə), La Guaira.

Guam (gwäm), an island in the W Pacific, the largest of the Mariana Islands: an unincorporated U.S. territory. 120,000; 212 sq. mi. (549 sq. km). *Cap.:* Agaña. *Abbr.:* GU —**Gua·ma/ni·an** (-mä/nē-ən), *n., adj.*

Gua·na·ba·ra Bay, (gwä/nə bär/ə, gwä/-), an inlet of the Atlantic in SE Brazil.

Gua·na·jua·to (gwä/nä hwä/tō), **1.** a state in central Mexico. 3,542,103; 11,805 sq. mi. (30,575 sq. km). **2.** the capital of this state: center of the silver-mining region. 65,258.

Guang·dong (gwäng/dông/) also **Kwangtung,** a province in SE China. 63,640,000; 89,344 sq. mi. (231,401 sq. km). *Cap.:* Guangzhou.

Guang·xi Zhuang (gwäng/shē/ jwäng/) also **Kwangsi Chuang,** an autonomous region in S China. 39,460,000; 85,096 sq. mi. (220,399 sq. km). *Cap.:* Nanning.

Guang·zhou or **Kwang·chow** or **Kuang·chou** (gwäng/jō/), the capital of Guangdong province, in SE China, on the Zhu Jiang. 3,290,000. Also called **Canton.**

Guan·tá·na·mo (gwän tä/nə mō/), a city in SE Cuba: U.S. naval base. 174,400.

Guantá/namo Bay, a bay on the SE coast of Cuba.

Gua·po·ré (gwä/pŏŏ rä/), **1.** a river forming part of the boundary between Brazil and Bolivia, flowing NW to the Mamoré River. 950 mi. (1530 km) long. **2.** former name of RONDÔNIA.

Gua·ra·pua·va (gwä′rä pwä′vä), a city in S Brazil. 126,080.

Guar·da·fui (gwär′də fwē′), **Cape,** a cape at the E extremity of Africa.

Gua·ru·lhos (gwä rōōl′yəs), a city in SE Brazil, NE of São Paulo. 426,693.

Gua·te·ma·la (gwä′tə mä′lə), **1.** a republic in N Central America. 8,990,000; 42,042 sq. mi. (108,889 sq. km). **2.** Also called **Gua′tema′la Cit′y.** the capital of this republic. 1,500,000. —**Gua′te·ma′lan,** *adj., n.*

Guay·a·quil (gwī′ə kēl′), **1.** a seaport in W Ecuador, on the Gulf of Guayaquil. 1,300,868. **2. Gulf of,** an arm of the Pacific in SW Ecuador.

Guay·mas (gwī′mäs), a seaport in NW Mexico. 84,730.

Guay·na·bo (gwī nä′bō), a city in N Puerto Rico, SE of Bayamón. 65,075.

Guel·ders (gel′dərz), GELDERLAND.

Guelph (gwelf), a city in SE Ontario, in S Canada. 78,235.

Guer·ni·ca (gwâr′ni kə, gär′-), Basque town in N Spain: bombed and destroyed 1937 by German planes helping the insurgents in the Spanish Civil War.

Guern·sey (gûrn′zē), **Isle of,** one of the Channel Islands, in the English Channel. 55,482; 25 sq. mi. (65 sq. km).

Guer·re·ro (gə râr′ō), a state in S Mexico. 2,560,262; 24,885 sq. mi. (64,452 sq. km). *Cap.:* Chilpancingo.

Gui·an·a (gē an′ə, -ä′nə, gi an′ə), **1.** a vast tropical region in NE South America, bounded by the Orinoco, Negro, and Amazon rivers and the Atlantic. **2.** the coastal portion of this region, which includes Guyana, French Guiana, and Suriname. —**Gui·an·a, Gui′a·nese′** (-ə nēz′, -nēs′), *adj., n., pl.* **-an·ans, -a·nese.**

Gui·enne or **Guy·enne** (gwē yen′), a former province in SW France.

Gui·lin (gwē′lin′) also **Kweilin,** a city in the NE Guangxi Zhuang region, in S China. 235,000.

Guin·ea (gin′ē), **1.** a coastal region in W Africa, extending from the Gambia River to the Gabon estuary. **2.** Formerly, **French Guinea.** an independent republic in W Africa, on the Atlantic coast. 6,530,000; ab. 96,900 sq. mi. (251,000 sq. km). *Cap.:* Conakry. **3. Gulf of,** a part of the Atlantic Ocean that projects into the W coast of Africa and extends from Ivory Coast to Gabon. —**Guin′e·an,** *adj., n.*

Guin′ea-Bissau′, a republic on the W coast of Africa, between Guinea and Senegal: formerly a Portuguese overseas province; gained independence in 1974. 932,000; 13,948 sq. mi. (36,125 sq. km). *Cap.:* Bissau. Formerly, **Portuguese Guinea.**

Gui·yang (gwē′yäng′) also **Kweiyang,** the capital of Guizhou province, in S China. 1,380,000.

Gui·zhou (gwē′jō′), **1.** Also, **Kweichow.** a province in S China. 30,080,000; 67,181 sq. mi. (173,999 sq. km). *Cap.:* Guiyang. **2.** former name of FENGJIE.

Gu·ja·rat (gŏŏj′ə rät′, gōō′jə-), **1.** a region in W India, N of the Narbada River. **2.** a state in W India, on the Arabian Sea. 33,960,905; 72,138 sq. mi. (186,837 sq. km). *Cap.:* Gandhinagar.

Guj·ran·wa·la (gŏŏj′rən wä′lə, gōōj′-), a city in NE Pakistan. 597,000.

Gü·lek Bo·ğaz (gy lek′ bō äz′), Turkish name of the CILICIAN GATES.

Gulf·port (gulf′pôrt′, -pōrt′), a city in SE Mississippi, on the Gulf of Mexico. 39,676.

Gulf′ States′, 1. the states of the U.S. bordering on the Gulf of Mexico: Florida, Alabama, Mississippi, Louisiana, and Texas. **2.** Also called **Persian Gulf States.** the oil-producing countries on or near the Persian Gulf: Bahrain, Iran, Iraq, Kuwait, Oman, Qatar, Saudi Arabia, and the United Arab Emirates.

Gum·ri (gōōm rē′), a city in NW Armenia, NW of Yerevan. 120,000. Formerly, **Leninakan.**

Gun·tur (gŏŏn tŏŏr′), a city in E Andhra Pradesh, in SE India. 367,000.

Gu·ryev (gŏŏr′yəf), a port in W Kazakhstan, at the mouth of the Ural River on the Caspian Sea. 142,000.

Gus·ta·vo A. Ma·de·ro (gōōs tä′vō ä′ mä the′-rō), official name of GUADALUPE HIDALGO.

Gut′ of Can·so (gut), CANSO (def. 2).

Guy·a·na (gi an′ə, -ä′nə), an independent republic on the NE coast of South America: a former British protectorate; gained independence 1966; member of the Commonwealth of Nations. 812,000; 82,978 sq. mi. (214,913 sq. km). *Cap.:* Georgetown. Formerly, **British Guiana. —Guy′a·nese′** (-ə nēz′, -nēs′), *n., pl.* **-nese,** *adj.*

Guy·enne (gwē yen′), Guienne.

Gwa·li·or (gwä′lē ôr′), **1.** a former state in central India, now part of Madhya Pradesh. **2.** a city in N Madhya Pradesh. 560,000.

Gwent (gwent), a county in S Wales. 440,100; 531 sq. mi. (1376 sq. km).

Gwe·ru (gwā′roō), a city in central Zimbabwe. 79,000. Formerly, **Gwe·lo** (gwā′lō).

Gwyn·edd (gwin′eth), a county in NW Wales. 236,000; 1493 sq. mi. (3866 sq. km).

Györ (dyœr), a city in NW Hungary. 131,000.

Haag (häкн), **Den** (den), a Dutch name of The HAGUE.

Haar·lem (här′ləm), a city in the W Netherlands, W of Amsterdam. 157,556.

Ha·ba·na (ä vä′nä), Spanish name of HAVANA.

Ha·chi·o·ji (hä′chē ô′jē), a city on SE Honshu, in Japan, W of Tokyo. 425,000.

Hack·ney (hak′nē), a borough of Greater London. 187,400.

Ha·dhra·maut or **Ha·dra·maut** (hä′drə mōt′), a region on the S coast of Arabia, on the Arabian Sea, in the Republic of Yemen.

Ha·gen (hä′gən), a city in North Rhine-Westphalia, in W Germany. 209,200.

Ha·gers·town (hā′gərz toun′), a city in NW Maryland. 33,670.

Hague (hāg), **The,** a city in the W Netherlands, near the North Sea: site of the government, the royal residence, and the International Court of Justice. 444,313. Dutch, **Den Haag, 's Graven-hage.**

Hai·fa (hi′fə), a seaport in NW Israel. 230,000.

Haight-Ash·bur·y (hāt′ash′ber ē, -bə rē), a district of San Francisco: a center for hippies and the drug culture in the 1960s.

Hai·kou (hi′kō′), the capital of Hainan province, on N Hainan island, in SE China. 266,303.

Hai·nan (hi′nän′), an island in the South China Sea, separated from the mainland by Qiongzhou Strait: constitutes a province in S China. 6,000,000; 12,430 sq. mi. (32,200 sq. km). *Cap.:* Haikou.

Hai′nan′ Strait′, QIONGZHOU STRAIT.

Hai·naut (e nō′), **1.** a medieval county in territory now in SW Belgium and N France. **2.** a province in SW Belgium. 1,271,649; 1437 sq. mi. (3722 sq. km). *Cap.:* Mons.

Hai·phong (hi′fong′), a seaport in N Vietnam, near the Gulf of Tonkin. 1,190,900.

Hai·ti (hā′tē), **1.** a republic in the West Indies occupying the W part of the island of Hispaniola. 5,300,000; 10,714 sq. mi. (27,750 sq. km). *Cap.:* Port-au-Prince. **2.** a former name of HISPANIOLA. —**Hai·tian** (hā′shən, -tē ən), *adj., n.*

Ha·ko·da·te (hä′kə dä′tē), a seaport on S Hokkaido, in N Japan. 320,152.

Ha·le·a·ka·la′ Na′tional Park′ (hä′le ä′kä lä′), a national park on the island of Maui, Hawaii: site of dormant volcano **(Haleakala)**, 10,023 ft. (3055 m) high. 45 sq. mi. (116 sq. km).

Hal·i·car·nas·sus (hal′ə kär nas′əs), an ancient city of Caria, in SW Asia Minor: site of the Mausoleum, one of the seven wonders of the ancient world. —**Hal′i·car·nas′si·an, Hal′i·car·nas′se·an,** *adj.*

Hal·i·fax (hal′ə faks′), **1.** the capital of Nova Scotia, in SE Canada. 113,577. **2.** a city in West Yorkshire, in N central England. 91,171. —**Hal·i·go·ni·an** (hal′i gō′nē ən), *adj., n.*

Hal·le (häl′ə), a city in central Germany, NW of Leipzig. 236,044.

Hal·ma·he·ra (hal′mə her′ə, häl′-), an island in NE Indonesia: the largest of the Moluccas. 6928 sq. mi. (17,944 sq. km).

Halm·stad (hälm′städ′), a seaport in SW Sweden. 76,042.

Häl·sing·borg (hel′sing bôr′y°), a seaport in SW Sweden, opposite Helsingør, Denmark. 106,982.

Ha·ma (hä′mä, hä mä′), a city in W Syria, on the Orontes River. 176,640. Biblical name, **Ha·math** (hä′mäth, hä mäth′).

Ham·a·dan (ham′ə dan′; *Pers.* ha ma dän′), a city in W Iran. 165,785. Ancient, **Ecbatana.**

Ha·ma·mat·su (hä′mä mä′tsoō), a city on S central Honshu, in central Japan. 518,000.

Ham·burg (ham′bûrg, häm′boŏrg), a seaport and state in N Germany, on the Elbe River. 1,593,600; 292 sq. mi. (755 sq. km).

Ham·den (ham′dən), a town in S Connecticut. 51,071.

Ha·meln (hä′məln), a city in N central Germany, on the Weser River: scene of the legend of the Pied Piper of Hamelin. 55,580. English, **Ham·e·lin** (ham′ə lin).

Ham·hung (häm′hoŏng′), a city in central North Korea. 775,000.

Ham·il·ton (ham′əl tən), **1.** former name of CHURCHILL RIVER. **2. Mount,** a mountain in W California, near San Jose: site of Lick Observatory. 4209 ft. (1283 m). **3.** a seaport in SE Ontario, in SE Canada, on Lake Ontario. 306,728. **4.** a city on central North Island, in New Zealand. 154,606. **5.** a city in S Scotland, SE of Glasgow. 51,529. **6.** a city in SW Ohio. 61,368. **7.** the capital of Bermuda. 3000.

Ham·il·ton In·let, an arm of the Atlantic in SE Labrador, Newfoundland, in E Canada, an estuary of the Churchill River. 150 mi. (240 km) long.

Hamm (häm), a city in North Rhine–Westphalia, in W Germany. 171,100.

Ham·mer·fest (hä′mər fest′), a seaport in N Norway: the northernmost town in Europe. 7062.

Ham·mer·smith (ham′ər smith′), a borough of Greater London, England. 172,300.

Ham·mond (ham′ənd), a city in NW Indiana, near Chicago. 84,236.

Hamp·shire (hamp′shēr, -shər), a county in S England. 1,537,000; 1460 sq. mi. (3780 sq. km). Also called **Hants.**

Hamp·stead (hamp′stid, -sted), a former borough of London, England, now part of Camden.

Hamp·ton (hamp′tən), a city in SE Virginia, on Chesapeake Bay. 133,793.

Hamp′ton Roads′, a channel in SE Virginia between the mouth of the James River and Chesapeake Bay.

Han (hän), a river flowing from central China into the Chang Jiang at Wuhan. 900 mi. (1450 km) long.

Han′ Cit′ies, Wuhan.

Han·ford (han′fərd), a locality in SE Washington, on the Columbia River: site of an atomic energy plant.

Hang·zhou or **Hang·chow** (häng′jō′), the capital of Zhejiang province, in E China, on Hangzhou Bay. 1,250,000.

Hang′zhou (or **Hang′chow**) **Bay′,** a bay of the East China Sea.

Han·kou or **Han·kow** (hang′kou′; *Chin.* hän′kō′), a former city in E Hubei province, in E China: now part of Wuhan.

Han·ni·bal (han′ə bəl), a port in NE Missouri, on the Mississippi: Mark Twain's boyhood home. 18,811.

Ha·noi (ha noi′, hə-), the capital of Vietnam, in the N part, on the Red River. 2,000,000.

Han·o·ver (han′ō vər), **1.** a former province in NW Germany: now a district in Lower Saxony. **2.**

the capital of Lower Saxony, in N central Germany. 495,300. German, **Han·no·ver** (hä nō′-vər).

Hants (hants), HAMPSHIRE.

Han·yang (hän′yäng′), a former city in E Hubei province, in E China: now part of Wuhan.

Ha·rap·pa (hə rap′ə), a village in Pakistan: site of successive cities of the Indus valley civilization. —**Ha·rap′pan,** *adj.*

Ha·rar or **Har·rar** (här′ər), a city in E Ethiopia. 62,160.

Ha·ra·re (hə rär′ā), the capital of Zimbabwe, in the NE part. 656,100. Formerly, **Salisbury.**

Har·bin (här′bin′), the capital of Heilongjiang province, in NE China. 2,630,000.

Har·gei·sa (här gā′sə), a city in NW Somalia. 400,000.

Ha·ri·a·na (hur′ē ä′nə), HARYANA.

Har·in·gey (har′ing gā′), a borough of Greater London, England. 232,800.

Har·lem (här′ləm), **1.** a section of New York City, in the NE part of Manhattan. **2.** a tidal river in New York City, between the boroughs of Manhattan and the Bronx, which, with Spuyten Duyvil Creek, connects the Hudson and East rivers. 8 mi. (13 km) long. —**Har′lem·ite′,** *n.*

Har′ley Street′ (här′lē), a street in London, England: noted for the eminent doctors who have offices there.

Har·lin·gen (här′lin jən), a city in S Texas. 48,735.

Har′ney Peak′ (här′nē), a mountain in SW South Dakota: the highest peak in the Black Hills. 7242 ft. (2207 m).

Har′pers (or **Har′per's**) **Fer′ry** (här′pərz), a town in NE West Virginia at the confluence of the Shenandoah and Potomac rivers: site of John Brown's raid 1859. 361.

Har·rar (här′ər), HARAR.

Har·ris·burg (har′is bûrg′), the capital of Pennsylvania, in the S part, on the Susquehanna River. 52,376.

Har·row (har′ō), a borough of Greater London, in SE England. 201,300.

Hart·ford (härt′fərd), the capital of Connecticut, in the central part, on the Connecticut River. 139,739.

Ha·ry·a·na or **Ha·ri·a·na** (hur/ē ä/nə), a state in NW India, formed in 1966 from the S part of Punjab. 12,850,000; 17,074 sq. mi. (44,222 sq. km). *Cap.* (shared with Punjab): Chandigarh.

Harz/ Moun/tains (härts), a range of low mountains in central Germany between the Elbe and Weser rivers. Highest peak, Brocken, 3745 ft. (1141 m).

Ha·sa (hä/sə), a region in E Saudi Arabia, on the Persian Gulf. Also called **El Hasa.**

Hash/e·mite King/dom of Jor/dan (hash/ə- mit/), official name of JORDAN.

Has·tings (hā/stingz), a seaport in E Sussex, in SE England: William the Conqueror defeated the Saxons near here 1066. 74,600.

Hat·ter·as (hat/ər əs), **Cape,** a promontory on an island off the E coast of North Carolina.

Hat·ties·burg (hat/ēz bûrg/), a city in SE Mississippi. 40,865.

Hat·tu·sas (hät/tŏŏ säs/), the capital of the ancient Hittite empire in Asia Minor: site of modern Bogazkoy, Turkey.

Haute-Nor·man·die (ōt nôr män dē/) a metropolitan region in NW France. 1,692,800; 4700 sq. mi. (12,317 sq. km).

Ha·van·a (hə van/ə), a seaport in and the capital of Cuba, on the NW coast. 2,014,800. Spanish, **Habana.**

Ha·vel (hä/fəl), a river in NE Germany, flowing S through Berlin, and W into the Elbe. 212 mi. (341 km) long.

Hav·er·ford (hav/ər fərd), a township in SE Pennsylvania, near Philadelphia. 52,349.

Ha·ver·hill (hā/vər əl, -vrəl), a city in NE Massachusetts, on the Merrimack River. 51,418.

Ha·ver·ing (hā/vər ing), a borough of Greater London, England. 237,200.

Ha·vre (hä/vrə, -vər; *Fr.* A/vR°), LE HAVRE.

Haw., Hawaii.

Ha·wai·i (hə wī/ē, -wä/-, -wä/yə, hä vä/ē), **1.** a state of the United States comprising the Hawaiian Islands in the N Pacific: a U.S. territory 1900–59; admitted to the Union 1959. 1,108,229; 6424 sq. mi. (16,638 sq. km). *Cap.:* Honolulu. *Abbr.:* HI, Haw. **2.** the largest island of Hawaii, in the SE part. 117,500; 4035 sq. mi. (10,415 sq. km). —**Ha·wai/ian,** *n., adj.*

Hawai·ian Is·lands, a group of islands in the N Pacific, 2090 mi. (3370 km) SW of San Francisco: includes the islands of Hawaii, Maui, Oahu, Kauai, Molokai, Lanai, Niihau, Kahoolawe, and other islands and islets. Formerly, **Sandwich Islands.**

Hawai·i Volca·noes Na·tional Park', a national park on the island of Hawaii that includes the active volcanoes Kilauea and Mauna Loa. 358 sq. mi. (927 sq. km).

Haw·thorne (hô/thôrn'), a city in SW California, SW of Los Angeles. 71,349.

Hay·ward (hā/wərd), a city in central California, SE of Oakland. 111,498.

He·bei (hœ/bā/) also **Hopeh** or **Hopei,** a province in NE China. 56,310,000; 78,200 sq. mi. (202,700 sq. km). *Cap.*: Shijiazhuang.

Heb·ri·des (heb/ri dēz'), a group of islands (**Inner Hebrides** and **Outer Hebrides**) off the W coast of and belonging to Scotland. 29,615; ab. 2900 sq. mi. (7500 sq. km). Also called **Western Isles.** —**Heb/ri·de/an,** *adj., n.*

He·bron (hē/brən), a city in W Jordan: occupied by Israel 1967. 38,348. Arabic, **El Khalil.**

Heer·len (hāR/lən), a city in the SE Netherlands. 94,321.

He·fei (hœ/fā/) also **Hofei,** the capital of Anhui province, in E China. 821,812.

Hei·del·berg (hid/l bûrg'), a city in NW Baden-Württemberg, in SW Germany: university, founded 1386. 127,500.

Heil·bronn (hil/bron, -brôn), a city in N Baden-Württemberg, in SW Germany. 110,900.

Hei·long Jiang (hā/lông/ jyäng'), Chinese name of the **Amur.**

Hei·long·jiang (hā/lông/jyäng/) also **Hei·lung·kiang** (-lŏŏng/gyäng/), a province in NE China, S of the Amur River. 33,320,000; 179,000 sq. mi. (463,600 sq. km). *Cap.*: Harbin.

He·jaz (hi jaz/), **Hijaz.**

Hel·e·na (hel/ə nə), the capital of Montana, in the W part. 23,938.

Hel·go·land (hel/gō länt/) a German island in the North Sea. ¼ sq. mi. (0.6 sq. km).

Hel·i·con (hel/i kon', -kən), a mountain in S central Greece: regarded as the abode of Apollo and the Muses. 5738 ft. (1749 m).

He·li·op·o·lis (hē/lē op/ə lis), **1.** Biblical name,

On. an ancient ruined city in N Egypt, on the Nile delta. **2.** ancient Greek name of BAALBEK.

Hel·las (hel′əs), ancient Greek name of GREECE.

Hel·les (hel′is), **Cape,** a cape in European Turkey at the S end of Gallipoli Peninsula.

Hel·les·pont (hel′ə spont′), ancient name of the DARDANELLES. —**Hel′les·pont′ine** (-spon′tin, -tin), adj.

Hell′ Gate′, a narrow channel in the East River, in New York City.

Hel·mand (hel′mənd), a river in S Asia, flowing SW from E Afghanistan to a lake in E Iran. 650 mi. (1045 km) long.

Hel·sing·ør (hel′sing œr′), a seaport on NE Zealand, in NE Denmark: the scene of Shakespeare's *Hamlet.* 56,607. Also called **Elsinore.**

Hel·sin·ki (hel′sing kē, hel sing′-), the capital of Finland, on the S coast. 490,034. Swedish, **Hel·sing·fors** (hel′sing fôrz′).

Hel·ve·tia (hel vē′shə), Latin name of SWITZERLAND. —**Hel·ve′tian,** adj., n.

He·nan (hœ′nän′) also **Honan,** a province in E China. 78,080,000; 64,479 sq. mi. (167,000 sq. km). *Cap.*: Zhengzhou.

Hen·der·son (hen′dər sən), a city in SE Nevada, near Las Vegas. 64,942.

Hen·don (hen′dən), a former urban district in Middlesex, in SE England: now part of Barnet.

Heng·e·lo (heng′ə lō′), a city in the E Netherlands. 75,990.

Heng·yang (hœng′yäng′), a city in E central Hunan province, in E China. 240,000. Formerly, **Heng·chow** (hœng′jō′).

Hen·ley-on-Thames (hen′lē), a city in SE Oxfordshire, in S England: annual rowing regatta. 31,744. Also called **Hen′ley.**

Hen·ry (hen′rē), **Cape,** a cape in SE Virginia at the mouth of the Chesapeake Bay.

Her·a·cle·a (her′ə klē′ə), an ancient Greek city in S Italy, near the Gulf of Taranto: Roman defeat 280 B.C.

He·rak·li·on (i rak′lē ən, -rä′klē-), IRAKLION.

He·rat (he rät′), a city in NW Afghanistan. 140,323.

Her·ce·go·vi·na (*Serbo-Croatian.* her′tse gô′vinä), HERZEGOVINA.

Her·cu·la·ne·um (hûr′kyə lā′nē əm), an ancient

city in SW Italy, on the Bay of Naples: buried along with Pompeii by the eruption of Mount Vesuvius in A.D. 79; partially excavated. —**Her′cu·la/ne·an,** *adj.*

Her·e·ford (her′ə fərd), **1.** a city in Hereford and Worcester, in W England. 47,300. **2.** HEREFORD-SHIRE.

Her′eford and Worces/ter, a county in W England. 665,100; 1516 sq. mi. (3926 sq. km).

Her·e·ford·shire (her′ə fərd shēr′, -shər), a former county in W England, now part of Hereford and Worcester.

Her·mon (hûr′mən), **Mount,** a mountain in SW Syria, in the Anti-Lebanon range. 9232 ft. (2814 m).

Her·mo·si·llo (eʀ/mô sē′yô), the capital of Sonora, in NW Mexico. 340,779.

Herne (hûrn), a city in W Germany, in the Ruhr region. 174,200.

Hert·ford (här′fərd, härt′fərd), **1.** the county seat of Hertfordshire, in SE England. 20,379. **2.** HERT-FORDSHIRE.

Hert·ford·shire (här′fərd shēr′, -shər, härt′-), a county in SE England. 986,800; 631 sq. mi. (1635 sq. km). Also called **Hertford, Herts** (härts, hûrts).

Her·to·gen·bosch, 's (*Du.* seʀ′tō ᴋʜən bôs′), 's HERTOGENBOSCH.

Her·ze·go·vi·na (her′tsə gō vē′nə), a historic region in SE Europe: a former Turkish province; a part of Austria-Hungary (1878–1914); now part of Bosnia and Herzegovina. Serbo-Croatian, **Herce-govina.** —**Her′ze·go·vi/ni·an,** *adj., n.*

Hes·pe·ri·a (he sper′ē ə), a city in S California. 50,418.

Hesse (hes), a state in central Germany. 5,508,000; 8150 sq. mi. (21,110 sq. km). *Cap.:* Wiesbaden. German, **Hes·sen** (hes′ən). —**Hes·sian** (hesh′ən), *adj., n.*

Hes·ton and I·sle·worth (hes′tən; i′zəl-wûrth′), a former borough, now part of Hounslow, in SE England, near London.

HI Hawaii.

Hi·a·le·ah (hi′ə lē′ə), a city in SE Florida, near Miami: racetrack. 188,004.

Hi·ber·ni·a (hi bûr′nē ə), Latin name of IRELAND. —**Hi·ber′ni·an,** *adj., n.*

Hi·dal·go (hi dal′gō; *Sp.* ē ᵗħäl′gô), a state in central Mexico. 1,822,296; 8057 sq. mi. (20,870 sq. km). *Cap.:* Pachuca.

Hi·ga·shi·o·sa·ka (hi gä′shē ō sä′kə), a city on S Honshu, in Japan, E of Osaka. 503,000.

High·land (hī′lənd), a region in N Scotland, including a number of the Inner Hebrides. 200,608; 9710 sq. mi. (25,148 sq. km).

High·lands (hī′ləndz), **the**, a mountainous region in N Scotland, N of a line connecting Dumbarton and Aberdeen.

High′ Point′, a city in central North Carolina. 69,496.

Hii·u·maa (hē′ōō mä′), an island in the Baltic, E of and belonging to Estonia. 373 sq. mi. (965 sq. km).

Hi·jaz or **He·jaz** (hi jaz′), a region in Saudi Arabia bordering on the Red Sea, formerly an independent kingdom: contains the Islamic holy cities of Medina and Mecca. ab. 150,000 sq. mi. (388,500 sq. km). *Cap.:* Mecca.

Hil·des·heim (hil′des him′), a city in N central Germany. 103,400.

Hil·ling·don (hil′ing dən), a borough of Greater London, England. 232,200.

Hill′ of Tar′a, See under TARA.

Hi·lo (hē′lō), a seaport on E Hawaii island, in SE Hawaii. 35,269.

Hil·ver·sum (hil′vər səm), a city in central Netherlands. 92,141.

Hi·ma·chal Pra·desh (hi mä′chəl prə dāsh′), a state in N India. 4,237,569; 10,904 sq. mi. (28,241 sq. km). *Cap.:* Shimla.

Him·a·la·yas (him′ə lā′əz, hi mäl′yəz), **the**, a mountain range extending about 1500 mi. (2400 km) along the border between India and Tibet. Highest peak, Mt. Everest, 29,028 ft. (8848 m). Also called **Him′ala′ya Moun′tains**. —**Him′a·la′yan**, *adj., n.*

Hi·me·ji (hē′me jē′), a city on SW Honshu, in S Japan, W of Kobe. 451,000.

Hin·du Kush (hin′dōō kōōsh′, kush′), a mountain range in S Asia, mostly in NE Afghanistan, extending W from the Himalayas. Highest peak, Tirich Mir, 25,230 ft. (7690 m).

Hin·du·stan (hin′dōō stän′, -stan′), **1.** a region of N India, esp. the part N of the Deccan. **2.** the predominantly Hindu areas of India, as contrasted with the predominantly Muslim areas of Pakistan. —**Hin′du·stan′i,** *adj.*

Hip·po Re·gi·us (hip′ō rē′jē əs), a seaport of ancient Numidia: St. Augustine was bishop here A.D. 395–430; the site of modern Annaba, in Algeria. Also called **Hippo.**

Hi·ra·ka·ta (hē RÄ′kä tä′), a city on S Honshu, in Japan, NE of Osaka. 383,000.

Hi·ro·shi·ma (hēr′ō shē′mə, hi rō′shə mə), a seaport on SW Honshu, in SW Japan: first military use of atomic bomb Aug. 6, 1945. 1,034,000.

His·pa·ni·a (hi spā′nē ə, -spän′yə), Latin name of Iberian Peninsula.

His·pan·io·la (his′pən yō′lə), an island in the West Indies, comprising the republic of Haiti and the Dominican Republic. 30,285 sq. mi. (78,460 sq. km). Formerly, **Haiti, San Domingo, Santo Domingo.**

His·sar·lik (hi sär lik′), the modern name of the site of ancient Troy.

Ho·bart (hō′bərt, -bärt), the capital of Tasmania, in SE Australia. 175,082.

Ho·bo·ken (hō′bō kən), a seaport in NE New Jersey, opposite New York City. 42,460.

Ho′ Chi′ Minh′ Cit′y (hō′ chē′ min′), a seaport in S Vietnam. 4,000,000. Formerly, **Saigon.**

Ho·dei·da (hŏŏ dā′dä), a seaport in the Republic of Yemen, on the Red Sea. 155,110.

Hoek van Hol·land (*Du.* hōōk vän hôl′änt), HOOK OF HOLLAND.

Ho·fei (*Chin.* hu′fā′), HEFEI.

Ho·fuf (hŏŏ fōōf′), a city in E Saudi Arabia. 100,000.

Hoh·hot (hō′hōt′) also **Huhehot,** the capital of Inner Mongolia, in N China. 700,000.

Hok·kai·do (ho ki′dō), an island in N Japan, N of Honshu. 30,303 sq, mi. (78,485 sq. km). Formerly, **Yezo.**

Hol·guín (ōl gēn′), a city in NE Cuba. 194,700.

Hol·land (hol′ənd), **1.** the NETHERLANDS. **2.** a medieval county and province on the North Sea, corresponding to the modern North and South Holland provinces of the Netherlands. —**Hol′land·er,** *n.*

Hol·lan·di·a (ho lan′dē ə), former name of JAYAPURA.

Hol·ly·wood (hol′ē wŏŏd′), **1.** the NW part of Los Angeles, Calif.: center of the American motion-picture industry. **2.** a city in SE Florida near Miami. 121,697.

Ho·lon (ĸнô lôn′), a city in W central Israel: a suburb of Tel Aviv. 143,600.

Hol·stein (hōl′stīn, -stēn), a region in N Germany, at the base of the peninsula of Jutland: a former duchy. Compare SCHLESWIG-HOLSTEIN.

Ho·ly Cross′, Mount of the, a peak in central Colorado, in the Sawatch Range, in the Rocky Mountains. 14,005 ft. (4269 m).

Hol·y·head (hol′ē hed′), a seaport on Holy Island in NW Wales. 10,940.

Ho·ly Is′land, 1. Also called **Lindisfarne.** an island off the E coast of Northumberland, England. 3 mi. (4.8 km) long. **2.** Formerly, **Hol′yhead Is′land.** an island off the W coast of Anglesey, in NW Wales. 7 mi. (11 km) long.

Ho·ly Land′, PALESTINE (def. 1).

Hol·yoke (hōl′yōk, hō′lē ōk′), a city in SW Massachusetts. 44,678.

Ho·ly Ro′man Em′pire, a Germanic empire located chiefly in central Europe (800–1806).

Homs (hôms), a city in W Syria. 354,508.

Ho·nan (hō′nan′), HENAN.

Hon·do (hon′dō), a river flowing NE from Guatemala along the border of Belize and Mexico to the Caribbean Sea. 150 mi. (240 km) long.

Hon·du·ras (hon dŏŏr′əs, -dyŏŏr′-), **1.** a republic in NE Central America. 4,300,000; 43,277 sq. mi. (112,087 sq. km). *Cap.*: Tegucigalpa. **2. Gulf of,** an arm of the Caribbean Sea, bordered by Belize, Guatemala, and Honduras. —**Hon·du′ran,** *n.,* *adj.*

Hong Kong (hong′ kong′), **1.** a British crown colony comprising Hong Kong island (29 sq. mi.; 75 sq. km), Kowloon peninsula, nearby islands, and the adjacent mainland bordering SE China: reverting to Chinese sovereignty in 1997. 5,660,000; 404 sq. mi. (1046 sq. km). *Cap.*: Victoria. **2.** VICTORIA (def. 1). —**Hong′ Kong′er, Hong′kong′ite,** *n.*

Ho·ni·a·ra (hō′nē är′ə), the capital of the Solomon Islands, on N Guadalcanal. 26,000.

Hon·o·lu·lu (hon′ə lōō′lōō), the capital of Hawaii, on S Oahu. 365,272.

Hon·shu (hon′shōō), an island in central Japan: chief island of the country. 88,851 sq. mi. (230,124 sq. km).

Hood (hŏŏd), **Mount,** a volcanic peak in N Oregon, in the Cascade Range. 11,253 ft. (3430 m).

Hoogh·ly or **Hug·li** (hōōg′lē), a river in NE India, in West Bengal: the westernmost channel by which the Ganges enters the Bay of Bengal. 120 mi. (195 km) long.

Hook′ of Hol′land, a cape and the harbor it forms in the SW Netherlands. Dutch, **Hoek van Holland.**

Hoo′ver Dam′ (hōō′vər), official name of BOULDER DAM.

Ho·peh or **Ho·pei** (hō′pā′), HEBEI.

Hor·muz (hôr mōōz′, hôr′muz) also **Ormuz, Strait of,** a strait between Iran and the United Arab Emirates, connecting the Persian Gulf and the Gulf of Oman.

Horn (hôrn), **Cape,** CAPE HORN.

Hos·pi·ta·let (ôs′pē tä let′), a city in NE Spain, near Barcelona. 276,865.

Hot′ Springs′, a city in central Arkansas: adjoins a national park (**Hot′ Springs′ Na′tional Park′**) noted for its thermal mineral springs. 35,166.

Houns·low (hounz′lō), a borough of Greater London, England. 203,300.

Hou·sa·ton·ic (hōō′sə ton′ik), a river flowing S from NW Massachusetts to Long Island Sound near Stratford, Connecticut. 148 mi. (240 km) long.

Hous·ton (hyōō′stən), a city in SE Texas: a port on a ship canal, connected with the Gulf of Mexico. 1,630,553. —**Hous·to′ni·an** (-stō′nē ən), *adj., n.*

How·rah (hou′rä), a city in West Bengal, in E India, on the Hooghly River opposite Calcutta. 599,740.

Hra·dec Krá·lo·vé (hrä′dets krä′lə vā), a town in NW Czechoslovakia, on the Elbe River: Austrians defeated by Prussians in Battle of Sadowa 1866. 100,000. German, **Königgrätz.**

Hsia·men (*Chin.* shyä′mun′), XIAMEN.

Hsin·hsiang (*Chin.* shin′shyäng′), XINXIANG.

Hsi·ning (shē′ning′), XINING.

Huai·nan (hwi′nän′), a city in central Anhui province, in E China. 1,070,000.

Huam·bo (*Port.* wäm′bô), a city in central Angola. 67,000. Formerly, **Nova Lisboa.**

Huan·ca·yo (wäng kä′yô), a city in central Peru. 199,200.

Huang Hai (hwäng′ hī′), YELLOW SEA.

Huang He (hwäng′ hœ′) also **Hwang Ho,** a river flowing from W China into the Gulf of Bohai. 2800 mi. (4510 km) long. Also called **Yellow River.**

Huas·ca·ran (wäs′kä rän′), a mountain in W Peru, in the Andes. 22,205 ft. (6768 m).

Hub (hub), **the,** Boston, Mass. (used as a nickname).

Hu·bei (hy′bā′) also **Hupeh** or **Hupei,** a province in central China. 49,890,000; 72,394 sq. mi. (187,500 sq. km). *Cap.:* Wuhan.

Hub·li-Dar·war (hōōb′lē där′wär), a city in W Karnataka, in SW India. 526,000.

Hud·ders·field (hud′ərz fēld′), a town in West Yorkshire, in N central England. 130,060.

Hud·son (hud′sən), a river in E New York, flowing S to New York Bay. 306 mi. (495 km) long.

Hud′son Bay′, a large inland sea in N Canada. 850 mi. (1370 km) long; 600 mi. (965 km) wide; 400,000 sq. mi. (1,036,000 sq. km).

Hud′son Strait′, a strait connecting Hudson Bay and the Atlantic. 450 mi. (725 km) long; 100 mi. (160 km) wide.

Hué (hwā), a seaport in central Vietnam: former capital of Annam. 200,000.

Huel·va (wel′vä), a seaport in SW Spain, near the Gulf of Cádiz. 135,427.

Hug·li (hōōg′lē), HOOGHLY.

Hu·he·hot (hōō′hä′hōt′), HOHHOT.

Hui·la (wē′lä), **Mount,** a volcano in central Colombia. 18,700 ft. (5700 m).

Hull (hul), **1.** Official name, **Kingston upon Hull.** a seaport in Humberside, in E England, on the Humber River. 279,700. **2.** a city in SW Quebec, in SE Canada, on the Ottawa River opposite Ottawa. 58,722.

Hum·ber (hum′bər), an estuary of the Ouse and Trent rivers in E England. 37 mi. (60 km) long.

Hum·ber·side (hum′bər sid′), a county in NE England. 848,200; 1356 sq. mi. (3525 sq. km).

Hu·nan (hōō′nän′), a province in S China.

56,960,000; 81,274 sq. mi. (210,500 sq. km). *Cap.:* Changsha.

Hun·ga·ry (hung′gə rē), a republic in central Europe. 10,604,000; 35,926 sq. mi. (93,050 sq. km). *Cap.:* Budapest. Hungarian, **Magyarország.** —**Hun·gar′i·an** (-gâr′ē ən), *n., adj.*

Hung·nam (hōōng′näm′), a seaport in W North Korea. 150,000.

Hun·ting·don·shire (hun′ting dən shēr′, -shər), a former county in E England, now part of Cambridgeshire. Also called **Hun′ting·don, Hunts** (hunts).

Hun·ting·ton (hun′ting tən), a city in W West Virginia, on the Ohio River. 54,844.

Hun′tington Beach′, a city in SW California, SE of Los Angeles. 181,519.

Hun′tington Park′, a city in SW California, near Los Angeles. 56,065.

Hunts·ville (hunts′vil), a city in N Alabama. 159,789.

Hu·peh or **Hu·pei** (hōō′pā′, -bā′), HUBEI.

Hu·ron (hyŏŏr′ən, -on; *often* yŏŏr′-), **Lake,** a lake between the U.S. and Canada: second largest of the Great Lakes. 23,010 sq. mi. (59,595 sq. km).

Hwang Hai (hwäng′ hī′), YELLOW SEA.

Hwang Ho (hwäng′ hō′), HUANG HE.

Hyde′ Park′ (hīd), **1.** a public park in London, England. **2.** a village in SE New York, on the Hudson: site of the estate and burial place of Franklin D. Roosevelt and Eleanor Roosevelt. 2550.

Hy·der·a·bad (hī′dər ə bäd′, -bad′), **1.** a former state in S India, now part of Andhra Pradesh. **2.** the capital of Andhra Pradesh, India, in the W part. 2,528,000. **3.** a city in SE Pakistan, on the Indus River. 795,000.

Hy·met·tus (hī met′əs), a mountain in SE Greece, near Athens. 3370 ft. (1027 m). —**Hy·met′ti·an, Hy·met′tic,** *adj.*

Hyr·ca·ni·a (hər kā′nē ə), a province in an ancient Persian empire, SE of the Caspian Sea. —**Hyr·ca′ni·an,** *adj., n.*

I

IA or **Ia.**, Iowa.

I·a·și (yäsh, yá′shē), Romanian name of Jassy.

I·ba·dan (ē bäd′n), a city in SW Nigeria. 1,060,000.

I·ba·gué (ē′vä ge′), a city in W central Colombia. 292,965.

I·be·ri·a (ī bēr′ē ə), **1.** Also called **Ibe′rian Penin′sula.** a peninsula in SW Europe, comprising Spain and Portugal. **2.** an ancient region S of the Caucasus in what is now the Georgian Republic. —**I·be′ri·an,** adj., n.

I·bi·za (i bē′zə) also **Iviza,** a Spanish island in the SW Balearic Islands, in the W Mediterranean Sea. 209 sq. mi. (541 sq. km). —**I·bi′zan,** adj.

I·çá (ē′sä), Portuguese name of Putumayo.

I·car·i·a or **I·kar·i·a** (i kâr′ē ə, ī kâr′-, ē′kä rē′-ə), a Greek island in the Aegean Sea: part of the Southern Sporades group. 7702; 99 sq. mi. (256 sq. km).

Ice·land (īs′lənd), **1.** a large island in the N Atlantic between Greenland and Scandinavia. 39,698 sq. mi. (102,820 sq. km). **2.** a republic including this island and several smaller islands: formerly Danish; independent since 1944. 247,357. *Cap.:* Reykjavik. —**Ice′land·er** (-lan′dər, -lən dər), n. —**Ice·land′ic,** adj.

I·chi·ka·wa (ē chē′kä wä′), a city on E Honshu, in Japan, NE of Tokyo. 405,000.

I·chi·no·mi·ya (ē′chē nō′mē yə), a city on central Honshu, in central Japan. 257,000.

I·co·ni·um (ī kō′nē əm), ancient name of Konya.

ID or **Id.,** Idaho.

I·da (ī′də), **Mount, 1.** Turkish, **Kazdaği.** a mountain in W Turkey, in NW Asia Minor, SE of ancient Troy. 5810 ft. (1771 m). **2.** the highest mountain in Crete. 8058 ft. (2456 m). —**I·dae·an** (ī dē′ən), adj.

Ida., Idaho.

I·da·ho (ī′də hō′), a state in the NW United States. 1,006,749; 83,557 sq. mi. (216,415 sq. km). *Cap.:* Boise. *Abbr.:* ID, Id., Ida. —**I′da·ho·an,** adj., n.

I′daho Falls′, a city in E Idaho. 41,774.

Id·u·mae·a or **Id·u·me·a** (id′yŏŏ mē′ə), Greek name of Edom. —**Id′u·mae′an,** *adj., n.*

Ie·per (ē′pər), Flemish name of Ypres.

I·fe (ē′fā), a town in SW Nigeria. 214,500.

If·ni (ēf′nē), a former Spanish enclave on the W coast of Morocco, ceded to Morocco 1969.

I·gua·çú or **I·gua·zú** (ē′gwä sŏŏ′), a river in S Brazil, flowing W to the Paraná River. 380 mi. (610 km) long.

I′guaçú Falls′, a waterfall on the Iguaçú River, on the boundary between Brazil and Argentina. 210 ft. (64 m) high.

IJ or **Ij** (ī), an inland arm of the IJsselmeer in the Netherlands: Amsterdam located on its S side.

IJs·sel or **Ijs·sel** (ī′səl), a river in the central Netherlands flowing N to the IJsselmeer. 70 mi. (110 km) long.

IJs·sel·meer or **Ijs·sel·meer** (ī′səl mâr′), a lake in the NW Netherlands created by the diking of the Zuider Zee. 465 sq. mi. (1204 sq. km).

I·kar·i·a (i kâr′ē ə, i kâr′-, ē′kä rē′ə), Icaria.

IL, Illinois.

Île-de-France (ēl də fRÄNs′), **1.** a historic region and former province in N central France, including Paris and the region around it. **2.** a metropolitan region in N central France. 10,250,900; 4637 sq. mi. (12,012 sq. km).

Île du Dia·ble (ēl dv dyA′bl°), French name of Devil's Island.

I·le·sha (i lā′shə), a town in SW Nigeria. 273,400.

I·lhé·us (ē lye′ŏōs), a seaport in E Brazil. 100,687.

I·li·gan (ē′lē gän′), a city in the Philippines, on NW Mindanao. 227,000.

Il·i·on (il′ē ən, -on′), *n.* Greek name of ancient Troy.

Il·i·um (il′ē əm), *n.* Latin name of ancient Troy.

Ill., Illinois.

I·llam·pu (ē yäm′pŏō), a peak of Mount Sorata, in W Bolivia.

I·lli·ma·ni (ē′yē mä′nē), a mountain in W Bolivia, in the Andes, near La Paz. 21,188 ft. (6458 m).

Il·li·nois (il′ə noi′; *sometimes* -noiz′), **1.** a state in the central United States. 11,430,602; 56,400 sq. mi. (146,075 sq. km). *Cap.:* Springfield. *Abbr.:* IL, Ill. **2.** a river flowing SW from NE Illinois to the

Mississippi River: connected by a canal with Lake Michigan. 273 mi. (440 km) long.

Il·lyr·i·a (i lēr/ē ə), an ancient country along the E coast of the Adriatic.

Il·lyr·i·cum (i lēr/i kəm), a Roman province in ancient Illyria.

I·lo·i·lo (ē/lō ē/lō), a seaport on S Panay, in the central Philippines. 311,000.

I·lo·rin (i lôr/in), a town in W central Nigeria. 343,900.

Impe/rial Val/ley, an irrigated agricultural region in SE California, adjacent to Mexico, formerly a part of the Colorado Desert: it is largely below sea level and contains the Salton Sink.

Imp·hal (imp/hul), the capital of Manipur state, in NE India. 155,639.

IN, Indiana.

I·na·ri (in/ə rē, -är ē), **Lake,** a lake in NE Finland. ab. 500 sq. mi. (1295 sq. km).

In·chon (in/chon/), a seaport in W South Korea. 1,387,475. Formerly, **Chemulpo.**

Ind., Indiana.

In·de·pend·ence (in/di pen/dəns), a city in W Missouri: starting point of the Santa Fe and Oregon trails. 112,301.

In·di·a (in/dē ə), **1.** a republic in S Asia: formerly a British colony; gained independence in 1947; became a republic within the Commonwealth of Nations in 1950. 844,000,000; 1,246,880 sq. mi. (3,229,419 sq. km). *Cap.:* New Delhi. **2.** a subcontinent in S Asia, S of the Himalayas, occupied by Bangladesh, Bhutan, India, Nepal, and Pakistan. —**In/di·an,** *n., adj.*

In·di·an·a (in/dē an/ə), a state in the central United States. 5,544,159; 36,291 sq. mi. (93,995 sq. km). *Cap.:* Indianapolis. *Abbr.:* IN, Ind. —**In/di·an/an, In/di·an/i·an,** *adj., n.*

In·di·an·ap·o·lis (in/dē ə nap/ə lis), the capital of Indiana, in the central part. 731,327.

In/dian Des/ert, THAR DESERT.

In/dian O/cean, an ocean S of Asia, E of Africa, and W of Australia. 28,357,000 sq. mi. (73,444,630 sq. km).

In/dian States/ and A/gencies, the 560 former semidependent states and agencies in India and Pakistan: all except Kashmir incorporated into the republics of India and Pakistan (1947–49). Also called **Native States.**

In·dies (in/dēz), **the, 1.** WEST INDIES (def. 1). **2.** EAST INDIES.

In·do·chi·na (in/dō chi/nə), a peninsula in SE Asia, between the Bay of Bengal and the South China Sea, comprising Vietnam, Cambodia, Laos, Thailand, W Malaysia, and Burma. Compare FRENCH INDOCHINA. —**In/do·chi·nese/** (-nēz/, -nēs/), *adj., n., pl.* **-nese.**

In·do·ne·sia (in/də nē/zhə, -shə), **Republic of,** a republic in the Malay Archipelago, consisting of Sumatra, Java, Sulawesi, the S part of Borneo, Irian Jaya, and about 3000 small islands: received independence from the Netherlands in 1949. 172,000,000; ab. 741,100 sq. mi. (1,919,400 sq. km). *Cap.:* Jakarta. Formerly, **Netherlands East Indies, Dutch East Indies.** —**In/do·ne/sian,** *n., adj.*

In·dore (in dôr/), **1.** a former state in central India: now part of Madhya Pradesh. **2.** a city in W Madhya Pradesh, in central India. 827,000.

In·dus (in/dəs), a river in S Asia, flowing from W Tibet through India and Pakistan to the Arabian Sea. 1900 mi. (3060 km) long.

In·gle·wood (ing/gəl wŏŏd/), a city in SW California, near Los Angeles. 109,602.

In/land Sea/, a sea in SW Japan, enclosed by the islands of Honshu, Shikoku, and Kyushu. 240 mi. (385 km) long.

Inn (in), a river in central Europe, flowing from S Switzerland through Austria and Germany into the Danube. 320 mi. (515 km) long.

In/ner Heb/rides, See under HEBRIDES.

In/ner Mongo/lia, an autonomous region in NE China, adjoining the Mongolian People's Republic. 20,290,000; 454,600 sq. mi. (1,177,400 sq. km). *Cap.:* Hohhot. Official name, **In/ner Mongo/lia Auton/omous Re/gion.**

Inns·bruck (inz/brŏŏk), a city in W Austria, on the Inn river. 117,287.

In·ter·la·ken (in/tər lä/kən, in/tər lä/kən), a town in central Switzerland between the lakes of Brienz and Thun: tourist center. 4852.

In/tra·coast/al Wa/terway (in/trə kō/stəl, in/-), a system of canals and naturally sheltered bays and channels, extending 2666 mi. (4300 km) along the Atlantic and Gulf coasts of the U.S.: maintained to protect small craft from the open sea.

In·ver·ness (in/vər nes/, in/vər nes/), **1.** Also called **In/ver·ness/shire** (-shēr, -shər). a historic county in NW Scotland, in the Highland region, in N Scotland. 61,077; 1080 sq. mi. (2797 sq. km).

Io., Iowa.

Io·an·ni·na (yô ä/nē nä, yä/nē nä), a city in NW Greece. 44,362.

I·o·na (ī ō/nə), an island in the Hebrides, off the W coast of Scotland: center of early Celtic Christianity.

I·o·ni·a (ī ō/nē ə), an ancient region on the W coast of Asia Minor and on adjacent islands in the Aegean: colonized by the ancient Greeks. —**I·o/ni·an,** n., adj.

Io/nian Is/lands, a group of Greek islands including Corfu, Levkas, Ithaca, Cephalonia, and Zante off the W coast of Greece, and Kithira off the S coast.

Io/nian Sea/, an arm of the Mediterranean between S Italy, E Sicily, and Greece.

I·o·wa (ī/ə wə), **1.** a state in the central United States. 2,776,755; 56,290 sq. mi. (145,790 sq. km). Cap.: Des Moines. Abbr.: IA, Ia., Io. **2.** a river flowing SE from N Iowa to the Mississippi River. 291 mi. (470 km) long. —**I/o·wan,** adj., n.

I/owa Cit/y, a city in SE Iowa. 59,738.

I·poh (ē/pō), capital of Perak state, in W Malaysia. 300,727.

Ips·wich (ip/swich), a city in SE Suffolk, in E England. 116,500.

I·qui·que (ē kē/ke), a seaport in N Chile. 132,948.

I·qui·tos (ē kē/tôs), a city in NE Peru, on the upper Amazon. 247,900.

I·rak·li·on or **He·rak·li·on** (i rak/lē ən, -rä/klē-), a seaport on the N coast of Crete, in Greece. 243,622. Also called **Candia.**

I·ran (i ran/, i rän/, ī ran/), **1.** Formerly (until 1935), **Persia.** a republic in SW Asia: an Islamic republic since 1979. 53,920,000; ab. 635,000 sq. mi. (1,644,650 sq. km). Cap.: Teheran. **2.** Plateau of, a plateau in SW Asia, mostly in Iran and Afghanistan, extending from the Tigris to the Indus rivers. —**I·ra·ni·an** (i rā′nē ən, i rä′-, i rā′-), *n., adj.*

I·ra·pua·to (ēr′ə pwä′tō), a city in Guanajuato state, in central Mexico. 246,308.

I·raq (i rak′, i räk′), a republic in SW Asia, N of Saudi Arabia and W of Iran, centering in the Tigris-Euphrates basin of Mesopotamia. 17,060,000; 172,000 sq. mi. (445,480 sq. km). *Cap.:* Baghdad. —**I·ra′qi, I·ra′ki,** *n., pl.* **-qis, -kis,** *adj.*

I·ra·zu (ē′rä sōō′), **Mount,** a volcano in central Costa Rica. 11,200 ft. (3414 m).

Ire·land (ī°r′lənd), **1.** Latin, **Hibernia.** an island of the British Isles, W of Great Britain, comprising Northern Ireland and the Republic of Ireland. 32,375 sq. mi. (83,850 sq. km). **2. Republic of.** Formerly, **Irish Free State** (1922–37), **Eire** (1937–49). a republic occupying most of the island of Ireland. 3,540,000; 27,137 sq. mi. (70,285 sq. km). *Cap.:* Dublin. Irish, **Eire.** —**Ire′land·er,** *n.* —**I·rish** (ī′rish), *adj., n.*

I·ri·an Ja·ya (ēr′ē än′ jä′yä), a province of Indonesia, in the W part of the island of New Guinea: a Dutch territory until 1963. 1,173,875; ab. 159,000 sq. mi. (411,810 sq. km). *Cap.:* Jayapura. Also called **West Irian.** Formerly, **Netherlands New Guinea, Dutch New Guinea.**

I′rish Free′ State′, a former name of the Republic of IRELAND.

I′rish Sea′, a part of the Atlantic between Ireland and England.

Ir·kutsk (ēr kōōtsk′), a city in the S Russian Federation in Asia, on the Angara, W of Lake Baikal. 626,000.

I·ron·de·quoit (i ron′di kwoit′), a city in W New York. 57,648.

I′ron Gate′ (or **Gates′**), a gorge cut by the Danube through the Carpathian Mountains, between Yugoslavia and SW Romania. 2 mi. (3.2 km) long.

Ir·ra·wad·dy (ir′ə wod′ē, -wô′dē), a river flowing S through Burma to the Bay of Bengal. 1250 mi. (2015 km) long.

Ir·tysh or **Ir·tish** (ir tish′), a river in central Asia, flowing NW from the Altai Mountains in China through NE Kazakhstan and the Russian Federation to the Ob River. ab. 1840 mi. (2960 km) long.

Ir·vine (ûr′vīn), a city in SW California. 110,330.

Ir·ving (ûr′ving), a city in NE Texas, near Dallas. 155,037.

Ir·ving·ton (ûr′ving tən), a town in NE New Jersey, near Newark. 61,493.

I·sar (ē′zär), a river in central Europe, flowing NE from W Austria through S Germany to the Danube River. 215 mi. (345 km) long.

Is·chia (ē′skyä), an Italian island in the Tyrrhenian Sea, W of Naples. 18 sq. mi. (47 sq. km).

I·sère (ē zâr′), a river in SE France, flowing from the Alps to the Rhone River. 150 mi. (240 km) long.

Is·fa·han (is′fə hän′) also **Ispahan,** a city in central Iran: the capital of Persia from the 16th into the 18th century. 986,753.

I·sis (ī′sis), the local name of the Thames River at Oxford.

Is·ken·de·run (is ken′də rōōn′), **1.** Formerly, **Alexandretta.** a seaport in S Turkey, on the Gulf of Iskenderun. 124,900. **2. Gulf of,** an inlet of the Mediterranean, off the S coast of Turkey.

Is·la de la Ju·ven·tud (ēz′lä ᵭe lä hōō′ven tōōᵭ′), Spanish name of the Isle of Youth.

Is·la de Pas·cua (ēz′lä ᵭe päs′kwä), Spanish name of Easter Island.

Is·lam·a·bad (is lä′mə bäd′, -lam′ə bad′), the capital of Pakistan, in the N part, near Rawalpindi. 201,000.

Is·las Ca·na·rias (ēz′läs kä nä′ryäs), Spanish name of Canary Islands.

Is·las Mal·vi·nas (ēz′läz mäl vē′näs), Spanish name of Falkland Islands.

Isle′ Roy′ale (roi′əl), an island in Lake Superior: a part of Michigan; a national park (**Isle′ Roy′ale Na′tional Park′**). 208 sq. mi. (540 sq. km).

Is·ling·ton (iz′ling tən), a borough of N London, England. 168,700.

Is·ma·i·li·a or **Is·ma·′i·li·ya** (is′mä ə lē′ə, -mī-ə-), a seaport at the midpoint of the Suez Canal, in NE Egypt. 236,300. —**Is·ma·′il′i** (-il′ē), **Is′ma·′il′i·an** (-il′ē ən), *n., adj.*

Is·pa·han (is′pə hän′), Isfahan.

Is·ra·el (iz′rē əl, -rā-), a republic in SW Asia, on the Mediterranean: formed as a Jewish state in 1948. 4,440,000; 7984 sq. mi. (20,679 sq. km).

Cap.: Jerusalem. —**Is·rae/li** (-rā/lē), *n., pl.* **-lis,** (*esp. collectively*) **-li,** *adj.*

Is·sus (is/əs), an ancient town in Asia Minor, in Cilicia: victory of Alexander the Great over Darius III, 333 B.C.

Is·syk-Kul (is/ik kŏŏl/, -kŏŏl/), a mountain lake in NW Kyrgyzstan. 2250 sq. mi. (5830 sq. km).

Is·tan·bul (is/tän bŏŏl/, -tan-, -täm-), a seaport in NW Turkey, on both sides of the Bosporus: site of capital of Byzantine and Ottoman empires. 5,494,900. Formerly (A.D. 330–1930), **Constantinople.**

Is·tri·a (is/trē ə), a peninsula in W Croatia, projecting into the N Adriatic. Also called **Is/trian Penin/sula.** —**Is/tri·an,** *adj., n.*

I·ta·bu·na (ē/tə bŏŏ/nə), a city in E Brazil. 129,938.

Ital/ian East/ Af/rica, a former Italian territory in E Africa, formed in 1936 by the merging of Eritrea, Italian Somaliland, and Ethiopia.

Ital/ian Soma/liland, a former Italian colony in E Africa: now part of Somalia.

It·a·ly (it/l ē), a republic in S Europe, comprising a peninsula S of the Alps, and Sicily, Sardinia, Elba, and other smaller islands: a kingdom 1870–1946. 57,400,000; 116,294 sq. mi. (301,200 sq. km). *Cap.:* Rome. Italian, **I·ta·lia** (ē tä/lyä). —**I·tal·ian** (i tal/yən), *n., adj.*

I·tas·ca (i tas/kə), **Lake,** a lake in N Minnesota: one of the sources of the Mississippi River.

Ith·a·ca (ith/ə kə), **1.** one of the Ionian Islands, off the W coast of Greece: legendary home of Ulysses. 4156; 37 sq. mi. (96 sq. km). **2.** a city in S New York at the S end of Cayuga Lake. 28,732. —**Ith/a·can,** *adj., n.*

I·va·no-Fran·kovsk (i vä/nō fräng kôfsk/, -kofsk/), a city in W Ukraine, S of Lvov. 150,000.

I·va·no·vo (ē vä/nə və), a city in the W Russian Federation in Europe, NE of Moscow. 479,000.

I·vi·za (*Sp.* ē vē/sä), IBIZA.

I/vory Coast/, a republic in W Africa: formerly part of French West Africa; gained independence 1960. 11,630,000; 127,520 sq. mi. (330,275 sq. km). *Cap.:* Abidjan. French, **Côte d'Ivoire.** —**I·vo·ri·an** (ī vôr/ē ən, i vōr/-), *adj., n.*

I·wa·ki (ē wä/kē), a city on NE Honshu, in Japan. 356,000.

I·wo (ē′wō), a city in SW Nigeria. 261,600.

I·wo Ji·ma (ē′wə jē′mə, ē′wō), one of the Volcano Islands, in the N Pacific, S of Japan: under U.S. administration after 1945; returned to Japan 1968.

Ix·elles (ēk sel′), a city in central Belgium, near Brussels. 75,723. Flemish, **Elsene.**

Ix·tac·cí·huatl or **Iz·tac·cí·huatl** (ēs′täk sē′wät′l), also **Ix·ta·cí·huatl** (-tä sē′-), an extinct volcano in S central Mexico, SE of Mexico City. 17,342 ft. (5286 m).

I·za·bal (ē′zə bäl′, -sə-), **Lake,** a lake in E Guatemala: the largest in the country. ab. 450 sq. mi. (1165 sq. km).

I·zhevsk (ē′zhifsk), the capital of the Udmurt Autonomous Republic, in the Russian Federation in Europe, NE of Kazan. 635,000.

Iz·mir (iz′mēr), **1.** Formerly, **Smyrna.** a seaport in W Turkey on the Gulf of Izmir: important city of Asia Minor from ancient times. 1,489,817. **2. Gulf of.** Formerly, **Gulf of Smyrna.** an arm of the Aegean Sea in W Turkey. 35 mi. (56 km) long; 14 mi. (23 km) wide.

Iz·mit (iz mit′), a seaport in NW Turkey, on the Sea of Marmara. 236,144.

J

Jab·al·pur (jub/əl pŏŏr/), a city in central Madhya Pradesh, in central India. 758,000.

Jack·son (jak/sən), **1.** the capital of Mississippi, in the central part. 196,637. **2.** a city in W Tennessee. 53,320.

Jack/son Hole/, a valley in NW Wyoming, near the Teton Range.

Jack·son·ville (jak/sən vil/), a seaport in NE Florida, on the St. Johns River. 635,230.

Ja·dot·ville (zhad/ō vēl/), former name of LIKASI.

Ja·én (hä en/), a city in S Spain, NNW of Granada. 102,826.

Jaf·fa (jaf/ə, jä/fə; *locally* yä/fä) also **Yafo**, a former seaport in W Israel, part of Tel Aviv-Jaffa since 1950: ancient Biblical town. Ancient, **Joppa**.

Jaff·na (jäf/nə), a seaport in N Sri Lanka. 118,224.

Jai·pur (jī/pŏŏr), **1.** a former state in NW India, now part of Rajasthan. **2.** the capital of Rajasthan, in NW India. 1,105,000.

Ja·kar·ta or **Dja·kar·ta** (jə kär/tə), the capital of Indonesia, on the NW coast of Java. 6,503,449. Formerly, **Batavia.**

Ja·la·pa (hä lä/pä), the capital of Veracruz, in E Mexico. 212,769.

Ja·lis·co (hə lis/kō, -lēs/-), a state in W Mexico. 5,198,374; 31,152 sq. mi. (80,685 sq. km). *Cap.:* Guadalajara.

Ja·mai·ca (jə mā/kə), **1.** an island in the West Indies, S of Cuba. 4413 sq. mi. (11,430 sq. km). **2.** a republic coextensive with this island: formerly a British colony; became independent in 1962; a member of the Commonwealth of Nations. 2,300,000. *Cap.:* Kingston. —**Ja·mai/can,** *adj., n.*

Jam·bi or **Djam·bi** (jäm/bē), **1.** a province on SE Sumatra, in W Indonesia. **2.** Formerly, **Telanaipura.** a river port in and the capital of this province. 230,373.

James (jāmz), **1.** a river flowing E from the W part of Virginia to Chesapeake Bay. 340 mi. (547 km) long. **2.** a river flowing S from central North Dakota to the Missouri River. 710 mi. (1143 km) long.

James/ Bay/, the S arm of Hudson Bay, in E

Canada between Ontario and Quebec provinces. 300 mi. (483 km) long; 160 mi. (258 km) wide.

James·town (jāmz′toun′), **1.** a village in E Virginia: first permanent English settlement in North America 1607; restored 1957. **2.** the capital of St. Helena, in the S Atlantic Ocean. 1516.

Jam·mu (jum′ŏŏ), the winter capital of Jammu and Kashmir, in the SW part, in N India. 206,135.

Jam′mu and Kash′mir, official name of KASHMIR (def. 2).

Jam·na·gar (jäm nug′ər), a city in W Gujarat, in W central India. 317,000.

Jam·shed·pur (jäm′shed pŏŏr′), a city in SE Bihar, in NE India. 670,000.

Janes·ville (jānz′vil), a city in S Wisconsin. 52,133.

Ja·nic·u·lum (jə nik′yə ləm), a ridge near the Tiber in Rome, Italy. —**Ja·nic′u·lan,** adj.

Jan May·en (yän′ mī′ən), a volcanic island in the Arctic Ocean between Greenland and Norway: a possession of Norway. 144 sq. mi. (373 sq. km).

Ja·pan (jə pan′), **1.** a constitutional monarchy on a chain of islands off the E coast of Asia: main islands, Hokkaido, Honshu, Kyushu, and Shikoku. 122,260,000; 141,529 sq. mi. (366,560 sq. km). *Cap.:* Tokyo. Japanese, **Nihon, Nippon. 2. Sea of,** the part of the Pacific Ocean between Japan and mainland Asia. —**Jap·a·nese** (jap′ə nēz′, -nēs′), n., pl. **-nese,** adj.

Ja·pu·rá (zhä′pŏŏ rä′), a river flowing E from the Andes in SW Colombia to the Amazon. 1750 mi. (2820 km) long.

Jas′per Na′tional Park′ (jas′pər), a national park in the Rocky Mountains in W Alberta, in SW Canada. 4200 sq. mi. (10,878 sq. km).

Jas·sy (yä′sē), a city in NE Romania. 314,156. Romanian, Iaşi.

Ja·va (jä′və), the main island of Indonesia. 91,269,528 (with Madura); 51,032 sq. mi. (132,173 sq. km). —**Ja′va·nese′,** n., pl. **-nese,** adj.

Ja·va·ri (zhä′və rē′), a river in E South America, flowing NE from Peru to the upper Amazon, forming part of the boundary between Peru and Brazil. 650 mi. (1045 km) long. Spanish, **Yavarí.**

Ja′va Sea′, a sea between Java and Borneo.

Ja/va Trench/, a trench in the Indian Ocean, S of Java: deepest known part of Indian Ocean. 25,344 ft. (7725 m) deep.

Jax·ar·tes (jak sär/tēz), ancient name of SYR DARYA.

Ja·ya·pu·ra or **Dja·ja·pu·ra** (jä/yə pŏŏr/ə), the capital of Irian Jaya, on the NE coast, in Indonesia. 149,618. Formerly, **Hollandia.**

Jeb·el ed Druz (jeb/əl ed drōōz/), a mountainous region in S Syria: inhabited by Druzes. ab. 2700 sq. mi. (6995 sq. km). Also called **Jeb/el Druz/** (drōōz/).

Jeb/el Mu/sa (mōō/sä), a mountain in NW Morocco, opposite Gibraltar: one of the Pillars of Hercules. 2775 ft. (846 m). Ancient, **Abyla.**

Jed·da (jed/ə), JIDDA.

Jef·fer·son Cit/y (jef/ər sən), the capital of Missouri, in the central part, on the Missouri River. 33,619.

Je·hol (jə hōl/), **1.** a region and former province in NE China: incorporated into Manchukuo by the Japanese 1932–45. 74,297 sq. mi. (192,429 sq. km). **2.** former name of CHENGDE.

Je·mappes (zhə map/), a town in SW Belgium. 12,455.

Je·na (yā/nä), a city in central Germany: Napoleon decisively defeated the Prussians here in 1806. 108,010.

Jer·ba (jer/bə), DJERBA.

Je·rez (hə rās/, -rez/), a city in SW Spain. 180,444. Also called **Jerez/ de la Fron·te/ra** (frun târ/ə). Formerly, **Xeres.**

Jer·i·cho (jer/i kō/), an ancient city of Palestine, N of the Dead Sea.

Jer·sey (jûr/zē), **1.** a British island in the English Channel: the largest of the Channel Islands. 80,212; 44 sq. mi. (116 sq. km). *Cap.:* St. Helier. **2.** NEW JERSEY. —**Jer/sey·an,** *n., adj.* —**Jer/sey·ite/,** *n.*

Jer/sey Cit/y, a seaport in NE New Jersey, opposite New York City. 228,537.

Je·ru·sa·lem (ji rōō/sə ləm, -zə-), an ancient holy city for Jews, Christians, and Muslims: divided between Israel and Jordan 1948–67; Jordanian sector annexed by Israel 1967; capital of Israel since 1950. 482,700. —**Je·ru/sa·lem·ite/,** *n., adj.*

Jes·sel·ton (jes′əl tən), former name of KOTA KINABALU.

Jew′ish Auton′omous Re′gion, an autonomous region in the Khabarovsk territory of the Russian Federation in E Siberia. 216,000; 13,900 sq. mi. (36,000 sq. km). *Cap.:* Birobidzhan.

Jez·re·el (jez′rē əl, -el′), **Plain of,** ESDRAELON. —**Jez′re·el·ite′,** *n.*

Jhan·si (jän′sē), a city in SW Uttar Pradesh, in N central India. 281,000.

Jhe·lum (jā′ləm), a river in S Asia, flowing from S Kashmir into the Chenab River in Pakistan. 450 mi. (725 km) long.

Jia·mu·si (jyä′my′sē′) also **Chiamussu, Kiamusze,** a city in E Heilongjiang province, in NE China. 300,000.

Jiang·su (jyäng′sy′) also **Kiangsu,** a maritime province in E China. 62,130,000; 39,460 sq. mi. (102,200 sq. km). *Cap.:* Nanjing.

Jiang·xi (jyäng′shē′) also **Kiangsi,** a province in SE China. 35,090,000; 63,629 sq. mi. (164,729 sq. km). *Cap.:* Nanchang.

Jid·da (jid′ə) also **Jedda,** the seaport of Mecca, in W Saudi Arabia, on the Red Sea. 1,500,000.

Ji·hla·va (yi′hlä vä), a city in W Moravia, in central Czechoslovakia. 53,074.

Ji·lin (jē′lin′) also **Kirin, 1.** a province in NE China, N of the Yalu River. 23,150,000; 72,201 sq. mi. (187,001 sq. km). *Cap.:* Changchun. **2.** a port city in this province, on the Songhua River: a former provincial capital. 1,140,000.

Ji·long (jē′lông′), CHILUNG.

Ji·nan or **Tsi·nan** (jē′nän′), the capital of Shandong province, in E China. 1,430,000.

Jin·ja (jin′jä), a city in SE Uganda, on Lake Victoria. 47,300.

Jin·zhou or **Chin·chow** (jin′jō′), a city in S Liaoning province, in NE China. 750,000.

Jiu·long (*Chin.* jyʊ′lông′), KOWLOON.

Jo·ão Pes·so·a (zhŏŏ oʊɴ′ pe sō′ə), the capital of Paraíba, in NE Brazil. 290,247.

Jodh·pur (jod′pər, jōd′pŏŏr), **1.** Also called **Marwar.** a former state in NW India, now part of Rajasthan. **2.** a city in central Rajasthan, in NW India. 494,000.

Jod·rell Bank′ (jod′rəl), the site of a radio astronomy observatory in NE Cheshire, England, that operates a 250-ft. (76-m) radio telescope.

Jog·ja·kar·ta (jog/jə kär/tə, jōg/-) also **Jokjakarta,** a city in central Java, in S Indonesia. 398,727. Dutch, **Djokjakarta.**

Jo·han·nes·burg (jō han/is bûrg/, -hä/nis-, yō-), a city in S Transvaal, in the NE Republic of South Africa. 1,609,408. Dutch.

John/ o'Groats/ House/ (jon/ ə grōts/), a locality at the N tip of Scotland, near Duncansby Head, N Caithness, traditionally thought of as the northernmost point of Britain. Also called **John/ o'Groat's/.**

Johns·town (jonz/toun/), a city in SW Pennsylvania: disastrous flood 1889. 35,496.

Jo·hore (jə hôr/, -hōr/), a state in Malaysia, on S Malay Peninsula. 1,638,229; 7330 sq. mi. (18,985 sq. km).

Johore/ Bah/ru (bä/rōō), the capital of Johore state, Malaysia, in the S part. 249,880.

Join·vi·le (zhoin vē/lē), a seaport in S Brazil. 216,986. Formerly, **Join·vil/le.**

Jok·ja·kar·ta (jok/jə kär/tə, jōk/-), JOGJAKARTA.

Jo·li·et (jō/lē et/), a city in NE Illinois. 76,836.

Jo·lo (hô lô/), an island in the SW Philippines: the main island of the Sulu Archipelago. 237,683; 345 sq. mi. (894 sq. km).

Jön·kö·ping (yœn/chœ ping), a city in S Sweden. 108,962.

Jon·quière (Fr. zhôn kyer/), a city in S Quebec, in E Canada. 58,467.

Jop·lin (jop/lin), a city in SW Missouri. 38,893.

Jop·pa (jop/ə), ancient name of JAFFA.

Jor·dan (jôr/dn), **1.** Official name, **Hashemite Kingdom of Jordan.** a kingdom in SW Asia, consisting of the former Transjordan and a part of Palestine that, since 1967, has been occupied by Israel. 2,970,000; 37,264 sq. mi. (96,514 sq. km). *Cap.:* Amman. **2.** a river in SW Asia, flowing from S Lebanon through the Sea of Galilee, then S between Israel and Jordan through W Jordan into the Dead Sea. 200 mi. (320 km) long. —**Jor·da/ni·an** (-dā/nē ən), *n., adj.*

Jos (jôs), a city in central Nigeria. 149,000.

Juan de Fu·ca (wän/ di fyōō/kə, fōō/-, hwän/), **Strait of,** a strait between Vancouver Island and NW Washington. 100 mi. (160 km) long; 15–20 mi. (24–32 km) wide. Also called **Juan′ de Fu′ca Strait′.**

Juan Fer·nán·dez (wän′ fər nan′diz, hwän′), a group of three islands in the S Pacific, 400 mi. (645 km) W of and belonging to Chile: Alexander Selkirk, the alleged prototype of Robinson Crusoe, marooned here 1704.

Juá·rez (wär′ez, hwär′-), **Ciudad,** Ciudad Juárez.

Ju·ba (jōō′bä), a river in E Africa, flowing south from S Ethiopia through Somalia to the Indian Ocean. 1000 mi. (1609 km).

Ju·dae·a (jōō dē′ə), Judea. —**Ju·dae′an,** adj., n.

Ju·dah (jōō′də), the Biblical kingdom of the Hebrews in S Palestine, including the tribes of Judah and Benjamin. —**Ju′dah·ite′,** n.

Ju·de·a (jōō dē′ə), the S region of ancient Palestine: existed under Persian, Greek, and Roman rule; divided between Israel and Jordan in 1948; occupied by Israel since 1967. —**Ju·de′an,** n., adj.

Ju·go·sla·vi·a (yōō′gō slä′vē ə), Yugoslavia. —**Ju′go·slav′** (-släv′, -slav′), n. —**Ju′go·sla′vi·an,** adj., n. —**Ju′go·slav′ic,** adj.

Juiz de Fo·ra (zhwēz′ də fôr′ə), a city in SE Brazil, N of Rio de Janeiro. 299,432.

Ju·juy (hōō hwē′), a city in NW Argentina. 124,487.

Jul′ian Alps′ (jōōl′yən), a mountain range in N Slovenia. Highest peak, 9394 ft. (2863 m).

Jul·lun·dur (jul′ən dər), a city in N Punjab, in NW India. 405,700.

Jum·na (jum′nə), a river in N India, flowing SE from the Himalayas to the Ganges at Allahabad. 860 mi. (1385 m) long.

Jun·di·aí (zhōōn′dyä ē′), a city in SE Brazil, NW of São Paulo. 221,888.

Ju·neau (jōō′nō), the capital of Alaska, in the SE part. 19,528.

Jung·frau (yōōng′frou′), a mountain in S Switzerland, in the Bernese Alps. 13,668 ft. (4166 m).

Ju·ra (jōōr′ə), **1.** Also called **Ju′ra Moun′tains.** a mountain range in W central Europe, between France and Switzerland, extending from the Rhine to the Rhone. Highest peak, 5654 ft. (1723 m). **2.** a canton in W Switzerland. 64,800; 323 sq. mi. (837 sq. km).

Ju·ru·á (zhōōr′ōō ä′), a river in E and W South

America, flowing NE from E Peru through W Brazil to the Amazon. 1200 mi. (1930 km) long.

Jut·land (jut/lənd), a peninsula comprising the continental portion of Denmark. 11,441 sq. mi. (29,630 sq. km). Danish, **Jyl·land** (yyl/län).
—**Jut/land·er,** *n.* —**Jut/land·ish,** *adj.*

Jy·väs·ky·lä (yy/vas ky/la), a city in S central Finland. 65,719.

K

K2 (kä/tōō/), a mountain in N Kashmir, in the Karakoram range: second highest peak in the world. 28,250 ft. (8611 m). Also called **Godwin Austen, Dapsang.**

Kab·ar·di·no-Bal/kar Auton/omous Repub/-lic (kab/ər dē/nō bôl/kär, -bal/-), an autonomous republic in the Russian Federation in N Caucasia, N of the Georgian Republic. 760,000; 4825 sq. mi. (12,500 sq. km). *Cap.:* Nalchik.

Ka·bul (kä/bōōl, -bəl, kə bōōl/), **1.** the capital of Afghanistan, in the NE part. 913,164. **2.** a river flowing E from NE Afghanistan to the Indus River in Pakistan. 360 mi. (580 km) long.

Kab·we (käb/wā), a city in central Zambia: old mining town. 190,752. Formerly, **Broken Hill.**

Ka·di·yev·ka (kə dē/yəf kə), former name of **STAKHANOV.**

Ka·du·na (kə dōō/nə), a city in central Nigeria. 247,100.

Kaf·frar·i·a (kə frâr/ē ə), a region in the S Republic of South Africa that is inhabited mostly by the Xhosa. —**Kaf·frar/i·an,** *adj., n.*

Ka·fi·ri·stan (kä/fi ri stän/, kaf/ər ə stan/), former name of **NURISTAN.**

Ka·fu·e (kə fōō/ā, kä-), a river in Zambia, flowing into the Zambezi River above Kariba Lake. ab. 600 mi. (965 km) long.

Ka·ge·ra (kä gâr/ə), a river in equatorial Africa flowing into Lake Victoria from the west: the most remote headstream of the Nile. 430 mi. (690 km) long.

Ka·go·shi·ma (kä/gə shē/mə), a seaport on S Kyushu, in SW Japan. 525,000.

Ka·ho·o·la·we (kä hō/ō lä/wā, -vä), an island in central Hawaii, S of Maui: uninhabited. 45 sq. mi. (117 sq. km).

Kai·e·teur Falls/ (ki/ē tōōr/, ki/ē tōōr/), a waterfall in central Guyana, on a tributary of the Essequibo River. 741 ft. (226 m) high.

Kai·feng (ki/fung/), a city in NE Henan province, in E China: a former provincial capital. 330,000.

Kai·lu·a (ki lōō/ə), a city on SE Oahu, in Hawaii. 35,812.

Kair·ouan or **Kair·wan** (ker wän′, ki°r-), a city in NE Tunisia: a holy city of Islam. 72,254.

Kai·sers·lau·tern (kī′zərz lou′tərn, -zərs-), a city in S Rhineland-Palatinate, in SW Germany. 97,664.

Ka·la·ha·ri (kä′lə här′ē, kal′ə-), a desert region in SW Africa, largely in Botswana. 100,000 sq. mi. (259,000 sq. km).

Ka·lakh (kä′läкн), an ancient Assyrian city on the Tigris River: its ruins are near Mosul in N Iraq.

Kal·a·ma·zoo (kal′ə mə zōō′), a city in SW Michigan. 80,277.

Ka·lat or **Khe·lat** (kə lät′), a region in S Baluchistan, in SW Pakistan.

Ka·le·mie (kə lā′mē), a city in E Zaire, on Lake Tanganyika. 172,297. Formerly, **Albertville.**

Kal·gan (käl′gän′), former name of ZHANGJIAKOU.

Kal·goor·lie (kal gŏŏr′lē), a city in SW Australia: chief center of gold-mining industry in Australia. 10,087, with suburbs 19,848.

Ka·li·man·tan (kä′lē män′tän), Indonesian name of BORNEO, esp. referring to the southern, or Indonesian, part.

Ka·li·nin (kə lē′nin), former name (1934–90) of TVER.

Ka·li·nin·grad (kə lē′nin grad′), a seaport in the W Russian Federation, on the Baltic Sea. 394,000. German, **Königsberg.**

Ka·lisz (kä′lish), a city in central Poland. 100,340.

Kal·mar (käl′mär), a seaport in SE Sweden, on Kalmar Sound. 54,915. —**Kal·mar′i·an** (-mär′ē-en), adj.

Kal′mar Sound′, a strait between SE Sweden and Öland Island. 85 mi. (137 km) long; 14 mi. (23 km) wide.

Kal′myk Auton′omous Repub′lic (kal′mik), an autonomous republic in the Russian Federation in Europe, on the NW shore of the Caspian Sea. 322,000; 29,300 sq. mi. (75,000 sq. km). Also called **Kal·myk·i·a** (kal mik′ē ə).

Ka·lu·ga (ku lōō′gə), a city in the W Russian Federation in Europe, SW of Moscow. 307,000.

Ka·ma (kä′mə), a river in the E Russian Federation in Europe, flowing SW from the central Ural Mountains region into the Volga River S of Kazan. 1200 mi. (1930 km) long.

Ka·ma·ku·ra (kä/mə kŏŏr/ə), a city on SE Honshu, in central Japan: great bronze statue of Buddha. 172,612.

Kam·chat·ka (kam chät/kə, -chat/-), a peninsula in the NE Russian Federation in Asia, between the Bering Sea and the Sea of Okhotsk. 750 mi. (1210 km) long; 104,200 sq. mi. (269,880 sq. km) wide. —**Kam·chat/kan,** adj.

Ka·mensk-U·ral·ski (kä/mənsk yŏŏ ral/skē), a city in the Russian Federation in Asia, near the Ural Mountains SE of Ekaterinburg. 204,000.

Ka·mi·na (kä mē/nə), a city in S Zaire. 160,020.

Kam·loops (kam/lōōps), a city in S British Columbia, in SW Canada. 61,773.

Kam·pa·la (käm pä/lə, kam-), the capital of Uganda, in the S part. 458,423.

Kam·pu·che·a (kam/pōō chē/ə), **People's Republic of,** a former official name of CAMBODIA. —**Kam/pu·che/an,** adj., n.

Kan., Kansas.

Ka·nan·ga (kə näng/gə), a city in central Zaire. 290,898. Formerly, **Luluabourg.**

Ka·na·ra (kə när/ə, kä/nər ə), a region in SW India, on the Deccan Plateau. ab. 60,000 sq. mi. (155,400 sq. km).

Ka·na·za·wa (kä/nə zä/wə), a seaport on W Honshu, in central Japan. 421,000.

Kan·chen·jun·ga (kän/chən jōōng/gə), a mountain in S Asia, between NE India and Nepal, in the E Himalayas: third highest in the world. 28,146 ft. (8579 m).

Kan·da·har (kun/də här/), a city in S Afghanistan. 178,409.

Kan·dy (kan/dē, kän/-), a city in central Sri Lanka: famous Buddhist temples. 97,872.

Ka·ne·o·he (kä/nā ō/hā), a town on E Oahu, in Hawaii. 29,919.

Ka·no (kä/nō), a city in N Nigeria. 487,100.

Kan·pur (kän/pŏŏr), a city in S Uttar Pradesh, in N India, on the Ganges River. 1,688,000.

Kans., Kansas.

Kan·sas (kan/zəs), **1.** a state in the central United States. 2,477,574; 82,276 sq. mi. (213,094 sq. km). Cap.: Topeka. Abbr.: KS, Kans., Kan., Kas. **2.** a river in NE Kansas, flowing E to the Missouri River. 169 mi. (270 km) long. —**Kan/san,** adj., n.

Kan·sas Cit·y, 1. a city in W Missouri, at the confluence of the Kansas and Missouri rivers. 435,146. **2.** a city in NE Kansas, adjacent to Kansas City, Mo. 149,767.

Kan·su (*Chin.* kän′sy′, gän′-), GANSU.

Kao·hsiung (gou′shyŏŏng′), a seaport on SW Taiwan. 1,300,000.

Ka·o·lack or **Ka·o·lak** (kä′ō lak′, kou′lak), a city in W Senegal. 115,679.

Ka·ra·chai′-Cher·kess′ Auton′omous Region (kär′ə chi′chər kes′), an autonomous region in the Russian Federation in Europe, in the Caucasus. 418,000; 5442 sq. mi. (14,100 sq. km).

Ka·ra·chi (kə rä′chē), a seaport in S Pakistan, near the Indus delta: former national capital; now capital of Sind province. 5,208,170.

Ka·ra·fu·to (kär′ə fōō′tō), Japanese name of SAKHALIN.

Ka·ra·gan·da (kar′ə gən dä′), a city in central Kazakhstan. 614,000.

Ka·ra·kal·pak′ Auton′omous Repub′lic (kar′ə kal pak′), an autonomous republic in NW Uzbekistan. 1,214,000; 63,938 sq. mi. (165,600 sq. km).

Ka·ra·ko·ram (kär′ə kôr′əm, -kōr′-, kar′-), **1.** a mountain range in NW India, in N Kashmir. Highest peak, K2, 28,250 ft. (8611 m). **2.** a pass traversing this range, on the route from NE Kashmir in India to Xinjiang Uygur in China. 18,300 ft. (5580 m).

Ka·ra·ko·rum (kär′ə kôr′əm, -kōr′-, kar′-), a ruined city in central Mongolian People's Republic: ancient capital of the Mongol Empire.

Ka·ra Kum (kar′ə kōōm′, kŏŏm′, kär′ə), a desert S of the Aral Sea, largely in Turkmenistan. ab. 110,000 sq. mi. (284,900 sq. km).

Ka′ra Sea′ (kär′ə), an arm of the Arctic Ocean between Novaya Zemlya and the N Russian Federation.

Kar·ba·la (kär′bə lə), KERBELA.

Ka·re·lia (kə rēl′yə), **1.** a region in the NW Russian Federation in Europe, comprising Lake Ladoga and Lake Onega and the adjoining area along the E border of Finland. **2.** KARELIAN AUTONOMOUS REPUBLIC. —**Ka·re′lian**, *adj., n.*

Kare′lian Auton′omous Repub′lic, an autonomous republic in the NW Russian Federation in Europe. 792,000; 66,500 sq. mi. (172,240 sq. km). *Cap.:* Petrozavodsk.

Kare′lian Isth′mus, a narrow strip of land between Lake Ladoga and the Gulf of Finland, in the NW Russian Federation.

Ka·ri·ba (kə rē′bə), an artificial lake on the border of Zimbabwe and Zambia, formed by a dam (**Kari′ba Dam′**): site of hydroelectric power project. ab. 2000 sq. mi. (5200 sq. km).

Kar·kheh (kär kā′, -кнā′), a river in SW Iran, flowing SW to marshes along the Tigris River. ab. 350 mi. (565 km) long.

Karl-Marx-Stadt (kärl′märks′shtät′), former name (1953–90) of CHEMNITZ.

Kar·lo·vy Va·ry (kär′lə vē vär′ē), a city in W Czechoslovakia: hot mineral springs. 58,541. German, **Karls·bad** (kärlz′bät′).

Karls·ruh·e (kärlz′rōō′ə), a city in SW Germany, in Baden-Württemberg, on the Rhine. 260,500.

Karl·stad (kärl′städ, -stä), a city in S Sweden. 74,892.

Kar·nak (kär′nak), a village in E Egypt, on the Nile: the N part of the ruins of ancient Thebes.

Kar·na·ta·ka (kär nä′tə kə), a state in S India. 37,043,451; 74,326 sq. mi. (192,504 sq. km). *Cap.:* Bangalore. Formerly, **Mysore.**

Kar·roo (kə rōō′), a vast plateau in the S Republic of South Africa, in Cape of Good Hope province. 100,000 sq. mi. (260,000 sq. km); 3000–4000 ft. (900–1200 m) above sea level.

Kars (kärs), a city in NE Turkey. 58,799.

Ka·run (kə rōōn′, kä-), a river in SW Iran, flowing SW to the Shatt-al-Arab. ab. 515 mi. (830 km) long.

Kas., Kansas.

Ka·sai (kə sī′, kä-), a river in SW Africa, flowing from Angola through Zaire to the Congo (Zaire) River. ab. 1338 mi. (2154 km) long.

Kas·bah or **Cas·bah** (kaz′bä, käz′-), the older, native Arab quarter of a North African city, esp. Algiers.

Ka·shi (kä′shē′, kash′ē), a city in W Xinjiang Uygur, in extreme W China. 274,128. Also called **Kash·gar** (kash′gär, käsh′-).

Kash·mir (kash′mēr, kazh′-, kash mēr′, kazh-), **1.** Also, **Cashmere.** a region in SW Asia, in N India: sovereignty in dispute between India and Pakistan since 1947. **2.** Official name, **Jammu and Kashmir.** the part of this region occupied by India, forming a state in the Indian union. 5,981,600; ab. 53,500 sq. mi. (138,000 sq. km). *Cap.:* Srinagar (summer); Jammu (winter). —**Kash·mir′i,** *n., pl.* **-mir·is,** *adj.*

Kas·sa·la (kä′sä lä′, kas′ə lə), a city in the E Sudan. 99,652.

Kas·sel (kas′əl, kä′səl), a city in central Germany. 188,500.

Ka·strop-Rau·xel (kä′strəp rouk′səl, kas′trəp-), CASTROP-RAUXEL.

Ka·tah·din (kə tä′dn), **Mount,** the highest peak in Maine, in the central part. 5273 ft. (1607 m).

Ka·tan·ga (kə täng′gə, -tang′-), former name of SHABA.

Ka·thi·a·war (kä′tē ə wär′), a peninsula on the W coast of India.

Kat·mai (kat′mī), **Mount,** an active volcano in SW Alaska. 6715 ft. (2047 m).

Kat′mai Na′tional Park′, a national park in SW Alaska including Mt. Katmai and the Valley of Ten Thousand Smokes. 5806 sq. mi. (l5,038 sq. km).

Kat·man·du or **Kath·man·du** (kät′män dōō′, kat′man-), the capital of Nepal, in the central part. 235,160.

Ka·to·wi·ce (kä′tə vēt′sə, kat′ə-), a city in S Poland. 363,000.

Kat·rine (ka′trin), **Loch,** a lake in central Scotland. 8 mi. (13 km) long.

Kat·te·gat (kat′i gat′, kä′ti gät′), a strait between Jutland and Sweden. 40–70 mi. (64–113 km) wide.

Ka·u·a·i (kä′ōō ä′ē, kou′ī), an island in NW Hawaii. 49,100; 558 sq. mi. (1445 sq. km).

Kau·nas (kou′näs), a city in S central Lithuania. 423,000. Russian, **Kovno.**

Ka·val·la (kə val′ə, -vä′lə), a seaport in E Greece. 46,000.

Ka·ver·i (kô′və rē, kä′-), CAUVERY.

Ka·vir′ Des′ert (kə vēr′), DASHT-I-KAVIR.

Ka·wa·gu·chi (kä′wə gōō′chē), a city on SE Honshu, in central Japan, N of Tokyo. 379,357.

Ka·wa·sa·ki (kä′wə sä′kē), a seaport on SE Honshu, in central Japan, SW of Tokyo. 1,040,698.

Kay·se·ri (ki/se rē/, -zə-), a city in central Turkey. 207,039. Ancient, **Caesarea.**

Ka·zakh·stan (kä/zäk stän/), a republic in central Asia, NE of the Caspian Sea and W of China. 16,538,000; 1,049,155 sq. mi. (2,717,300 sq. km). *Cap.:* Alma-Ata. Formerly (1936–91), **Kazakh/ So/viet So/cialist Repub/lic. —Ka·zakh/,** *n.*

Ka·zan (kə zän/), the capital of the Tatar Autonomous Republic in the SE Russian Federation in Europe, near the Volga River. 1,094,000.

Kaz·bek (käz bek/), **Mount,** an extinct volcano in the central Caucasus Mountains between the Georgian Republic and the Russian Federation. 16,541 ft. (5042 m).

Kaz·da·ği (käz/dä gē/), Turkish name of Mount IDA (def. 1).

Kaz·vin (kaz vēn/), QAZVIN.

K.C., Kansas City.

Ke·a (*Gk.* ke/ä), KEOS.

Kecs·ke·mét (kech/kə mät/), a city in central Hungary. 105,000.

Ke·dah (kā/dä), a state in Malaysia, on the W central Malay Peninsula. 1,116,140; 3660 sq. mi. (9480 sq. km).

Ke·dron (kē/drən), KIDRON.

Kee/ling Is/lands (kē/ling), COCOS ISLANDS.

Kee·lung (kē/lŏŏng/), CHILUNG.

Kee·wa·tin (kē wāt/n), a district in the Northwest Territories, in N Canada. 228,160 sq. mi. (590,935 sq. km).

Ke·fal·li·ni·a (ke/fä lē nē/ä), Greek name of CEPHALONIA.

Kef·la·vík (kyep/lə vēk/, -vik, kef/-), a town in SW Iceland: international airport. 7133.

Ke·lan·tan (kə lan/tan, -län tän/), a state in Malaysia, on the central Malay Peninsula. 893,753; 5750 sq. mi. (14,893 sq. km). *Cap.:* Kota Bharu.

Ke·low·na (ki lō/nə), a city in S British Columbia, in SW Canada. 61,213.

Ke·me·ro·vo (kem/ə rō/və, -ər ə və), a city in the S Russian Federation in Asia, NE of Novosibirsk. 520,000.

Ken., Kentucky.

Ke/nai Fjords/ Na/tional Park/ (kē/ni), a national park in S Alaska: ice field and coastal fjords. 1047 sq. mi. (2711 sq. km).

Ke·nai Penin·sula, a peninsula in S Alaska, between Cook Inlet and Prince William Sound.

Ken·il·worth (ken'l wûrth'), a town in central Warwickshire, in central England, SE of Birmingham. 20,121.

Ke·ni·tra (kə nē'trə), a port in NW Morocco, NE of Rabat. 188,194.

Ken·ne·bec (ken'ə bek', ken'ə bek'), a river flowing S through W Maine to the Atlantic. 164 mi. (264 km) long.

Ken·ne·bunk·port (ken'ə bungk'pôrt', -pōrt'), a town in SW Maine: summer resort. 2952.

Ken·ne·dy (ken'i dē), **Cape,** former name (1963–73) of Cape CANAVERAL.

Ken·ner (ken'ər), a city in SE Louisiana, near New Orleans. 72,033.

Ken'ne·saw Moun'tain (ken'ə sô'), a mountain in N Georgia, near Atlanta: battle 1864. 1809 ft. (551 m).

Ke·no·sha (kə nō'shə), a port in SE Wisconsin, on Lake Michigan. 80,352.

Ken'sing·ton and Chel'sea (ken'zing tən), a borough of Greater London, England. 133,100.

Kent (kent), **1.** a county in SE England. 1,510,500; 1442 sq. mi. (3735 sq. km). **2.** an early English kingdom in SE Britain. **—Kent'ish,** adj.

Ken·tuck·y (kən tuk'ē), **1.** a state in the E central United States. 3,685,296; 40,395 sq. mi. (104,625 sq. km). Cap.: Frankfort. Abbr.: KY, Ken., Ky. **2.** a river flowing NW from E Kentucky to the Ohio River. 259 mi. (415 km) long. **—Ken·tuck'i·an,** adj., n.

Ken·ya (ken'yə, kēn'-), **1.** a republic in E Africa: a member of the Commonwealth of Nations and formerly a British crown colony and protectorate. 22,800,000; 223,478 sq. mi. (578,808 sq. km). Cap.: Nairobi. **2. Mount,** an extinct volcano in central Kenya. 17,040 ft. (5194 m). **—Ken'yan,** adj., n.

Ke·os (kē'os), a Greek island in the Aegean, off the SE coast of the Greek mainland. 1666; 56 sq. mi. (145 sq. km). Also called **Kea.**

Ke·ra·la (kā'rə lə, ker'ə-), a state in SW India on the Arabian Sea. 25,403,217; 15,035 sq. mi. (38,940 sq. km). Cap.: Trivandrum.

Ker·be·la (kûr′bə lə) also **Karbala,** a town in central Iraq: holy city of the Shi'ite sect. 184,574.

Kerch (kerch), **1.** a seaport in E Crimea, in SE Ukraine, on Kerch Strait. 173,000. **2.** a strait connecting the Sea of Azov and the Black Sea. 25 mi. (40 km) long.

Ker·gue·len (kûr′gə len′, -lən), an archipelago in the S Indian Ocean: a possession of France. 2700 sq. mi. (7000 sq. km).

Ker·ky·ra (ker′kē rä), Greek name of CORFU.

Ker·man (ker män′, ker-), a city in SE Iran. 257,284.

Ker·man·shah (ker′män shä′, -shô′, kûr′-), former name of BAKHTARAN.

Ker·ry (ker′ē), a county in W Munster province, in the SW Republic of Ireland. 122,734; 1815 sq. mi. (4700 sq. km). *Co. seat:* Tralee.

Ket·ter·ing (ket′ər ing), a city in SW Ohio. 60,569.

Kew (kyōō), a part of Richmond, in Greater London, England: famous botanical gardens **(Kew′ Gar′dens).**

Key′ Lar′go (lär′gō), the largest island of the Florida Keys. 30 mi. (48 km) long; 2 mi. (3.2 km) wide.

Key′ West′, 1. the westernmost island of the Florida Keys, in the Gulf of Mexico. 4 mi. (6.4 km) long; 2 mi. (3.2 km) wide. **2.** a seaport on this island: the southernmost city in the U.S. 24,292.

Kha·ba·rovsk (kə bär′əfsk), **1.** a territory of the Russian Federation in NE Asia. 1,565,000; 965,400 sq. mi. (2,500,400 sq. km). **2.** the capital of this territory, in the SE part, on the Amur River. 601,000.

Kha·kass′ Auton′omous Re′gion (kə käs′, кнə-), an autonomous region in the Russian Federation, in S Siberia. 569,000; 23,855 sq. mi. (61,900 sq. km). *Cap.:* Abakan.

Khal·ki·di·ki (кнäl′kē ᵵнē′kē), Greek name of CHALCIDICE.

Khal·kis (кнäl kēs′), Greek name of CHALCIS.

Kha·nia (кнä nyä′), Greek name of CANEA.

Khan Ten·gri (kän′ teng′grē, кнän′), a mountain in the Tien Shan range, on the boundary between Kyrgyzstan and China. 22,949 ft. (6995 m). Also called **Tengri Khan.**

Khar·kov (kär′kôf, -kof), a city in NE Ukraine: former capital of Ukraine. 1,611,000.

Khar·toum or **Khar·tum** (kär toom′), the capital of the Sudan, at the junction of the White and Blue Nile rivers. 476,218.

Khartoum′ North′, a city in E central Sudan, on the Blue Nile River, opposite Khartoum. 341,146.

Khe·lat (kə lät′), KALAT.

Kher·son (ker sôn′), a port in S Ukraine, on the Dnieper River, near the Black Sea. 358,000.

Khi·os (KHē′ôs), Greek name of CHIOS.

Khir·bet Qum·ran (kēr′bet koom′rän), an archaeological site in W Jordan, near the NW coast of the Dead Sea: Dead Sea Scrolls found here 1947.

Khi·va (kē′və), a former Asian khanate on the Amu Darya River, S of the Aral Sea: now divided between Uzbekistan and Turkmenistan.

Khmel·nit·sky (kmel nit′skē, kə mel-), a city in W Ukraine, SW of Kiev. 210,000.

Khmer′ Repub′lic (kmâr, kə mâr′), a former official name of CAMBODIA.

Kho·dzhent (kō jent′, КHə-) also **Khu·dzhand** (kə jänt′, КHə-), a city in N Tajikistan, on the Syr Darya. 153,000. Formerly (1936–91), **Leninabad.**

Khul·na (kool′nə), a city in S Bangladesh, on the delta of the Ganges. 646,359.

Khy′ber Pass′ (kī′bər), the chief mountain pass between Pakistan and Afghanistan, W of Peshawar. 33 mi. (53 km) long; 6825 ft. (2080 m) high.

Kia·mu·sze (*Chin.* jyä′mōō′su′), JIAMUSI.

Kiang·si (kyang′sē′), JIANGXI.

Kiang·su (kyang′sōō′), JIANGSU.

Ki·dron (kē′drən, kid′rən) also **Kedron,** a ravine E of Jerusalem, a traditional site of judgment.

Kiel (kēl), the capital of Schleswig-Holstein in N Germany, at the Baltic end of the Kiel Canal. 237,800.

Kiel′ Canal′, a canal in N Germany, connecting the North and Baltic seas. 61 mi. (98 km) long.

Kiel·ce (kyel′tse), a city in S Poland. 201,000.

Ki·ev (kē′ef, -ev), the capital of Ukraine, on the Dnieper River. 2,587,000. —**Ki′ev·an,** *adj., n.*

Ki·ga·li (kē gä′lē), the capital of Rwanda, in the central part. 156,650.

Ki·lau·e·a (kē′lou ā′ä, -ā′ə, kil′ō-), an active volcanic crater on the E slope of Mauna Loa on the island of Hawaii. 4090 ft. (1247 m).

Kil·dare (kil dâr′), a county in Leinster, in the E Republic of Ireland. 116,015; 654 sq. mi. (1695 sq. km).

Kil·i·man·ja·ro (kil′ə mən jär′ō), a volcanic mountain in NE Tanzania: highest peak in Africa. 19,321 ft. (5889 m).

Kil·ken·ny (kil ken′ē), **1.** a county in Leinster, in the SE Republic of Ireland. 73,094; 796 sq. mi. (2060 sq. km). **2.** its county seat. 9466.

Kil·lar·ney (ki lär′nē), **Lakes of,** three lakes in the SW Republic of Ireland.

Kil·leen (ki lēn′), a city in central Texas. 63,535.

Kil·lie·cran·kie (kil′ē krang′kē), a mountain pass in central Scotland, in the Grampians.

Kil·mar·nock (kil mär′nək), a city in the Strathclyde region, in SW Scotland. 52,080.

Kim·ber·ley (kim′bər lē), a city in E Cape of Good Hope province, in the central Republic of South Africa: diamond mines. 149,667.

Kin·a·ba·lu or **Kin·a·bu·lu** (kin′ə bə loo′), a mountain in N Sabah, in Malaysia: highest peak on the island of Borneo. 13,455 ft. (4101 m).

Kin·car·dine (kin kär′dn), a former county in E Scotland. Also called **Kin·car′dine·shire** (-shēr′, -shər).

Ki·nesh·ma (kē′nish mə), a city in the NW Russian Federation in Europe, NW of Nizhni Novgorod. 101,000.

Kings′ Can′yon Na′tional Park′, a national park in E California in the Sierra Nevada: mountains, deep granite gorges; giant sequoias. 708 sq. mi. (1835 sq. km).

Kings·ton (kingz′tən, king′stən), **1.** the capital of Jamaica. 600,000. **2.** a port in SE Ontario, in SE Canada, on Lake Ontario. 55,050.

Kings′ton upon′ Hull′, official name of HULL (def. 1).

Kings′ton upon′ Thames′ (temz), a borough of Greater London, England. 135,900.

Kings·town (kingz′toun′), the capital of St. Vincent and the Grenadines, on SW St. Vincent island. 28,942.

Kin·ross (kin rôs′, -ros′), a historic county in E Scotland. Also called **Kin·ross′shire** (-shēr, -shər).

Kin·sha·sa (kin shä′sə, kin′shä sə), the capital of Zaire, in the NW part, on the Zaire (Congo) River. 2,653,558. Formerly, **Léopoldville.**

Ki·o·ga (kyō′gə), **Lake,** KYOGA, Lake.

Kir·ghi·zia (kir gē′zhə, -zhē ə), KYRGYZSTAN.

Kir·ghiz′ So′viet So′cialist Repub′lic (kir-gēz′), former name (1936–91) of KYRGYZSTAN.

Ki·ri·ba·ti (kēr′ē bä′tē, kēr′ə bas′), a republic in the central Pacific Ocean, on the equator, comprising 33 islands. 66,250; 263 sq. mi. (681 sq. km). *Cap.:* Tarawa. Formerly, **Gilbert Islands.**

Ki·rin (kē′rin′), JILIN.

Ki·riti·mati (kə ris′məs), one of the Line Islands belonging to Kiribati, in the central Pacific: largest atoll in the Pacific. 1737; ab. 220 sq. mi. (575 sq. km). Formerly, **Christmas Island.**

Kirk·cal·dy (kər kôl′dē, -kô′dē, -kä′-), a city in SE Fife, in E Scotland, on the Firth of Forth. 147,963.

Kirk·cud·bright (kər kōō′brē), a historic county in SW Scotland. Also called **Kirk·cud′bright·shire′** (-shēr′, -shər).

Kirk·pat·rick (kûrk pa′trik), **Mount,** a mountain in Antarctica, near Ross Ice Shelf. ab. 14,855 ft. (4528 m).

Kir·kuk (kir kōōk′), a city in N Iraq. 207,852.

Ki·rov (kēr′ôf, -of), a city in the E Russian Federation in Europe, N of Kazan. 421,000. Formerly, **Vyatka.**

Ki·ro·va·bad (ki rō′və bad′), a city in NW Azerbaijan. 270,000.

Ki·ro·vo·grad (ki rō′və grad′), a city in S central Ukraine. 269,000.

Ki·san·ga·ni (ki zäng′gä nē, kē′säng gä′-), a city in N Zaire, on the Zaire (Congo) River. 282,650. Formerly, **Stanleyville.**

Kish (kish), an ancient Sumerian and Akkadian city: its site is 8 mi. (13 km) east of the site of Babylon in S Iraq.

Ki·shi·nev (kish′ə nef′, -nôf′, -nof′), the capital of Moldova, in the central part. 565,000. Romanian, **Chişinău.**

Kist·na (kist′nə), former name of KRISHNA.

Ki·su·mu (kē sōō′mōō), a city in W Kenya. 152,643.

Ki·ta·kyu·shu (ki tä′kē ōō′shōō), a seaport on N Kyushu, in S Japan. 1,042,000.

Kitch·e·ner (kich′ə nər), a city in S Ontario, in SE Canada. 150,604.

Ki·thi·ra (kē′thər ə, -thə rä′) also **Cythera,** a Greek island in the Mediterranean, S of Peloponnesus: site in ancient times of temple of Aphrodite.

Kit′ty Hawk′, a village in NE North Carolina: Wright brothers' airplane flight 1903.

Ki·twe (kē′twā), a city in N Zambia. 449,442.

Ki·vu (kē′vōō), **Lake,** a lake in central Africa, between Zaire and Rwanda. 1100 sq. mi. (2849 sq. km).

Ki·zil Ir·mak (ki zil′ ēr mäk′), a river flowing N through central Turkey to the Black Sea. 600 mi. (965 km) long.

Kjö·len (chœ′lən), a mountain range on the border between Norway and Sweden. Highest peak, 7005 ft. (2135 m).

Kla·gen·furt (klä′gən fŏŏrt′), a city in S Austria. 87,321.

Klai·pe·da (klī′pi də), a seaport in NW Lithuania, on the Baltic. 204,000. German, **Memel.**

Klam·ath (klam′əth), a river flowing from SW Oregon through NW California into the Pacific. 250 mi. (405 km) long.

Klon·dike (klon′dīk), **1.** a region of the Yukon territory in NW Canada: gold rush 1897–98. **2.** a river in this region, flowing into the Yukon. 90 mi. (145 km) long.

Knos·sos or **Cnos·sus** (nos′əs), a ruined city in N central Crete: capital of the ancient Minoan civilization. —**Knos′si·an,** adj.

Knox·ville (noks′vil), a city in E Tennessee, on the Tennessee River. 165,121.

Ko·ba·rid (kō′bə rēd′), a village in W Slovenia, formerly in Italy: defeat of the Italians by the Germans and Austrians 1917. Italian, **Caporetto.**

Ko·be (kō′bē, -bā), a seaport on S Honshu, in S Japan. 1,413,000.

Kø·ben·havn (kœ′bən houn′), Danish name of **Copenhagen.**

Ko·blenz (kō′blents), **Coblenz.**

Ko·chi (kō′chē), a seaport on central Shikoku, in SW Japan. 310,000.

Ko·di·ak (kō′dē ak′), an island in the N Pacific, near the base of the Alaska Peninsula. 100 mi. (160 km) long.

Ko·dok (kō′dok), modern name of Fashoda.

Ko·fu (kō′fōō), a city on S Honshu, in central Japan. 199,272.

Ko·hi·ma (kō′hē mä′), the capital of Nagaland, in E India. 67,200.

Ko·kand (ko kand′), a city in NE Uzbekistan, SE of Tashkent. 153,000.

Ko·ko·mo (kō′kə mō′), a city in central Indiana. 47,808.

Ko·ko Nor (kō′kō′ nôr′), Qinghai (def. 2).

Ko′la Penin′sula (kō′lə), a peninsula in the NW Russian Federation in Europe, between the White and Barents seas.

Kol·ha·pur (kō′lə pŏŏr′), a city in S Maharashtra, in SW India. 351,000.

Köln (kœln), German name of Cologne.

Ko·lom·na (kə lôm′nə), a city in the W Russian Federation in Europe, SE of Moscow. 147,000.

Ko·lozs·vár (kô′lôzh vär′), Hungarian name of Cluj-Napoca.

Ko·ly·ma or **Ko·li·ma** (kə lē′mə), a river in the NE Russian Federation in Asia, flowing NE to the Arctic Ocean. 1000 mi. (1610 km) long.

Koly′ma Range′, a mountain range in NE Siberia in the NE Russian Federation.

Ko′mi Auton′omous Repub′lic (kō′mē), an autonomous republic in the NW Russian Federation in Europe. 1,263,000; 160,540 sq. mi. (415,900 sq. km). *Cap.:* Syktyvkar.

Kom·so·molsk (kom′sə môlsk′), a city in the E Russian Federation in Asia, on the Amur River. 316,000. Also called **Komsomolsk′-on-Amur′**.

Kö·nig·grätz (kœ′niкн grets′), German name of Hradec Králové.

Kö·nigs·berg (kœ′niкнs berк′), German name of Kaliningrad.

Kon·stanz (kôn′stänts), German name of Constance (def. 2).

Kon·ya (kôn′yä, kôn yä′), a city in S Turkey, S of Ankara. 438,859. Ancient, **Iconium**.

Koo·te·nay or **Koo·te·nai** (kōōt′n ā′, -n ē′), a river flowing from SW Canada through NW Montana and N Idaho, swinging back into Canada to the Columbia River. 400 mi. (645 km) long.

Koo′tenay Lake′, a lake in W Canada, in S British Columbia. 64 mi. (103 km) long.

Kor·do·fan (kôr′dō fän′), a province in the central Sudan. 3,103,000. ab. 147,000 sq. mi. (380,730 sq. km). *Cap.:* El Obeid.

Ko·re·a (kə rē′ə), **1.** a former country in E Asia, on a peninsula SE of Manchuria and between the Sea of Japan and the Yellow Sea: a kingdom prior to 1910; under Japanese rule 1910–45. **2. Democratic People's Republic of,** official name of NORTH KOREA. **3. Republic of,** official name of SOUTH KOREA. —**Ko·re′an,** *n., adj.*

Kore′a Strait′, a strait between Korea and Japan, connecting the Sea of Japan and the East China Sea. 120 mi. (195 km) long.

Ko·ri·ya·ma (kôr′ē ä′mə), a city on E central Honshu, in Japan. 302,000.

Kort·rijk (kôrt′rik), Flemish name of COURTRAI.

Kos or **Cos** (kos, kôs), one of the Greek Dodecanese Islands in the SE Aegean, off the SW coast of Turkey. 19,987; 111 sq. mi. (287 sq. km).

Kos·ci·us·ko (kos′kē us′kō, kos′ē-), **Mount,** the highest mountain in Australia, in SE New South Wales. 7316 ft. (2230 m).

Ko·ši·ce (kô′shi tse), a city in SE Slovakia, in SE Czechoslovakia. 232,000.

Ko·so·vo (kô′sə vō′, kos′ə-), an autonomous province within the Serbian republic, in S Yugoslavia. 1,800,000; 4203 sq. mi. (10,887 sq. km). *Cap.:* Priština.

Kos·rae (kos rī′), an island in the W Pacific: part of the Federated States of Micronesia. 4471; 42 sq. mi. (109 sq. km.)

Ko·stro·ma (kos′trə mä′), a city in the W Russian Federation in Europe, NE of Moscow, on the Volga. 276,000.

Ko·ta Bha·ru (or **Bah·ru**) (kō′tə bär′oo), the capital of Kelantan state, in Malaysia, on the E central Malay Peninsula. 170,559.

Ko′ta Ki·na·ba·lu′ (kin′ə bə loo′), the capital of the state of Sabah, in Malaysia, on the NW coast of Borneo. 55,997. Formerly, **Jesselton.**

Kov·no (kôv′nə), Russian name of KAUNAS.

Kow·loon (kou′loon′), **1.** a peninsula in SE China, opposite Hong Kong island: a part of the Hong Kong colony. 3 sq. mi. (7.8 sq. km). **2.** a seaport on this peninsula. 715,440. Also called **Jiulong.**

Ko·zhi·kode (kō′zhi kōd′), former name of CALICUT.

Kra (krä), **Isthmus of,** the narrowest part of the Malay Peninsula, between the Bay of Bengal and the Gulf of Thailand. 35 mi. (56 km) wide.

Kra·ka·tau or **Kra·ka·tao** (krak/ə tou/, krä/kə-) also **Kra·ka·to·a** (-tō/ə), a volcano and small island in Indonesia, between Java and Sumatra: violent eruption 1883.

Kra·ków or **Cra·cow** (krak/ou, krä/kou; *Pol.* krä/kŏŏf), a city in S Poland, on the Vistula: the capital of Poland 1320–1609. 716,000.

Kra·ma·torsk (krä/mə tôrsk/), a city in E Ukraine, in the Donets Basin. 198,000.

Kra·sno·dar (kras/nə där/), **1.** a territory of the Russian Federation in SE Europe. 4,814,000; 34,200 sq. mi. (88,578 sq. km). **2.** its capital, on the Kuban River, S of Rostov. 620,000.

Kra·sno·yarsk (kras/nə yärsk/), **1.** a territory of the Russian Federation in N and central Asia. 3,198,000; 827,507 sq. mi. (2,143,243 sq. km). **2.** its capital, on the Yenisei River. 912,000.

Kra·sny (krä/snē), Russian name of KYZYL.

Kre·feld (krä/feld, -felt/), a city in W North Rhine-Westphalia, in W Germany, NW of Cologne. 232,300.

Kre·men·chug (krem/ən chŏŏk/, -chŏŏg/), a city in central Ukraine, on the Dnieper River. 230,000.

Krish·na (krish/nə), a river in S India, flowing E from the Western Ghats to the Bay of Bengal. 800 mi. (1290 km) long. Formerly, **Kistna.**

Kris·tian·sand (kris/chən sand/, -sän/), a seaport in S Norway. 63,491.

Kri·voi Rog (kri voi/ rōg/, rôk/), a city in SE Ukraine, SW of Dnepropetrovsk. 713,000.

Kron·shtadt (krun shtät/, -stat/), a naval base in the NW Russian Federation, on an island in the Gulf of Finland.

Kru·gers·dorp (krōō/gərz dôrp/), a city in S Transvaal, in the NE Republic of South Africa, NW of Johannesburg. 91,202.

Krym or **Krim** (krīm), Russian name of the CRIMEA.

KS, Kansas.

Kua·la Lum·pur (kwä/lə lŏŏm pŏŏr/), the capital of Malaysia, in the SW Malay Peninsula. 937,875.

Kuang·chou (*Chin.* gwäng/jō/), GUANGZHOU.

Ku·ban (kŏŏ ban/, -bän/), a river flowing NW from

the Caucasus Mountains to the Black and the Azov seas. 512 mi. (825 km) long.

Ku·ching (kōō'ching), the capital of Sarawak state, in E Malaysia. 74,229.

Ku·fa (kōō'fə, -fa), a town in central Iraq: former seat of Abbasid caliphate; Muslim pilgrimage center.

Kui·by·shev (kwē'bə shef', -shev'), former name (1935–91) of SAMARA.

Ku·ma·mo·to (kōō'mə mō'tō), a city on W central Kyushu, in SW Japan. 550,000.

Ku·ma·si (kōō mä'sē), the capital of Ashanti district, in S Ghana. 376,246.

Kun·lun (kŏŏn'lŏŏn'), a mountain range in China, bordering on the N edge of the Tibetan plateau and extending W across central China: highest peak, 25,000 ft. (7620 m).

Kun·ming (kŏŏn'ming'), the capital of Yunnan province, in S China. 1,490,000. Formerly, **Yun·nan.**

Kun·san (kŏŏn'sän'), a seaport in W South Korea. 165,318.

Kuo·pio (kwô'pyô), a city in central Finland. 78,619.

Ku·ra (kōō rä'), a river flowing from NE Turkey, through the Georgian Republic and Azerbaijan, SE to the Caspian Sea. 950 mi. (1530 km) long.

Kur·di·stan (kûr'di stan', -stän'), a mountain and plateau region in SE Turkey, NW Iran, and N Iraq, inhabited largely by Kurds. 74,000 sq. mi. (191,660 sq. km). —**Kurd,** *n.* —**Kurd'ish,** *adj.*

Ku·re (kōōr'ē, kōōr'ā), a seaport on SW Honshu, in SW Japan. 234,550.

Kur·gan (kŏŏr gän', -gan') a city in the S Russian Federation in Asia, near the Ural Mountains. 354,000.

Ku·rile (or **Ku'ril**) **Is'lands** (kōōr'il, kōō rēl'), a chain of small islands off the NE coast of Asia, extending from N Japan to the S tip of Kamchatka: renounced by Japan in 1945; under Russian administration.

Kur·land (kōōr'lənd), COURLAND.

Kursk (kōōrsk), a city in the W Russian Federation in Europe, W of Voronezh. 434,000.

Kush (kŏŏsh, kush), CUSH.

Ku·sta·nai (kōō stu ni'), a city in N Kazakhstan, on the Tobol river. 212,000.

Ku·tai·si (ko͞o ti/sē) also **Ku·tais** (ko͞o tis/), a city in the W Georgian Republic. 235,000.

Kutch or **Cutch** (kuch), **1.** a former state in W India, now part of Gujarat state. **2. Rann of** (run), a salt marsh NE of this area. 9000 sq. mi. (23,310 sq. km).

Ku·wait (ko͞o wāt/), **1.** a sovereign monarchy in NE Arabia, on the NW coast of the Persian Gulf: formerly a British protectorate. 1,960,000; ab. 8000 sq. mi. (20,720 sq. km). **2.** the capital of this monarchy. 167,750. —**Ku·wai/ti** (-wā/tē), *n.*, *pl.* -tis, *adj.*

Kuz·netsk/ Ba/sin (ko͞oz netsk/), an industrial region in the S Russian Federation in Asia extending from Tomsk to Novokuznetsk: coal fields.

Kwa·ja·lein (kwä/jə lān/, -lən), an atoll in the Marshall Islands, in E Micronesia. 5064. ab. 78 mi. (126 km) long.

Kwang·chow (*Chin.* gwäng/jō/), GUANGZHOU.

Kwang·ju (gwäng/jo͞o/), a city in SW South Korea. 905,896.

Kwang·si Chuang (*Chin.* gwäng/sē/ chwäng/), GUANGXI ZHUANG.

Kwang·tung (*Chin.* gwäng/do͞ong/), GUANGDONG.

Kwa·zu·lu (kwä zo͞o/lo͞o), ZULULAND.

Kwei·chow (*Chin.* gwä/jō/), GUIZHOU.

Kwei·lin (*Chin.* gwä/lin/), GUILIN.

Kwei·yang (*Chin.* gwä/yäng/), GUIYANG.

KY or **Ky.,** Kentucky.

Kyo·ga or **Kio·ga** (kyō/gə), **Lake,** a lake in central Uganda. ab. 1000 sq. mi. (2600 sq. km).

Kyo·to (kē ō/tō, kyō/-), a city on S Honshu, in central Japan: the capital from 794–1868. 1,472,993.

Kyr·gyz·stan (kir/gi stän/), a republic in central Asia, S of Kazakhstan and N of Tajikistan. 4,291,000; 76,460 sq. mi. (198,500 sq. km). *Cap.:* Bishkek. Also called **Kirghizia.** Formerly (1936–91), **Kirghiz/ So/viet So/cialist Repub/-lic.**

Kyu·shu (kē o͞o/sho͞o, kyo͞o/-), an island in SW Japan. 15,750 sq. mi. (40,793 sq. km).

Ky·zyl (ki zil/), the capital of the Tuva Autonomous Republic, in the S Russian Federation in Asia. 80,000. Russian, **Krasny.**

Ky·zyl Kum (ki zil/ ko͞om/, ko͞om/), a desert, SE

of the Aral Sea, in Uzbekistan and Kazakhstan.
ab. 90,000 sq. mi. (233,100 sq. km).

L

LA or **La.**, Louisiana.

L.A., Los Angeles.

Laa·land (lol′ənd, lô′lən), LOLLAND.

La·be (lä′be), Czech name of the ELBE.

Lab·ra·dor (lab′rə dôr′), **1.** a peninsula in E Canada between Hudson Bay and the Atlantic, containing the provinces of Newfoundland and Quebec. **2.** the E portion of this peninsula, constituting the mainland part of Newfoundland. 113,641 sq. mi. (294,330 sq. km). —**Lab·ra·dor·e·an, Lab·ra·dor·i·an** (lab′rə dôr′ē ən), *adj., n.*

La·bu·an (lä′bōō än′), an island off the NW coast of Borneo: part of Sabah state; a free port. 14,904; 35 sq. mi. (91 sq. km).

Lac′ca·dive Is′lands (lä′kə dēv′, lak′ə-), a group of islands and coral reefs in the Arabian Sea, off the SW coast of India. ab. 7 sq. mi. (18 sq. km).

Lac′cadive, Min′i·coy, and A′min·di′vi Is′lands (min′i koi′; ä′min dē′vē, ä′min-), former name of LAKSHADWEEP.

Lac·e·dae·mon (las′i dē′mən), SPARTA. —**Lac′e·dae·mo′ni·an** (-di mō′nē ən), *adj., n.*

La·co·ni·a (lə kō′nē ə), an ancient country in the SE Peloponnesus, in S Greece. *Cap.:* Sparta. —**La·co′ni·an**, *adj., n.*

La Co·ru·ña (lä′ kə rōōn′yə), a seaport in NW Spain. 241,808. Also called **Coruña, Corunna.**

La Crosse (lə krôs′, kros′), a city in W Wisconsin, on the Mississippi River. 51,003.

La·do·ga (lä′də gə), **Lake**, a lake in the NW Russian Federation, NE of St Petersburg: largest lake in Europe. 7000 sq. mi. (18,000 sq. km).

La·e (lä′ā, lä′ē), a seaport in E Papua New Guinea. 79,600.

La·fa·yette (laf′ē et′, laf′ā-, lä′fē′-, -fā-), a city in S Louisiana. 94,440.

La·gash (lä′gash), an ancient Sumerian city between the Tigris and Euphrates rivers, at the modern village of Telloh in SE Iraq.

La·gos (lä′gōs, lä′gos), a seaport in SW Nigeria: former capital. 1,097,000.

La Guai·ra (lä gwī′rə), a seaport in N Venezuela: the port of Caracas. 26,154.

La·gu'na Beach' (lə gōō'nə), a town in SW California: resort. 23,170.

La Ha'bra (lə hä'brə), a city in SW California, near Los Angeles. 51,266.

La Hogue (lA ôg') also **La Houge** (lA ōōg'), a roadstead off the NW coast of France: naval battle, 1692.

La·hore (lə hôr', -hōr'), a city in NE Pakistan: the capital of Punjab province. 2,922,000.

Lah·ti (läʜ'tē), a city in S Finland, NNE of Helsinki. 94,948.

La Jol·la (lə hoi'ə), a resort area in NW San Diego, in S California.

Lake' Charles', a city in SW Louisiana. 70,580.

Lake' Dis'trict, a mountainous region in NW England containing many lakes. Also called **Lake' Coun'try.**

Lake·land (lāk'lənd), a city in central Florida. 70,576.

Lake' of the Woods', a lake in S Canada and the N United States, between N Minnesota and Ontario and Manitoba provinces. 1485 sq. mi. (3845 sq. km).

Lake' Plac'id, a town in NE New York, in the Adirondack Mountains: resort. 2490.

Lake·wood (lāk'wŏŏd'), **1.** a city in central Colorado, near Denver. 126,481. **2.** a city in SW California, near Los Angeles. 73,557. **3.** a city in NE Ohio, on Lake Erie, near Cleveland. 59,718.

Lak·sha·dweep (luk'shə dwēp'), a union territory of India comprising a group of islands and coral reefs in the Arabian Sea, off the SW coast of India. 40,237; ab. 12 sq. mi. (31 sq. km). Formerly, **Laccadive, Minicoy, and Amindivi Islands.**

La Lí·ne·a (lä lē'nē ə), a seaport in S Spain, near Gibraltar. 52,127.

La Man·cha (lä män'chə), a plateau region in central Spain.

Lam·beth (lam'bith), a borough of Greater London, England. 243,200.

La Me·sa (lä mā'sə), a city in SW California. 52,931.

Lam·pe·du·sa (lam'pi dōō'sə, -zə), an island in the Mediterranean, between Tunisia and Malta, belonging to Italy.

La·nai (lə ni′, lä nä′ē), an island in central Hawaii. 2204; 141 sq. mi. (365 sq. km).

Lan·ark (lan′ərk), a historic county in S Scotland. Also called **Lan′ark·shire′** (-shēr′, -shər).

Lan·ca·shire (lang′kə shēr′, -shər), a county in NW England. 1,381,300; 1174 sq. mi. (3040 sq. km). Also called **Lancaster.**

Lan·cas·ter (lang′kə stər; *for 2, 3 also* -kas tər), **1.** a city in Lancashire, in NW England. 130,400. **2.** a town in S California. 97,291. **3.** a city in SE Pennsylvania. 55,551. **4.** LANCASHIRE. —**Lan·cas′tri·an** (-kas′trē ən), *n.*

Land′ of the Mid′night Sun′, 1. any of those countries containing land within the Arctic Circle where there is a midnight sun in midsummer, esp. Norway, Sweden, or Finland. **2.** LAPLAND.

Land′ of the Ris′ing Sun′, JAPAN.

Land′s′ End′, a cape in Cornwall that forms the SW tip of England.

Lang·ley (lang′lē), a city in SW British Columbia, in SW Canada, near Vancouver. 53,434.

Langue·doc (läng dôk′), a former province in S France. *Cap.:* Toulouse. —**Langue·do·cian** (lang-dō′shən, lang′gwə dō′-), *adj., n.*

Langue·doc-Rous·sil·lon (läng dôk ROŌ sē-yôN′), a metropolitan region in S France. 2,011,900; 10,570 sq. mi. (27,376 sq. km).

Lan·sing (lan′sing), the capital of Michigan, in the S part. 127,321.

Lan·zhou or **Lan·chou** or **Lan·chow** (län′jō′), the capital of Gansu province, in N China, on the Huang He. 1,350,000.

La·od·i·ce·a (lā od′ə sē′ə, lā′ə də-), ancient name of LATAKIA.

Laoigh·is (lā′ish), a county in Leinster, in the central Republic of Ireland. 53,270; 623 sq. mi. (1615 sq. km). Also called **Leix.**

La·os (lā′ōs, lous, lā′os), a country in SE Asia: formerly part of French Indochina. 3,830,000; 91,500 sq. mi. (236,985 sq. km). *Cap.:* Vientiane. —**La·o·tian** (lā ō′shən), *n., adj.*

La Paz (lə päz′, päs′), **1.** the administrative capital of Bolivia, in the W part; Sucre is the official capital. 992,592; ab. 12,000 ft. (3660 m) above sea level. **2.** the capital of Baja California Sur, in NW Mexico. 75,000.

Lap·land (lap′land′), a region in N Norway, N

Sweden, N Finland, and the Kola Peninsula of the NW Russian Federation in Europe: inhabited by Lapps. —**Lap/land/er,** *n.*

La Pla·ta (lə plä/tə), a seaport in E Argentina. 455,000.

Lap/tev Sea/ (lap/tef, -tev), an arm of the Arctic Ocean N of the Russian Federation in Asia, between the Taimyr Peninsula and the New Siberian Islands. Formerly, **Nordenskjöld Sea.**

L'Aq·ui·la (lak/wə lə, lä/kwə-), AQUILA.

Lar·a·mie (lar/ə mē), a city in SE Wyoming. 24,410.

La·re·do (lə rā/dō), a city in S Texas, on the Rio Grande. 122,899.

Lar·go (lär/gō), a city in W Florida. 65,674.

La·ris·sa or **La·ri·sa** (lə ris/ə, lär/ə sə), a city in E Thessaly, in E Greece. 102,048.

La Ro·chelle (lä/ rō shel/), a seaport in W France. 77,494.

La Salle (lə sal/, säl/), a city in S Quebec, in E Canada: suburb of Montreal. 75,621.

Las Cru·ces (läs kroo/sis), a city in S New Mexico, on the Rio Grande. 62,126.

La Se·re·na (lä/ sə rā/nə), a seaport in central Chile. 106,617.

Lash·io (läsh/yō), a town in N Burma, NE of Mandalay: the SW terminus of the Burma Road.

Lash·kar (lush/kər), the modern part of Gwalior city in N India.

Las Pal·mas (läs päl/mäs), a seaport on NE Gran Canaria, in the central Canary Islands. 360,098.

La Spe·zia (lä spāt/sē ə), a seaport in NW Italy, on the Ligurian Sea. 107,435.

Las/sen Peak/ (las/ən), an active volcano in N California, in the S Cascade Range. 10,465 ft. (3190 m).

Las/sen Volcan/ic Na/tional Park/, a national park in N California, in the S Cascade Range, including Lassen Peak. 166 sq. mi. (430 sq. km).

Las Ve·gas (läs vā/gəs), a city in SE Nevada. 258,295.

Lat·a·ki·a (lat/ə kē/ə), a seaport in NW Syria, on the Mediterranean. 196,791. Ancient, **Laodicea.**

Lat/in Amer/ica, the part of the American continents south of the United States in which Span-

ish, Portuguese, or French is officially spoken.
—**Lat′in-Amer′ican,** *adj.* —**Lat′in Amer′ican,**
n.

Lat′in Quar′ter, a quarter of Paris on the south
side of the Seine, frequented for centuries by students, writers, and artists.

La·ti·um (lā′shē əm), a country in ancient Italy,
SE of Rome.

Lat·vi·a (lat′vē ə, lät′-), a republic in N Europe,
on the Baltic, S of Estonia: an independent state
1918–40; annexed by the Soviet Union 1940; regained independence 1991. 2,681,000; 25,395
sq. mi. (63,700 sq. km). *Cap.:* Riga. —**Lat′vi·an,**
n., adj.

Laun·ces·ton (lôn′ses/tən, län′-), a port in Australia, on N Tasmania. 88,486.

Lau·ren′tian Moun′tains (lô ren′shən), a
range of low mountains in E Canada, between the
St. Lawrence River and Hudson Bay. Also called
Lau·ren′tians.

Lau·sanne (lō zan′), the capital of Vaud, in W
Switzerland, on the Lake of Geneva. 123,700.

La·val (lə val′), a city in S Quebec, in E Canada,
NW of Montreal, on the St. Lawrence. 284,164.

Law·rence (lôr′əns, lor′-), **1.** a city in NE Massachusetts, on the Merrimack River. 70,207. **2.** a
city in E Kansas, on the Kansas River. 65,608.

Law·ton (lôt′n), a city in SW Oklahoma. 80,561.

Lead·ville (led′vil), a town in central Colorado:
historic mining town. 3879.

Leav·en·worth (lev′ən wûrth′, -wərth), a city in
NE Kansas: site of federal prison. 33,656.

Leb·a·non (leb′ə nən, -non′), a republic at the E
end of the Mediterranean, N of Israel. 3,500,000;
3927 sq. mi. (10,170 sq. km). *Cap.:* Beirut.
—**Leb′a·nese′** (-nēz′, -nēs′), *adj., n., pl.* **-nese.**

Leb′anon Moun′tains, a mountain range extending the length of Lebanon, in the central part.
Highest peak, 10,049 ft. (3063 m).

Le Bour·get (lə bŏŏr zhā′), a suburb of Paris:
former airport, landing site for Charles A. Lind-
bergh, May 1927.

Lec·ce (lech′ā), a city in SE Italy. 101,520.

Leeds (lēdz), a city in West Yorkshire, in N England. 749,000.

Lee·u·war·den (lā′wär′dn, -vär′-), a city in N
Netherlands. 85,435.

Lee·ward Is·lands (lē′wərd), a group of islands in the Lesser Antilles of the West Indies, extending from Puerto Rico SE to Martinique.

Left′ Bank′, a part of Paris, France, on the S bank of the Seine: frequented by artists, writers, and students.

Leg·horn (leg′hôrn′), English name of LIVORNO.

Leg·ni·ca (leg nēt′sə), a city in SW Poland. 98,600.

Le Ha·vre (lə hä′vrə, -vər; *Fr.* lə ʌ′vrᵉ), a seaport in N France, at the mouth of the Seine. 200,411. Also called **Havre.**

Le·high (lē′hī), a river in E Pennsylvania, flowing SW and SE into the Delaware River. 103 mi. (165 km) long.

Leh′man Caves′ (lē′mən), limestone caverns in E Nevada: part of Great Basin National Park.

Leices·ter (les′tər), **1.** a city in Leicestershire, in central England. 290,600. **2.** LEICESTERSHIRE.

Leices·ter·shire (les′tər shēr′, -shər), a county in central England. 836,500; 986 sq. mi. (2555 sq. km). Also called **Leicester.**

Lei·den or **Ley·den** (lid′n), a city in W Netherlands. 107,893.

Lein·ster (len′stər), a province in the E Republic of Ireland. 1,851,134; 7576 sq. mi. (19,620 sq. km).

Leip·zig (līp′sig, -sik), a city in E central Germany. 545,307.

Leith (lēth), a seaport in SE Scotland, on the Firth of Forth: now part of Edinburgh.

Lei·trim (lē′trim), a county in Connaught province, in N Republic of Ireland. 27,000; 589 sq. mi. (1526 sq. km).

Leix (lāks), LAOIGHIS.

Lei·zhou or **Lei·chou** (lā′jō′), also **Luichow,** a peninsula of SW Guangdong province, in SE China, between the South China Sea and the Gulf of Tonkin.

Le·man (lē′mən), **Lake,** GENEVA, Lake of.

Le Mans (lə män′), a city in NW France: auto racing. 155,245.

Lem·nos (lem′nos, -nōs) also **Limnos,** a Greek island in the NE Aegean. 186 sq. mi. (480 sq. km). *Cap.:* Mirina. —**Lem′ni·an,** *adj., n.*

Le·na (lē′nə, lā′-), a river in the Russian Federation in Asia, flowing NE from Lake Baikal into the Laptev Sea. 2800 mi. (4500 km) long.

Le·ni·na·bad (len′i nə bäd′), former (1936–91) name of KHODZHENT.

Le·ni·na·kan (len′i nə kän′), former name of GUMRI.

Le·nin·grad (len′in grad′), a former name (1924–91) of ST. PETERSBURG (def. 1).

Le′nin Peak′ (len′in), a peak in the Trans Alai range, in central Asia, between Kyrgyzstan and Tajikistan. 23,382 ft. (7127 m).

Len·ox (len′əks), a town in W Massachusetts, in the Berkshire Hills: a former estate (**Tangle-wood**) in the area is the site of music festivals. 6523.

Le·ón (lā ōn′), **1.** a province in NW Spain: formerly a kingdom. 598,721; 5936 sq. mi. (15,375 sq. km). **2.** the capital of this province. 137,414. **3.** a city in W Guanajuato, in central Mexico. 655,809. **4.** a city in W Nicaragua: the former capital. 100,982.

Lé·o·pold·ville (lē′ə pōld vil′, lā′-), former name of KINSHASA.

Le·pan·to (li pan′tō,), **1.** Greek, **Navpaktos**. a seaport in W Greece, on a strait between the Io-nian Sea and the Gulf of Corinth: site of naval battle (1571) in which the Turkish fleet was de-feated by allied European powers. **2. Gulf of,** CORINTH, Gulf of.

Le·pen·ski Vir (lep′ən skē vēr′), the site of an advanced Mesolithic fishing culture on the banks of the Danube in Yugoslavia.

Le·pon·tine Alps′ (li pon′tin), a central range of the Alps in S Switzerland and N Italy. Highest peak, 11,684 ft. (3561 m).

Lé·ri·da (ler′i də), a city in NE Spain. 111,507.

Les·bos (lez′bos, -bōs), a Greek island in the NE Aegean. 104,620; 836 sq. mi. (2165 sq. km). *Cap.:* Mytilene. Also called **Mytilene**. —**Les′bi·an** (-bē ən), *n., adj.*

Le·so·tho (lə sōō′tōō, -sō′tō), a monarchy in S Africa: formerly a British protectorate; gained in-dependence 1966; member of the Common-wealth of Nations. 1,670,000; 11,716 sq. mi. (30,344 sq. km). *Cap.:* Maseru. Formerly, **Basu-toland**.

Less′er Antil′les, See under ANTILLES.

Less·er Sun·da Is·lands, See under Sunda Islands.

Leth·bridge (leth′brij′), a city in S Alberta, in SW Canada. 58,841.

Leuc·tra (lōōk′trə), a town in ancient Greece, in Boeotia: Thebans defeated Spartans here 371 B.C.

Leu·ven (lœ′vən, lōō′-), a city in central Belgium. 84,180. French, **Louvain.**

Le·vant (li vant′), the lands bordering the E shores of the Mediterranean Sea. —**Lev·an·tine** (lev′ən tin′, -tēn′, li van′tin, -tīn′), *n., adj.*

Le·ven (lē′vən), **Loch,** a lake in E Scotland: ruins of a castle in which Mary Queen of Scots was imprisoned.

Le·ver·ku·sen (lā′vər kōō′zən), a city in North Rhine-Westphalia, in W Germany, on the Rhine. 154,700.

Lev·it·town (lev′it toun′), a town on W Long Island, in SE New York. 57,000.

Lev·kas (lef käs′), an island in the Ionian group, off the W coast of Greece. 114 sq. mi. (295 sq. km).

Lew·es (lōō′is), a city in East Sussex, in SE England: battle 1264. 84,400.

Lew·i·sham (lōō′ə shəm), a borough of Greater London, England. 231,600.

Lew·is with Har·ris (lōō′is; har′is), the northernmost island of the Outer Hebrides, in NW Scotland. 825 sq. mi. (2135 sq. km). Also called **Lew′is and Har′ris.**

Lex·ing·ton (lek′sing tən), **1.** a town in E Massachusetts, NW of Boston: first battle of the American Revolution fought here April 19, 1775. 29,479. **2.** a city in N Kentucky. 225,366.

Ley·den (lid′n), Leiden.

Ley·te (lā′tē), an island in the E central Philippines. 2786 sq. mi. (7215 sq. km).

Lha·sa (lä′sə, -sä, läs′ə), the capital of Tibet, in SW China: sacred city of Lamaism. 310,000.

Lho·tse (lōt sā′, hlōt-), a mountain peak in the Himalayas, on Nepal-Tibet border: fourth highest peak in the world. 27,890 ft. (8501 m).

L.I., Long Island.

Lian·yun·gang (lyän′yœn′gäng′) also **Lienyün-kang,** a city in NE Jiangsu province, in E China. 395,730.

Liao (lyou), a river in NE China, flowing through S Manchuria into the Gulf of Liaodong. 700 mi. (1125 km) long.

Liao·dong or **Liao·tung** (lyou′dông′), **1.** a peninsula in NE China, extending S into the Yellow Sea. **2. Gulf of,** a gulf W of this peninsula.

Liao·ning (lyou′ning′), a province in NE China. 37,260,000; 58,301 sq. mi. (151,000 sq. km). *Cap.:* Shenyang. Formerly, **Fengtien.**

Liao·yang (lyou′yäng′), a city in central Liaoning province, in NE China. 448,807.

Liao·yuan or **Liao·yüan** (lyou′ywän′), a city in SE Jilin province, in NE China. 759,587.

Li·ard (lē′ärd, lē ärd′, -är′), a river in W Canada, flowing from S Yukon through N British Columbia and the Northwest Territories into the Mackenzie River. 550 mi. (885 km) long.

Li·be·rec (lib′ə rets′), a city in NW Czechoslovakia. 104,000.

Li·be·ri·a (lī bēr′ē ə), a republic in W Africa: founded by freed American slaves 1822. 2,440,000; ab. 43,000 sq. mi. (111,370 sq. km). *Cap.:* Monrovia. —**Li·be′ri·an,** *adj., n.*

Lib′erty Is′land, a small island in upper New York Bay: site of the Statue of Liberty. Formerly, **Bedloe's Island.**

Li·bre·ville (*Fr.* lē brə vēl′), the capital of Gabon, in the W part, on the Gulf of Guinea. 350,000.

Lib·y·a (lib′ē ə), **1.** an ancient name of the part of N Africa W of Egypt. **2.** a republic in N Africa between Tunisia and Egypt: formerly a monarchy 1951–69. 3,960,000; 679,400 sq. mi. (1,759,646 sq. km). *Cap.:* Tripoli. —**Lib′y·an,** *n., adj.*

Lib′yan Des′ert, a desert in N Africa, in Libya, Egypt, and Sudan, W of the Nile: part of the Sahara. ab. 650,000 sq. mi. (1,683,500 sq. km).

Lich·field (lich′fēld′), a town in Staffordshire, in central England. 92,900.

Li·di·ce (lid′ə tsä′, -tsə), a village in W Czechoslovakia: destroyed by the Nazis in 1942 in reprisal for the assassination of a high Nazi official. 509.

Li·do (lē′dō), a chain of sandy islands in NE Italy, between the Lagoon of Venice and the Adriatic: resort.

Liech·ten·stein (lik′tən stin′, likн′-), a small principality in central Europe between Austria and

Switzerland. 27,700; 65 sq. mi. (168 sq. km). *Cap.:* Vaduz. —**Liech/ten·stein/er,** *n.*

Li·ège (lē äzh/, -ezh/), **1.** a province in E Belgium. 1,019,226; 1521 sq. mi. (3940 sq. km). *Cap.:* Liège. **2.** the capital of this province, on the Meuse River. 200,312. Flemish, **Luik.**

Lien·yün·kang (*Chin.* lyun/yyn/gäng/), LIANYUNGANG.

Lie·pā·ja or **Lie·pa·ja** (lē ep/ə yə, -ä yə), a seaport in W Latvia, on the Baltic. 114,900.

Lie·tu·va (lye/too vä), Lithuanian name of LITHUANIA.

Li·gu·ri·a (li gyoor/ē ə), a region in NW Italy. 1,749,572; 2099 sq. mi. (5435 sq. km). —**Li·gu/ri·an,** *adj., n.*

Ligu/rian Sea/, a part of the Mediterranean between Corsica and the NW coast of Italy.

Li·ka·si (li kä/sē), a city in S Zaire. 194,465. Formerly, **Jadotville.**

Lille (lēl), a city in N France. 177,218. Formerly, **Lisle.**

Li·long·we (li lông/wä), the capital of Malawi, in the SW part. 186,800.

Li·ma (lē/mə *for 1;* lī/mə *for 2*), **1.** the capital of Peru, near the Pacific coast. 4,605,043. **2.** a city in NW Ohio. 47,381.

Lim·bourg (*Fr.* lan boor/), See under LIMBURG.

Lim·burg (lim/bûrg; *Du.* lim/bœrkh), a medieval duchy in W Europe: now divided into a province in the SE Netherlands (**Limburg**) and a province in NE Belgium (**Limbourg**).

Lime·house (līm/hous/), a dock district in the East End of London, England, once notorious for its squalor: formerly a Chinese quarter.

Lim·er·ick (lim/ər ik), **1.** a county in N Munster, in the SW Republic of Ireland. 107,963; 1037 sq. mi. (2686 sq. km). **2.** its county seat: a seaport at the head of the Shannon estuary. 60,721.

Lim·nos (lēm/nôs), LEMNOS.

Li·moges (li mōzh/), a city in S central France. 147,406.

Li·món (lē mōn/), a seaport in E Costa Rica. 52,602. Also called **Puerto Limón.**

Li·mou·sin (lē moo zan/), **1.** a historic region and former province in central France. **2.** a metropolitan region in central France. 735,800; 6541 sq. mi. (16,942 sq. km).

Lim·po·po (lim pō′pō), a river in S Africa, flowing from the N Republic of South Africa through S Mozambique into the Indian Ocean. 1000 mi. (1600 km) long. Also called **Crocodile River.**

Lin·coln (ling′kən), **1.** the capital of Nebraska, in the SE part. 191,972. **2.** a city in Lincolnshire, in E central England. 73,200. **3.** LINCOLNSHIRE.

Lin·coln·shire (ling′kən shēr′, -shər), a county in E England. 574,600; 2272 sq. mi. (5885 sq. km). Also called **Lincoln.**

Lin·dis·farne (lin′dəs färn′), HOLY ISLAND (def. 1).

Line′ Is′lands, a group of 11 coral atolls in the central Pacific, S of Hawaii: eight of the islands belong to Kiribati, and three to the U.S.

Lin′ga·yén′ Gulf′ (ling′gä yen′), a gulf in the Philippines, on the NW coast of Luzon.

Lin·kö·ping (lēn′chœ′pĕng), a city in S Sweden. 118,602.

Lin·lith·gow (lin lith′gō), former name of WEST LOTHIAN.

Linz (lints), a port in N Austria, on the Danube River. 199,910.

Li·ons (lī′ənz), **Gulf of,** a wide bay of the Mediterranean off the coast of S France. French, **Golfe du Lion.**

Lip′a·ri Is′lands (lip′ə rē), a group of volcanic islands N of Sicily, belonging to Italy. 10,043; 44 sq. mi. (114 sq. km).

Li·petsk (lē′petsk), a city in the W Russian Federation in Europe, SSE of Moscow. 465,000.

Lip·pe (lip′ə), a former state in NW Germany: now part of North Rhine-Westphalia.

Lis·bon (liz′bən), the capital of Portugal, in the SW part, on the Tagus estuary. 807,937. Portuguese, **Lis·bo·a** (lēzh bô′ə).

Lisle (lēl), former name of LILLE.

Lith·u·a·ni·a (lith′ōō ā′nē ə), a republic in N Europe, on the Baltic, S of Latvia: an independent state 1918–40; annexed by Soviet Union 1940; regained independence 1991. 3,690,000; 25,174 sq. mi. (65,200 sq. km). *Cap.:* Vilnius. Lithuanian, **Lietuva.** —**Lith′u·a′ni·an,** *n., adj.*

Lit′tle Ab′aco, See under ABACO.

Lit′tle Amer′ica, a base in the Antarctic, on the Bay of Whales, S of the Ross Sea: established by Adm. Richard E. Byrd in 1929.

Lit′tle Big′horn, a river flowing N from N Wyoming to S Montana into the Bighorn River: General Custer and troops defeated by Indians 1876. 80 mi. (130 km) long.

Lit′tle Colora′do, a river flowing NW from E Arizona to the E edge of the Grand Canyon, where it flows into the Colorado River. 315 mi. (507 km) long.

Lit′tle Di′omede, See under DIOMEDE ISLANDS.

Lit′tle Missou′ri, a river in the NW United States, flowing from NE Wyoming NE into the Missouri in North Dakota. 560 mi. (900 km) long.

Lit′tle Rock′, the capital of Arkansas, in the central part, on the Arkansas River. 175,795.

Little St. Bernard, ST. BERNARD (def. 2).

Liu·zhou or **Liu·chou** or **Liu·chow** (lyōō′jō′), a city in central Guangxi Zhuang region, in S China. 190,000.

Liv·er·more (liv′ər môr′, -mōr′), a city in W California. 56,741.

Liv·er·pool (liv′ər pōōl′), a seaport in Merseyside, in W England, on the Mersey estuary. 476,000. —**Liv′er·pud′li·an** (-pud′lē ən), n., adj.

Liv·ing·stone (liv′ing stən), a town in SW Zambia, on the Zambezi River, near Victoria Falls: the former capital. 94,637.

Li·vo·ni·a (li vō′nē ə), **1.** a former Russian province on the Baltic: now part of Latvia and Estonia. **2.** a city in SE Michigan, near Detroit. 100,850. —**Li·vo′ni·an,** adj., n.

Li·vor·no (li vôr′nō), a seaport in W Italy on the Ligurian Sea. 173,114. English, **Leghorn.**

Liz′ard Head′, a promontory in SW Cornwall, in SW England: the southernmost point in England. Also called **The Lizard.**

Lju·blja·na (lōō′blē ä′nə, -nä), the capital of Slovenia, in the central part. 305,211.

Llan·el·ly (la nel′ē; Welsh. hla ne′hlē), a seaport in Dyfed, in S Wales. 76,800.

Lla·no Es·ta·ca·do (lä′nō es′tə kä′dō, lan′ō), a large plateau in the SW United States, in W Texas and SE New Mexico. Also called **Staked Plain.**

Lo·car·no (lō kär′nō), a town in S Switzerland, on Lake Maggiore: resort. 15,300.

Loch Ness (lok′ nes′, loкн′), NESS, Loch.

Lo·cris (lō′kris), either of two districts in the central part of ancient Greece. —**Lo′cri·an,** n., adj.

Lo·di (lō′dē for 1; lō′di for 2), **1.** a town in N Italy, SE of Milan: Napoleon's defeat of the Austrians 1796. 42,873. **2.** a city in central California, near Sacramento. 51,874.

Łódź (lŏŏj, lodz), a city in central Poland, SW of Warsaw. 849,000.

Lo·fo·ten Is·lands (lō′fōōt′n), a group of islands NW of and belonging to Norway: rich fishing grounds. 63,365; 474 sq. mi. (1228 sq. km).

Lo·gan (lō′gən), **Mount,** a mountain in NW Canada, in the St. Elias Mountains: second highest peak in North America. 19,850 ft. (6050 m).

Lo·gro·ño (lə grōn′yō), a city in N Spain. 118,770.

Loire (lwär), a river in France, flowing NW and W into the Atlantic. 625 mi. (1005 km) long.

Lol·land or **Laa·land** (lol′ənd, lô′lən), an island in SE Denmark, S of Zealand. 81,760; 495 sq. mi. (1280 sq. km).

Lom·bard Street (lom′bärd, -bərd, lum′-), a street in London, England: a financial center.

Lom·bard·y (lom′bər dē, lum′-), a region and former kingdom in N Italy. 8,886,402; 9190 sq. mi. (23,800 sq. km). —**Lom′bard** (-bärd, -bərd), n., adj.

Lom·bok (lom bok′), an island in Indonesia, E of Bali. 1,300,234; 1826 sq. mi. (4729 sq. km).

Lo·mé (lô mā′), the capital of Togo, on the Gulf of Guinea. 366,476.

Lo·mond (lō′mənd), **Loch,** a lake in W Scotland. 23 mi. (37 km) long.

Lon·don (lun′dən), **1.** a metropolis in SE England, on the Thames: capital of the United Kingdom. **2. City of,** an old city in the central part of the former county of London: the ancient nucleus of the modern metropolis. 4700; 1 sq. mi. (3 sq. km). **3. County of,** a former administrative county comprising the City of London and 28 metropolitan boroughs, now part of Greater London. **4. Greater,** an urban area comprising the city of London and 32 metropolitan boroughs. 6,770,400; 609 sq. mi. (1575 sq. km). **5.** a city in S Ontario, on the Thames River, in SE Canada. 269,140. —**Lon′don·er,** n.

Lon·don·der·ry (lun′dən der′ē), **Derry.**

Long′ Beach′, a city in SW California, S of Los Angeles. 429,433.

Long·ford (lông′fərd, long′-), a county in Leinster, in the N Republic of Ireland. 31,138; 403 sq. mi. (1044 sq. km).

Long′ Is′land, an island in SE New York: the New York City boroughs of Brooklyn and Queens are at its W end. 118 mi. (190 km) long.

Long′ Is′land Sound′, an arm of the Atlantic between Connecticut and Long Island. 90 mi. (145 km) long.

Long·mont (lông′mont, long′-), a city in N central Colorado. 51,555.

Longs′ Peak′ (lôngz, longz), a peak in N central Colorado, in Rocky Mountain National Park. 14,255 ft. (4345 m).

Lon·gueuil (lông gāl′, long-; *Fr.* lôn gœ′y°), a city in S Quebec, in E Canada, across from Montreal, on the St. Lawrence. 125,441.

Long·view (lông′vyoō′, long′-), a city in NE Texas. 70,311.

Look′out Moun′tain, a mountain ridge in Georgia, Tennessee, and Alabama: Civil War battle fought here, near Chattanooga, Tenn. 1863; highest point, 2126 ft. (648 m).

Loop (loōp), **the,** the main business district of Chicago.

Lo·rain (lə rān′, lô-, lō-), a port in N Ohio, on Lake Erie. 71,245.

Lo·rient (lô ryän′), a seaport in NW France, on the Bay of Biscay. 64,675.

Lor·raine (lə rān′, lô-, lō-), **1.** a medieval kingdom in W Europe along the Moselle, Meuse, and Rhine rivers. **2.** a historic region in NE France, once included in this kingdom: a former province. Compare ALSACE-LORRAINE. **3.** a metropolitan region in NE France. 2,313,200; 9092 sq. mi. (23,547 sq. km).

Los Al·a·mos (lôs al′ə mōs′, los), a town in central New Mexico, NW of Sante Fe: atomic research center. 11,039.

Los An·ge·les (lôs an′jə ləs, -lēz′, los; *sometimes* ang′gə-), a seaport in SW California: second largest city in the U.S. 3,485,398; with suburbs 6,997,000.

Lot (lôt), a river in S France, flowing W to the Garonne. 300 mi. (480 km) long.

Lo·thi·an (lō′thē ən), a region in SE Scotland. 743,700; 700 sq. mi. (1813 sq. km).

Lough Neagh (lok′ nā′, lokh′), NEAGH, Lough.

Lou·ise (lōō ēz′), **Lake,** a glacial lake in W Canada, in SW Alberta in Banff National Park. 5670 ft. (1728 m) above sea level.

Lou·i·si·an·a (lōō ē′/zē an′ə, lōō′ə zē-, lōō/ē-), a state in the S United States. 4,219,973; 48,522 sq. mi. (125,672 sq. km). *Cap.*: Baton Rouge. *Abbr.*: LA, La. —**Lou·i′/si·an′an, Lou·i′/si·an′i·an,** *adj., n.*

Lou·is·ville (lōō/ē vil′, -ə vəl), a port in N Kentucky, on the Ohio River: Kentucky Derby. 269,063. —**Lou′/is·vill′/ian,** *n.*

Lourdes (lŏōrd, lŏōrdz; *Fr.* lŏōrd), a city in SW France: Roman Catholic shrine famed for miraculous cures. 18,096.

Lou·ren·ço Mar·ques (lô ren′/sō mär′kes, lō-), former name of MAPUTO.

Louth (louth), a county in Leinster province, in the NE Republic of Ireland. 91,698; 317 sq. mi. (820 sq. km).

Lou·vain (lōō van′), French name of LEUVEN.

Low′ Coun′tries, the lowland region near the North Sea, forming the lower basin of the Rhine, Meuse, and Scheldt rivers, divided in the Middle Ages into numerous small states: corresponding to modern Belgium, Luxembourg, and the Netherlands.

Low·ell (lō/əl), a city in NE Massachusetts, on the Merrimack River. 103,439.

Low′er Califor′nia, BAJA CALIFORNIA.

Low′er Can′ada, former name of Quebec province 1791–1841.

Lower 48, the 48 contiguous states of the U.S.

Low′er Mer′i·on (mer/ē ən), a town in SE Pennsylvania, near Philadelphia. 59,651.

Low′er Pal′at·inate. See under PALATINATE.

Low′er Sax′ony, a state in NW Germany. 7,162,000; 18,294 sq. mi. (47,380 sq. km). *Cap.*: Hanover. German, **Niedersachsen.**

Low′er Tungu′ska. See under TUNGUSKA.

Lowes·toft (lōs/tôft, -toft, -təf), a seaport in NE Suffolk, in E England. 55,231.

Low·lands (lō/ləndz), **the,** a low, level region in S, central, and E Scotland.

Lo·yang (*Chin.* lô/yäng′), LUOYANG.

Lu·a·la·ba (lōō/ə lä/bə), a river in SE Zaire: a headstream of the Zaire (Congo) River. 400 mi. (645 km) long.

Lu·an·da (lŏŏ an′də, -än′-), the capital of Angola, in SW Africa. 1,200,000.

Lu·ang Pra·bang (lŏŏ äng′ prä bäng′), a city in N Laos, on the Mekong River: former royal capital. 44,244.

Lu·an·shya (lŏŏ än′shä, lwän′-), a town in central Zambia. 160,667.

Lu·a·pu·la (lŏŏ′ə pŏŏ′lə), a river in S central Africa, flowing E and N along the Zambia-Zaire border to Lake Mweru. ab. 300 mi. (485 km) long.

Lub·bock (lub′ək), a city in NW Texas. 186,206.

Lü·beck (lv′bek), a seaport in N Germany: important Baltic port in the medieval Hanseatic League. 210,500.

Lu·blin (lŏŏ′blin, -blēn), a city in E Poland. 324,000.

Lu·bum·ba·shi (lŏŏ′bŏŏm bä′shē), a city in S Zaire. 543,268.

Lu·ca·ni·a (lŏŏ kā′nē ə), **1.** an ancient region in S Italy, NW of the Gulf of Taranto. **2.** a modern region in S Italy, comprising most of the ancient region. 621,506; 3856 sq. mi. (9985 sq. km). Italian, **Basilicata.**

Luc·ca (lŏŏ′kə, -kä), a city in NW Italy, W of Florence. 91,656.

Lu·cerne (lŏŏ sûrn′), **1.** a canton in central Switzerland. 308,500; 576 sq. mi. (1490 sq. km). **2.** the capital of this canton, on Lake of Lucerne. 60,600. **3. Lake of,** a lake in central Switzerland. 24 mi. (39 km) long; 44 sq. mi. (114 sq. km). German, **Luzern.**

Lu·chow (lŏŏ′jō′), Luzhou.

Luck·now (luk′nou), the capital of Uttar Pradesh state, in N India. 1,007,000.

Lü·da or **Lü·ta** (lv′dä′), a municipality in S Liaoning province, in NE China, on the Liaodong peninsula: includes the seaports of Dalian and Lüshun.

Lü·der·itz (lŏŏ′dər its), a seaport in SW Namibia: diamond-mining center. 17,000.

Lu·dhi·a·na (lŏŏ′dē ä′nä), a city in central Punjab, in N India. 606,000.

Lud·wigs·ha·fen (lood′viKHs hä′fən, -viks-, lood′) a city in SW Germany, on the Rhine opposite Mannheim. 156,700.

Lu·gansk (lŏŏ gänsk′), a city in E Ukraine, in the

Donets Basin. 509,000. Formerly (1935–90), **Vo-roshilovgrad.**

Lu·go (lōō′gô), a city in NW Spain. 77,728.

Lui·chow (Chin. lwē′jō′), LEIZHOU.

Luik (loik, lōōk), Flemish name of LIÈGE.

Lu·le·å (lōō′lä ô′, -lē-), a seaport in NE Sweden, on the Gulf of Bothnia. 66,834.

Lu·lua·bourg (lōō′lwä bōōr′), former name of KANANGA.

Lu·né·ville (ly nä vēl′), a city in NE France, W of Strasbourg: treaty between France and Austria 1801. 24,700.

Luo·yang (lwô′yäng′) also **Loyang,** a city in N Henan province, in E China. 1,050,000.

Lu·ray (lōō rā′), a town in N Virginia: site of Luray Caverns. 3584.

Lu·ri·stan (lōōr′ə stän′, -stan′), a mountainous region in W Iran.

Lu·sa·ka (lōō sä′kə), the capital of Zambia, in the S central part. 818,994.

Lu·sa·ti·a (lōō sā′shē ə, -shə), a region in E Germany, between the Elbe and Oder rivers. —**Lu-sa′tian** (-shən), n., adj.

Lü·shun (ly′shyn′), a seaport in S Liaoning province, in NE China. 200,000. Also called **Port Arthur.** Compare LÜDA.

Lu·si·ta·ni·a (lōō′si tā′nē ə), an ancient region and Roman province in the Iberian Peninsula, corresponding generally to modern Portugal. —**Lu′si·ta′ni·an,** adj., n.

Lü·ta (Chin. ly′dä′), LÜDA.

Lu·te·tia (lōō tē′shə), ancient name of PARIS.

Lutsk (lōōtsk), a city in NW Ukraine, on the Styr River. 167,000.

Lux·em·bourg (luk′səm bûrg′), **1.** a grand duchy surrounded by Germany, France, and Belgium. 377,100; 999 sq. mi. (2585 sq. km). **2.** the capital of this grand duchy. 76,640. **3.** a province in SE Belgium: formerly a part of the grand duchy of Luxembourg. 226,452; 1706 sq. mi. (4420 sq. km). Also, **Luxemburg** (for defs. 1, 2). —**Lux′em·bourg′er,** n. —**Lux′em·bourg′i·an** (-bûr′jē-ən), adj.

Lux·or (luk′sôr), a town in S Egypt, on the Nile: ruins of ancient Thebes. 147,900.

Lu·zern (Ger. lōō tsern′), LUCERNE.

Lu·zhou or **Lu·chow** (lōō′jō′), a city in S Sichuan province, in central China, on the Chang Jiang. 360,000.

Lu·zon (lōō zon′), the chief island of the Philippines, in the N part of the group. 40,420 sq. mi. (104,688 sq. km).

Lvov (lə vôf′, -vof′), a city in W Ukraine: formerly in Poland. 790,000. Polish, **Lwów** (lvōōf).

Ly·all·pur (li′əl pŏŏr′), former name of FAISALA-BAD.

Lyc·a·o·ni·a (lik′ā ō′nē ə, -ōn′yə, li′kā-), an ancient country in S Asia Minor: later a Roman province.

Ly·ci·a (lish′ē ə), an ancient country in SW Asia Minor: later a Roman province.

Lyd·i·a (lid′ē ə), an ancient kingdom in W Asia Minor: under Croesus, a wealthy empire including most of Asia Minor. *Cap.:* Sardis.

Lynch·burg (linch′bûrg), a city in central Virginia. 66,049.

Lynn (lin), a seaport in E Massachusetts, on Massachusetts Bay. 81,245.

Lyn·wood (lin′wŏŏd′), a city in SW California. 61,945.

Ly·on·nais or **Ly·o·nais** (lē ə nā′), a former province in E France.

Ly·ons (lē ôn′, li′ənz), a city in E France at the confluence of the Rhone and Saône rivers. 418,476. French, **Lyon** (lyôN′).

Lys (lēs), a river in W Europe, in N France and W Belgium, flowing NE into the Scheldt River at Ghent. 120 mi. (195 km) long.

M

MA, Massachusetts.

Maas (mäs), Dutch name of the MEUSE.

Maas·tricht or **Maes·tricht** (mäs/trikt, -trikнt), a city in the SE Netherlands, on the Maas River. 115,782.

Ma·cao (mə kou/), **1.** a Portuguese overseas territory in S China, comprising a peninsula in the Zhu Jiang delta and two adjacent islands. 426,400; 6 sq. mi. (16 sq. km). **2.** the capital of this territory. Portuguese, **Ma·cáu/.** —**Mac·a·nese** (mak/ə nēz/, -nēs/), *n., pl.* **-nese**, *adj.*

Ma·ca·pá (mə kə pä/), the capital of Amapá, in NE Brazil, at the mouth of the Amazon. 89,081.

Ma·cas·sar or **Ma·kas·sar** (mə kas/ər), former name of UJUNG PANDANG.

Macas/sar Strait/, MAKASSAR STRAIT.

Mac·e·do·ni·a (mas/i dō/nē ə, -dōn/yə), **1.** Also, **Mac·e·don** (mas/i don/). an ancient kingdom in the Balkan Peninsula, in SE Europe: now a region including parts of Greece, the republic of Macedonia, and Bulgaria. **2.** a republic in SE Europe: formerly part of Yugoslavia. 2,040,000; 9928 sq. mi. (25,713 sq. km). *Cap.:* Skopje. —**Mac/e·do/ni·an,** *n., adj.*

Ma·cei·ó (mä/sā ō/), the capital of Alagoas, in E Brazil. 376,479.

Ma·chu Pic·chu (mä/chōō pēk/chōō, pē/chōō), the site of an ancient Incan city in the Andes, in S central Peru.

Ma·cí·as Ngue·ma Bi·yo·go (mə sē/əs əng-gwā/mə bē yō/gō), a former name of BIOKO.

Mac·ken·zie (mə ken/zē), **1.** a river in NW Canada, flowing NW from Great Slave Lake to the Arctic Ocean. 1120 mi. (1800 km) long; with tributaries, 2525 mi. (4065 km) long. **2.** a district in the SW Northwest Territories of Canada. 527,490 sq. mi. (1,366,200 sq. km).

Macken/zie Moun/tains, a mountain range in NW Canada. Highest peak, 9750 ft. (2971 m).

Mack·i·nac (mak/ə nô/), **1. Straits of,** a strait between Upper and Lower Michigan, connecting Lakes Huron and Michigan. **2.** an island at the entrance of this strait. 3 mi. (5 km) long.

Main (mān, min), a river in central Germany, flowing W from N Bavaria into the Rhine at Mainz. 305 mi. (490 km) long.

Maine (mān), **1.** a state in the NE United States, on the Atlantic coast. 1,227,928; 33,215 sq. mi. (86,027 sq. km). *Cap.*: Augusta. *Abbr.*: ME, Me. **2.** a former province in NW France. —**Main/er,** *n.*

Main·land (mān/land/, -lənd), **1.** the largest of the Shetland Islands. 18,268; ab. 200 sq. mi. (520 sq. km). **2.** POMONA (def. 2).

Mainz (mints), a port in SW central Germany, on the Rhine: capital of Rhineland-Palatinate. 172,400. French, **Mayence.**

Ma·jor·ca (mə jôr/kə, -yôr/-), a Spanish island in the W Mediterranean: the largest of the Balearic Islands. 534,511; 1405 sq. mi. (3640 sq. km). *Cap.*: Palma. Spanish, **Mallorca.** —**Ma·jor/can,** *adj., n.*

Ma·ka·lu (muk/ə lōō/), a mountain in the Himalayas, on the boundary between Nepal and Tibet. 27,790 ft. (8470 m).

Ma·kas·sar or **Ma·cas·sar** (mə kas/ər), former name of UJUNG PANDANG.

Makas/sar (or **Macas/sar**) **Strait/,** a strait between Borneo and Sulawesi.

Ma·ke·yev·ka (mə kā/əf kə), a city in SE Ukraine, NE of Donetsk. 455,000.

Ma·khach·ka·la (mə käch/kə lä/), a seaport and capital of Dagestan, in the SW Russian Federation, on the Caspian Sea. 315,000.

Mal/a·bar Coast/ (mal/ə bär/), a region along the SW coast of India extending inland to the Western Ghats. Also called **Malabar.**

Ma·la·bo (mə lä/bō), the capital of Equatorial Guinea, on N Bioko island. 40,000. Formerly, **Santa Isabel.**

Ma·lac·ca (mə lak/ə, -lä/kə), **1.** a state in Malaysia, on the SW Malay Peninsula. 464,754; 640 sq. mi. (1658 sq. km). **2.** the capital of this state. 88,073. **3. Strait of,** a strait between Sumatra and the Malay Peninsula. Also, **Melaka** (for defs. 1, 2). —**Ma·lac/can,** *adj., n.*

Má·la·ga (mal/ə gə, mä/lə-), **1.** a province in S Spain, in Andalusia. 1,215,479; 2813 sq. mi. (7285 sq. km). **2.** a seaport in S Spain, on the Mediterranean. 595,264.

Mal·a·gas'y Repub'lic (mal'ə gas'ē), former name of MADAGASCAR.

Ma·lang (mä läng'), a city on E Java, in S Indonesia. 511,780.

Mä·lar (mā'lər, -lär), **Lake,** a lake in S Sweden, extending W from Stockholm. 440 sq. mi. (1140 sq. km). Swedish, **Mä·lar·en** (me'lä rən).

Ma·la·tya (mä'lä tyä'), a city in central Turkey. 251,257. Ancient, **Melitene.**

Ma·la·wi (mə lä'wē), **1.** Formerly, **Nyasaland.** a republic in SE Africa, on Lake Malawi: formerly a British protectorate; became an independent member of the Commonwealth of Nations in 1964; a republic since 1966. 7,058,800; 49,177 sq. mi. (127,368 sq. km). *Cap.:* Lilongwe. **2. Lake.** Formerly, **Nyasa.** a lake in SE Africa, between Malawi, Tanzania, and Mozambique. 11,000 sq. mi. (28,500 sq. km). —**Ma·la'wi·an,** *adj., n.*

Ma·lay·a (mə lā'ə), **1.** MALAY PENINSULA. **2. Federation of,** a former federation of states in the S Malay Peninsula: a former British protectorate; now forms part of Malaysia. 50,690 sq. mi. (131,287 sq. km). —**Ma·lay'an,** *adj., n.*

Ma'lay Archipel'ago (mā'lā, mə lā'), an extensive island group in the Indian and Pacific oceans, SE of Asia, including the Sunda Islands, the Moluccas, and the Philippines. Also called **Malaysia.**

Ma'lay Penin'sula, a peninsula in SE Asia, consisting of W (mainland) Malaysia and the S part of Thailand. Also called **Malaya.**

Ma·lay·sia (mə lā'zhə, -shə), **1.** a constitutional monarchy in SE Asia: a federation, comprising Malaya, Sabah, and Sarawak; a member of the Commonwealth of Nations. 16,968,000; 127,317 sq. mi. (329,759 sq. km). *Cap.:* Kuala Lumpur. **2.** MALAY ARCHIPELAGO. —**Ma·lay'sian,** *adj., n.*

Mal·den (môl'dən), a city in E Massachusetts, near Boston. 53,884.

Mal·dives (môl'dēvz, mal'dīvz), **Republic of,** a republic in the Indian Ocean, SW of Sri Lanka, consisting of about 1200 islands: British protectorate 1887–1965. 214,139; 115 sq. mi. (298 sq. km). *Cap.:* Male. —**Mal·div'i·an** (-div'ē ən), *n., adj.*

Ma·le (mä'lā, -lē), the capital of the Maldives. 56,060.

Ma·le'bo Pool' (mä lä'bō), a widening of the Zaire (Congo) River on the boundary between Zaire and the People's Republic of the Congo, about 330 mi. (530 km) from its mouth. ab. 20 mi. (32 km) long. Also called **Stanley Pool**.

Ma·li (mä'lē), **Republic of,** a republic in W Africa: formerly a territory of France; gained independence 1960. 9,092,000; 478,841 sq. mi. (1,240,192 sq. km). *Cap.:* Bamako. Formerly, **French Sudan.** —**Ma'li·an,** *n., adj.*

Ma·lines (mȧ lēn'), French name of MECHLIN.

Ma·llor·ca (*Sp.* mä lyȯr'kä, -yȯr'-), MAJORCA. —**Ma·llor'can** (-yȯr'kən), *adj., n.*

Malm·ö (mal'mō, mäl'mœ), a seaport in S Sweden, on Øresund opposite Copenhagen, Denmark. 230,838.

Mal·ta (môl'tə), **1.** an island in the Mediterranean south of Sicily. 95 sq. mi. (246 sq. km). **2.** a republic consisting of this island and two adjacent islands: a former British colony; gained independence 1964; a member of the Commonwealth of Nations since 1974. 354,900; 122 sq. mi. (316 sq. km). *Cap.:* Valletta. —**Mal·tese'** *n., pl.* **-tese,** *adj.*

Mal'vern Hill' (mal'vərn), a plateau in E Virginia: battle 1862.

Mal'vern Hills' (môl'vərn, mô'-), a range of hills in W England, bisecting Hereford and Worcester: highest point, 1395 ft. (425 m).

Mam'moth Cave' Na'tional Park', a national park in central Kentucky: limestone caverns. 79 sq. mi. (205 sq. km).

Ma·mo·ré (mä'mə rā'), a river in Bolivia, flowing N to the Beni River on the border of Brazil to form the Madeira River. 700 mi. (1125 km) long.

Man (man), **Isle of,** an island of the British Isles, in the Irish Sea. 64,282; 227 sq. mi. (588 sq. km). *Cap.:* Douglas. —**Manx** (mangks), *adj., n.*

Man., Manitoba.

Ma·na·do (mə nä'dō), MENADO.

Ma·na·gua (mə nä'gwə), **1. Lake,** a lake in W Nicaragua. 390 sq. mi. (1010 sq. km). **2.** the capital of Nicaragua, in the W part. 682,111.

Ma·na·ma (mə nam'ə), the capital of Bahrain. 151,500.

Ma·nas·sas (mə nas'əs), a town in NE Virginia: battles of Bull Run 1861, 1862. 15,438.

Ma·naus (mä nous′), the capital of Amazonas, in NW Brazil, on the Negro River near its confluence with the Amazon. 611,763.

Man·ches·ter (man′ches/tər, -chə stər), **1.** a city in NW England. 451,000. **2.** a city in S New Hampshire. 99,567. —**Man·cu·ni·an** (man kyōō′nē ən), *n.*, *adj.*

Man·chu·kuo (man′chōō/kwō′), a former country (1932–45) in E Asia, under Japanese control: included Manchuria and parts of Inner Mongolia; now a part of China.

Man·chu·ri·a (man chŏŏr′ē ə), a historic region in NE China. ab. 413,000 sq. mi. (1,070,000 sq. km). —**Man·chu′ri·an,** *adj., n.*

Man·da·lay (man′dl ā′, man′dl ā′), a city in central Burma, on the Irrawaddy River. 532,985.

Man·hat·tan (man hat′n, mən-), **1.** Also called **Manhat′tan Is′land.** an island in New York City surrounded by the Hudson, East, and Harlem rivers. 13½ mi. (22 km) long. **2.** a borough of New York City approximately coextensive with Manhattan Island. 1,487,536. —**Man·hat′tan·ite′,** *n.*

Ma·nil·a (mə nil′ə), the capital of the Philippines, on W central Luzon. 1,587,000.

Manil′a Bay′, a bay in the Philippines, in W Luzon Island.

Ma·ni·pur (mun′i pŏŏr′), a state in NE India between Assam and Burma. 1,433,691; 8620 sq. mi. (22,326 sq. km). *Cap.:* Imphal.

Ma·ni·sa (mä′ni sä′), a city in W Turkey. 126,319. Ancient, **Magnesia.**

Man·i·to·ba (man′i tō/bə), **1.** a province in central Canada. 1,063,016; 250,946 sq. mi. (649,046 sq. km). *Abbr.:* Man. *Cap.:* Winnipeg. **2. Lake,** a lake in the S part of this province. 120 mi. (195 km) long; 1817 sq. mi. (4705 sq. km). —**Man′i·to′ban,** *adj., n.*

Man·i·tou·lin (man′i tōō′lin), an island in N Lake Huron belonging to Canada. 80 mi. (130 km) long. Also called **Man′itou′lin Is′land.**

Ma·ni·za·les (mä′nē sä′les), a city in W Colombia. 299,352.

Man·nar (mə när′), **Gulf of,** an inlet of the Indian Ocean bounded by W Sri Lanka, the island chain of Adam's Bridge, and S India.

Mann·heim (man′him, män′-), a city in SW Germany at the confluence of the Rhine and Neckar rivers. 295,200.

Mans·field (manz′fēld′), **1. Mount,** a mountain in N Vermont: highest peak of the Green Mountains, 4393 ft. (1339 m). **2.** a city in W Nottinghamshire, in central England. 100,000. **3.** a city in N Ohio. 50,627.

Man·su·ra (man sŏŏr′ə), EL MANSURA.

Man·tu·a (man′chŏŏ ə), a city in E Lombardy, in N Italy: birthplace of Virgil. 60,932. Italian, **Man·to·va** (män′tô vä). —**Man′tu·an,** *adj., n.*

Ma·nu′a Is′lands (mə nŏŏ′ə), a group of three small islands in the E part of American Samoa. 1700; ab. 5 sq. mi. (13 sq. km).

Man·za·nil·lo (män′sä nē′yō), a seaport in SE Cuba. 87,471.

Ma·pu·to (mə pŏŏ′tō), the capital of Mozambique, on Delagoa Bay. 491,800. Formerly, **Lourenço Marques.**

Mar·a·cai·bo (mar′ə ki′bō), **1.** a seaport in NW Venezuela. 890,553. **2. Lake,** a lake in NW Venezuela, an extension of the Gulf of Venezuela: the largest lake in South America. 6300 sq. mi. (16,320 sq. km).

Mar·a·can·da (mar′ə kan′də), ancient name of SAMARKAND.

Ma·ra·cay (mär′ə ki′), a city in NE Venezuela, SW of Caracas. 387,682.

Ma·ra·ñón (mär′ə nyōn′), a river in Peru, flowing N and then E, joining the Ucayali to form the Amazon. 1000 mi. (1600 km) long.

Ma·raş (mə räsh′), a city in S Turkey, NE of Adana. 212,206.

Mar·a·thon (mar′ə thon′), **1.** a plain in SE Greece, in Attica: the Athenians defeated the Persians here 490 B.C. **2.** an ancient village near this plain.

Mar·burg (mär′bŏŏrg, -bûrg), a city in central Germany. 75,092.

March·es (mär′chiz), **the, 1.** the border districts between England and Scotland or England and Wales. **2.** Italian, **Le Mar·che** (le mär′ke). a region in central Italy on the Adriatic. 1,412,404; 3743 sq. mi. (9695 sq. km).

Mar·cy (mär′sē), **Mount,** a mountain in NE New York: highest peak of the Adirondack Mountains, 5344 ft. (1629 m).

Mar del Pla·ta (mär′ ŧħel plä′tä, del), a city in E Argentina. 407,024.

Ma·ren·go (mə reng′gō), a village in Piedmont, in NW Italy: defeat of Austrians by Napoleon 1800.

Mar·gate (mär′git, -gāt), a city in NE Kent, in SE England: seaside resort. 122,500.

Mar′i·an′a Is′lands (mâr′ē an′ə, mar′-, mâr′-, mar′-), a group of 15 islands in the W Pacific, E of the Philippines: comprised of Guam, a U.S. possession, and the commonwealth of the Northern Mariana Islands. 396 sq. mi. (1026 sq. km). Also called **Mar′i·an′as.**

Ma·ri·a·na·o (mär′ē ə nä′ō), a city in NW Cuba, a suburb of Havana. 127,563.

Ma·rián·ské Láz·ně (mä′ryän ske läz′nye), a spa in W Bohemia, in W Czechoslovakia. 18,510. German, MARIENBAD.

Ma′ri Auton′omous Repub′lic (mär′ē), an autonomous republic in the Russian Federation in Europe. 750,000; 8994 sq. mi. (23,294 sq. km).

Ma·ri·bor (mär′i bôr′), a city in N Slovenia, on the Drava River. 185,699.

Ma·rie′ Byrd′ Land′ (mə rē′ bûrd′), a part of Antarctica, SE of the Ross Sea: discovered and explored by Adm. Richard E. Byrd.

Ma·riel (mär yel′), a seaport of Cuba, SW of Havana. 34,467.

Ma·ri·en·bad (mä Rē′ən bät′; *Eng.* mâr′ē ən bad′, mar′-), German name of MARIÁNSKÉ LÁZNĚ.

Ma·rin·du·que (mar′in dōō′kā, mär′-), an island of the Philippines, between Luzon and Mindoro islands. 347 sq. mi. (899 sq. km).

Mar′itime Alps′, a range of the Alps in SE France and NW Italy.

Mar′itime Prov′inces, the Canadian provinces of Nova Scotia, New Brunswick, and Prince Edward Island. Also called **Mar′i·times′.** —**Mar′i·tim′er,** *n.*

Ma·ri·tsa (mə rēt′sə), a river in S Europe flowing from S Bulgaria along the border between Greece and Turkey and into the Aegean. 300 mi. (485 km) long.

Ma·ri·u·pol (mar′ē ōō′pəl), a city in SE Ukraine, on the Sea of Azov. 503,000. Formerly (1948–89), **Zhdanov.**

Mark·ham (mär′kəm), **1. Mount,** a mountain in Antarctica, SW of the Ross Sea. 15,100 ft. (4600 m). **2.** a town in SE Ontario, in S Canada. 114,597.

Mar·ma·ra (mär′mər ə), **Sea of,** a sea in NW Turkey connected with the Black Sea by the Bosporus, and with the Aegean by the Dardanelles. 4300 sq. mi. (11,135 sq. km).

Mar·mo·la·da (mär′mə lä′də), a mountain in N Italy: highest peak of the Dolomites, 11,020 ft. (3360 m).

Marne (märn), a river in NE France flowing W to the Seine near Paris. 325 mi. (525 km) long.

Ma·roc (mA RÔK′), French name of Morocco.

Ma·ros (mo′RÔsh), Hungarian name of Mureş.

Mar·que′sas Is′lands (mär kā′zəz, -səz, -səs), a group of French islands in the S Pacific. 6000; 480 sq. mi. (1245 sq. km). **—Mar·que′san,** n., adj.

Mar·ra·kesh or **Mar·ra·kech** (mar′ə kesh′, mar′ə kesh′), a city in W Morocco. 439,728.

Mar·sa·la (mär sä′lə), a seaport in W Sicily. 46,300.

Mar·seilles (mär sā′), a seaport in SE France, on the Gulf of Lions. 1,110,511. French, **Mar·seille** (mAR se′y°).

Mar′shall Is′lands (mär′shəl), a group of 34 atolls in the W central Pacific: formerly a part of the Trust Territory of the Pacific Islands; since 1986 a self-governing area associated with the U.S. 40,609; 70 sq. mi. (181 sq. km). **—Mar′-shall·ese′** (-shə lēz′, -lēs′), n., pl. **-ese,** adj.

Mars′ton Moor′ (mär′stən), a former moor in NE England, west of York: Cromwell's victory over the Royalists 1644.

Mar·ta·ban (mär′tə bän′), **Gulf of,** an inlet of the Bay of Bengal, in Burma.

Mar′tha's Vine′yard (mär′thəz), an island off SE Massachusetts: summer resort. 6000; 108 sq. mi. (282 sq. km).

Mar·ti·nique (mär′tn ēk′), an island in the E West Indies: an overseas department of France. 336,000; 425 sq. mi. (1100 sq. km). Cap.: Fort-de-France. **—Mar′ti·ni′can,** n., adj.

Mar·war (mär′wär), Jodhpur (def. 1).

Mar·y·land (mer′ə lənd), a state in the E United States, on the Atlantic coast. 4,781,468; 10,577 sq. mi. (27,395 sq. km). Cap.: Annapolis. Abbr.: MD, Md. **—Mar′y·land·er,** n.

Ma·sa·da (mə säˈdə), an ancient fortress in Israel on the SW shore of the Dead Sea.

Ma·san (mäˈsän), a seaport in SE South Korea. 449,236. Formerly, **Ma·sam·po** (mə sämˈpō).

Mas·ba·te (mäs bäˈtē), one of the central islands of the Philippines. 1262 sq. mi. (3269 sq. km).

Ma·se·ru (mäˈsə rōṓ, mazˈə rōṓ), the capital of Lesotho, in the NW part. 109,382.

Mash·had (mash hadˈ) also **Meshed,** a city in NE Iran: Muslim shrine. 1,463,508.

Mas·qat (mus katˈ), MUSCAT.

Mass., Massachusetts.

Mas·sa·chu·setts (masˈə chōōˈsits), a state in the NE United States, on the Atlantic coast. 6,016,425; 8257 sq. mi. (21,385 sq. km). *Cap.:* Boston. *Abbr.:* MA, Mass.

Masˈsachuˈsetts Bayˈ, an inlet of the Atlantic, off the E coast of Massachusetts.

Mas·sif Cen·tral (mA sēf sän trAlˈ), a plateau in S central France.

Ma·su·ri·a (mə zōōrˈē ə), a region in NE Poland, formerly in East Prussia, Germany. German, **Ma·su·ren** (mä zōōˈRən).

Ma·ta·di (mə täˈdē), a seaport in W Zaire, near the mouth of the Zaire (Congo) River. 144,742.

Mat·a·mo·ros (matˈə môrˈəs, -ōs, -mōrˈ-), a seaport in NE Mexico, on the Rio Grande, opposite Brownsville, Tex. 238,840.

Ma·tan·zas (mə tanˈzəs), a seaport on the NW coast of Cuba. 105,400.

Mat·a·pan (matˈə panˈ), **Cape,** a cape in S Greece, at the S tip of the Peloponnesus.

Ma·thu·ra (mutˈōō rə), a city in W Uttar Pradesh, in N India: Hindu shrine and holy city; reputed birthplace of Krishna. 161,000. Formerly, **Muttra.**

Ma·to Gros·so (matˈə grōˈsō), **1.** a plateau in SW Brazil. **2.** a state in SW Brazil. 1,580,900; 340,155 sq. mi. (881,000 sq. km). *Cap.:* Cuiabá.

Maˈto Grosˈso do Sulˈ (dō sōōlˈ), a state in SW Brazil. 1,673,500; 135,347 sq. mi. (350,548 sq. km). *Cap.:* Campo Grande.

Ma·tsu (mätˈsōō, matˈ-), an island off the SE coast of China: administered by Taiwan. 11,000; 17 sq. mi. (44 sq. km).

Ma·tsu·ya·ma (mäˈtsōō yäˈmä), a seaport on NW Shikoku, in SW Japan. 430,000.

Mat·ter·horn (matˈər hôrnˈ), a mountain on the

border of Switzerland and Italy, in the Pennine Alps. 14,780 ft. (4505 m). French, **Mont Cervin.**

Ma·tu·rín (mä′tə rēn′, mat′ə-), a city in NE Venezuela. 154,976.

Mau·i (mou′ē), an island in central Hawaii. 84,000; 728 sq. mi. (1886 sq. km).

Mau·mee (mô mē′, mô′mē), a river in E Indiana and W Ohio flowing NE to Lake Erie. 175 mi. (280 km) long.

Mau·na Ke·a (mou′nə kā′ə, mô′nə kē′ə), a dormant volcano on the island of Hawaii. 13,784 ft. (4201 m).

Mau′na Lo′a (lō′ə), an active volcano on the island of Hawaii, in Hawaii Volcanoes National Park. 13,680 ft. (4170 m).

Mau·re·ta·ni·a or **Mau·ri·ta·ni·a** (môr′i tā′-nē ə), an ancient kingdom in NW Africa: it included the territory that is modern Morocco and part of Algeria. **—Mau′re·ta′ni·an,** adj.

Mau·ri·ta·ni·a (môr′i tā′nē ə), **1.** Official name, **Islamic Republic of Mauritania.** a republic in NW Africa: formerly a French colony; independent since 1960. 1,894,000; 398,000 sq. mi. (1,030,700 sq. km). Cap.: Nouakchott. **2.** MAURE-TANIA. **—Mau′ri·ta′ni·an,** adj., n.

Mau·ri·tius (mô rish′əs, -rish′ē əs), **1.** an island in the Indian Ocean, E of Madagascar. 720 sq. mi. (1865 sq. km). **2.** a republic consisting of this island and several other islands: formerly a British colony; independent since 1968. 1,075,000; 788 sq. mi. (2040 sq. km). Cap.: Port Louis. **—Mau·ri′tian,** adj., n.

May (mā), **Cape,** a cape at the SE tip of New Jersey, on Delaware Bay.

Ma·ya·güez (mä′yä gwes′), a seaport in W Puerto Rico. 96,193.

Ma·yence (mᴀ yäns′), French name of MAINZ.

May·o (mā′ō), a county in NW Connaught province, in the NW Republic of Ireland. 115,016; 2084 sq. mi. (5400 sq. km).

Ma·yon (mä yôn′), an active volcano in the Philippines, on SE Luzon Island. 7926 ft. (2415 m).

Ma·yotte (mä yŏt′), one of the Comoro Islands, in the Indian Ocean: a dependency of France. 77,300; 144 sq. mi. (373 sq. km).

Ma·za·tlán (mä′sä tlän′), a seaport in S Sinaloa, in W Mexico. 249,988.

Mba·bane (bä bän′, -bä′nē, əm bä-), the capital of Swaziland, in the NW part. 38,290.

Mban·da·ka (bän′dä kä′, əm bän′-), a city in W Zaire. 125,263. Formerly, **Coquilhatville.**

Mbi·ni (bē′nē, əm bē′-), the mainland portion of Equatorial Guinea, on the Gulf of Guinea. 10,040 sq. mi. (26,003 sq. km). Formerly, **Río Muni.**

Mbo·mu (bō′mōō, əm bō′-), Bomu.

Mbu·ji-Ma·yi (bōō′jē mi′-, -mä′yē, əm bōō′-), a city in S central Zaire. 423,363. Formerly, **Ba·kwanga.**

Mc·Al·len (mə kal′ən), a city in S Texas, on the Rio Grande. 84,021.

Mc·Kin·ley (mə kin′lē), **Mount,** a mountain in central Alaska, in Denali National Park: highest peak in North America, 20,320 ft. (6194 m).

Mc·Mur′do Sound′ (mək mûr′dō), an inlet of Ross Sea, in Antarctica, N of Victoria Land.

MD or **Md.,** Maryland.

ME or **Me.,** Maine.

Mead (mēd), **Lake,** a lake in NW Arizona and SE Nevada, formed by Hoover Dam on the Colorado River. 227 sq. mi. (588 sq. km).

Me·an·der (mē an′dər), ancient name of the MENDERES (def. 1).

Meath (mēth, mēŧħ), a county in Leinster, in the E Republic of Ireland. 95,602; 902 sq. mi. (2335 sq. km).

Mec·ca (mek′ə), a city in W Saudi Arabia: birthplace of Muhammad; spiritual center of Islam. 550,000. —**Mec′can,** adj., n.

Mech·lin (mek′lin), a city in N Belgium. 75,718. French, **Malines.** Flemish, **Mech·e·len** (meκμ′ə-lən).

Meck′len·burg–West′ern Pomera′nia (mek′lən bûrg′), a state in NE Germany. 2,100,000; 8842 sq. mi. (22,900 sq. km). *Cap.:* Schwerin. German, **Meck·len·burg–Vor·pom·mern** (mek′lən-bōōrk′fōr′pôm ərn).

Me·dan (me dän′), a city in NE Sumatra, in W Indonesia. 1,378,955.

Me·de·llín (med′l ēn′, mä′də yēn′), a city in W Colombia. 1,468,089.

Med·ford (med′fərd), **1.** a city in E Massachusetts, near Boston. 57,407. **2.** a city in SW Oregon. 45,290.

Me·di·a (mē′dē ə), an ancient country in W Asia,

S of the Caspian Sea, corresponding generally to NW Iran. *Cap.:* Ecbatana.

Me/dia Atropate/ne, an ancient region in NW Iran, formerly a part of Media. Also called **Atropatene.**

Med/icine Bow/ Range/ (bō), a range of the Rocky Mountains, in Wyoming and Colorado. Highest peak, 12,014 ft. (3662 m).

Med/icine Hat/, a city in SE Alberta, in SW Canada. 32,811.

Me·di·na (mə dē/nə), a city in W Saudi Arabia, where Muhammad was first accepted as the Prophet and where his tomb is located. 198,196.

Med/i·ter·ra/ne·an Sea/ (med/i tə rā/nē ən), a sea surrounded by Africa, Europe, and Asia. 2400 mi. (3865 km) long; 1,145,000 sq. mi. (2,965,550 sq. km). Also called **Mediterranean.**

Mee·rut (mēr/ət), a city in W Uttar Pradesh, in N India. 538,000.

Meg·a·ra (meg/ər ə), a city in ancient Greece: the chief city of Megaris. —**Me·gar·i·an, Me·gar·e·an** (mə gar/ē ən), **Me·gar/ic,** *adj.*

Meg·a·ris (meg/ər is), a district in ancient Greece between the Gulf of Corinth and the Saronic Gulf.

Me·gha·la·ya (mā/gə lā/ə), a state in NE India. 1,327,824; 8785 sq. mi. (22,489 sq. km). *Cap.:* Shillong.

Me·gid·do (mə gid/ō), an ancient city in N Israel, on the plain of Esdraelon: often identified with the Biblical Armageddon.

Meis·sen (mī/sən), a city in E central Germany on the Elbe River. 38,137.

Mé·ji·co (*Sp.* me/hē kô), Spanish name of MEXICO.

Mek·nès (mek nes/), a city in N Morocco: former capital of Morocco. 319,783.

Me·kong (mā/kông, -kong, mē/-), a river whose source is in SW China, flowing SE along most of the boundary between Thailand and Laos to the South China Sea. 2600 mi. (4200 km) long.

Me/kong Del/ta, the delta of the Mekong River in Vietnam.

Me·la·ka (mə lä/kə), MALACCA (defs. 1, 2).

Mel·a·ne·sia (mel/ə nē/zhə, -shə), one of the three principal divisions of Oceania, comprising

the island groups in the S Pacific NE of Australia.
—**Mel·a·ne′sian,** *adj., n.*

Mel·bourne (mel′bərn), **1.** the capital of Victoria, in SE Australia. 2,942,000. **2.** a city on the E coast of Florida. 59,646.

Me·lil·la (mā lēl′yä), a seaport belonging to Spain on the NE coast of Morocco, in NW Africa. 55,613.

Mel·i·te·ne (mel′i tē′nē), ancient name of MALA-TYA.

Me·li·to·pol (mel′ə tô′pəl), a city in SE Ukraine, NW of the Sea of Azov. 174,000.

Me·los (mē′los, -lōs) also **Milos,** a Greek island in the Cyclades, in the SW Aegean. 4560; 51 sq. mi. (132 sq. km). —**Me′li·an,** *adj., n.*

Mel·rose (mel′rōz′), a village in SE Scotland, on the Tweed River.

Mel·ville (mel′vil), **Lake,** a saltwater lake on the E coast of Labrador, Newfoundland, in E Canada. ab. 1133 sq. mi. (2935 sq. km).

Mel′ville Is′land, an island in the Arctic Ocean, N of Victoria Island, belonging to Canada. 16,141 sq. mi. (41,805 sq. km).

Mel′ville Penin′sula, a peninsula in N Canada, SE of the Gulf of Boothia. 250 mi. (405 km) long.

Me·mel (mā′məl, mem′əl), German name of KLAIPEDA.

Mem·phis (mem′fis), **1.** a port in SW Tennessee, on the Mississippi. 610,337. **2.** a ruined city in N Egypt, S of Cairo: the ancient capital of Egypt. —**Mem′phi·an,** *adj., n.* —**Mem′phite,** *adj., n.*

Me·na·do or **Ma·na·do** (mə nä′dō), a seaport on NE Sulawesi, in NE Indonesia. 169,684.

Men′ai Strait′ (men′ī), a strait between Angle-sey Island and the mainland of NW Wales. 14 mi. (23 km) long.

Me·nam (me näm′), former name of CHAO PHRAYA.

Men·de·res (men′de res′), **1.** Ancient, **Maeander, Meander.** a river in W Asia Minor, flowing into the Aegean. 240 mi. (385 km) long. **2.** Ancient, **Scamander.** a river in NW Asia Minor, flowing into the Dardanelles. 60 mi. (97 km) long.

Men·do·ci·no (men′də sē′nō), **Cape,** a cape in NW California: the westernmost point in California.

Men·do·za (men dō′zə), a city in W central Argentina. 596,796.

Men′lo Park′ (men′lō), a village in central New Jersey: site of Thomas Edison's laboratory, 1876–87.

Me·nor·ca (*Sp.* me nôr′kä), MINORCA.

Men·ton (men tôn′; *Fr.* män tôn′), a city in SE France, on the Mediterranean: resort. 25,072. Italian, **Men·to·ne** (men tô′ne).

Mer·ced (mər sed′), a city in central California. 56,216.

Mer·ci·a (mûr′shē ə, -shə), an early English kingdom in central Britain. —**Mer′ci·an,** *n., adj.*

Mé·ri·da (mer′i də, mā′rē-), **1.** the capital of Yucatán, in SE Mexico. 424,529. **2.** a city in W Venezuela. 143,805.

Mer·i·den (mer′i dn), a city in central Connecticut. 59,479.

Me·rid·i·an (mə rid′ē ən), a city in E Mississippi. 46,577.

Me·rín (me Rēn′), Spanish name of Lake MIRIM.

Mer·i·on·eth·shire (mer′ē on′ith sher′, -shər), a historic county in Gwynedd, in N Wales. Also called **Mer′i·on·eth.**

Mer·o·ë (mer′ō ē′), a ruined city in Sudan, on the Nile: a capital of ancient Ethiopia that was destroyed A.D. c350. —**Mer′o·ite′** (-ō it′), *n.*

Mer·ri·mack (mer′ə mak′), a river in central New Hampshire and NE Massachusetts, flowing S and NE to the Atlantic. 110 mi. (175 km) long.

Mer·sey (mûr′zē), a river in W England, flowing W from Derbyshire to the Irish Sea. 70 mi. (115 km) long.

Mer·sey·side (mûr′zē sid′), a metropolitan county in W England. 1,456,800; 250 sq. mi. (648 sq. km).

Mer·sin (mer sēn′), a seaport in S Turkey, on the NW coast of the Mediterranean Sea. 216,308.

Mer·thyr Tyd·fil (mûr′thər tid′vil), an administrative district in Mid Glamorgan, in S Wales. 58,500; 43 sq. mi. (113 sq. km).

Mer·ton (mûr′tn), a borough of Greater London, England. 164,500.

Me·sa (mā′sə), a city in central Arizona, near Phoenix. 288,091.

Me·sa′bi Range′ (mə sä′bē), a range of low

hills in NE Minnesota, noted for major iron-ore deposits.

Me·sa Ver·de Na·tional Park (vûr′dē, vûrd′), a national park in SW Colorado: ruins of prehistoric cliff dwellings. 80 sq. mi. (207 sq. km).

Me·shed (me shed′), MASHHAD.

Me·so·lon·gi·on (me′sô lông′gē ôn), Greek name of MISSOLONGHI.

Mes·o·po·ta·mi·a (mes′ə pə tā′mē ə), an ancient region in W Asia between the Tigris and Euphrates rivers: now part of Iraq. —**Mes′o·po·ta′-mi·an**, *adj.*, *n.*

Mes·quite (me skēt′, mi-), a city in NE Texas, E of Dallas. 101,484.

Mes·se·ne (me sē′nē), an ancient city in S Greece, in the SW Peloponnesus: capital of Messenia.

Mes·se·ni·a (mə sē′nē ə, -sēn′yə), a division of ancient Greece, in the SW Peloponnesus.

Mes·si·na (me sē′nə), **1.** a seaport in NE Sicily. 270,546. **2. Strait of,** a strait between Sicily and Italy. 2½ mi. (4 km) wide.

Metz (mets; *Fr.* mes), a city in NE France, on the Moselle River. 186,437.

Meuse (myo͞oz; *Fr.* mœz), a river in W Europe, flowing from NE France through E Belgium and S Netherlands into the North Sea. 575 mi. (925 km) long. Dutch, **Maas.**

Mex·i·cal·i (mek′si kal′ē), the capital of Baja California Norte, in NW Mexico, on the Mexican-U.S. border. 510,600.

Mex·i·co (mek′si kō′), **1.** a republic in S North America. 82,700,000; 756,198 sq. mi. (1,958,201 sq. km). *Cap.:* Mexico City. **2.** a state in central Mexico. 11,571,000; 8268 sq. mi. (21,415 sq. km). *Cap.:* Toluca. **3. Gulf of,** an arm of the Atlantic surrounded by the U.S., Cuba, and Mexico. 700,000 sq. mi. (1,813,000 sq. km). Mexican, **Mé·xi·co** (me′hē kô′); Spanish, **Méjico** (for defs. 1, 2). —**Mex′i·can**, *adj.*, *n.*

Mex′ico Cit′y, the capital of Mexico, in the Federal District, in the central part of Mexico. 18,748,000. Official name, **Mexico, D(istrito) F(ederal).**

MI, Michigan.

Mi·am·i (mi am′ē, -am′ə), **1.** a city in SE Florida, on Biscayne Bay. 358,548. **2.** a river in W Ohio, flowing S into the Ohio River. 160 mi. (260 km) long. —**Mi·am′i·an,** *n.*

Miam′i Beach′, a city in SE Florida on an island 2½ mi. (4 km) across Biscayne Bay from Miami: seaside resort. 92,639.

Mich., Michigan.

Mich·i·gan (mish′i gən), **1.** a state in the N central United States. 9,295,297; 58,216 sq. mi. (150,780 sq. km). *Cap.:* Lansing. *Abbr.:* MI, Mich. **2. Lake,** a lake in the N central U.S., between Wisconsin and Michigan: one of the five Great Lakes. 22,400 sq. mi. (58,015 sq. km). —**Mich′i·gan′der** (-gan′dər), **Mich′i·gan·ite′,** *n.*

Mi·cho·a·cán (mē′chō ä kän′), a state in SW Mexico. 2,868,824; 23,196 sq. mi. (60,080 sq. km). *Cap.:* Morelia.

Mi·cro·ne·sia (mī′krə nē′zhə, -shə), **1.** one of the three principal divisions of Oceania, comprising the small Pacific islands N of the equator and E of the Philippines, whose main groups are the Mariana Islands, the Caroline Islands, and the Marshall Islands. **2. Federated States of,** a group of islands in the W Pacific, in the Caroline Islands, comprising the islands of Pohnpei, Truk, Yap, and Kosrae: formerly a part of the Trust Territory of the Pacific Islands; now a self-governing area associated with the U.S. 108,600; 271 sq. mi. (701 sq. km). *Cap.:* Kolonia. —**Mi′cro·ne′sian,** *adj., n.*

Mid·del·burg (mid′l bûrg′), a city in the SW Netherlands. 39,152.

Mid′dle Atlan′tic States′, New York, New Jersey, and Pennsylvania.

Mid′dle Con′go, a former name of the People's Republic of the Congo.

Mid′dle East′, 1. Also called **Mideast.** the area from Libya east to Afghanistan, usu. including Egypt, Sudan, Israel, Jordan, Lebanon, Syria, Turkey, Iraq, Iran, Saudi Arabia, and the other countries of the Arabian peninsula. **2.** (formerly) the area including Iran, Afghanistan, India, Tibet, and Burma. —**Mid′dle East′ern,** *adj.*

Mid·dles·brough (mid′lz brə), a seaport in NE England, on the Tees estuary. 144,800.

Mid·dle·sex (mid′l seks′), a former county in SE England, now part of Greater London.

Mid·dle·town (mid′l toun′), a township in E New Jersey. 62,574.

Mid·dle West′, MIDWEST. **—Mid′dle West′ern,** *adj.* **—Mid′dle West′erner,** *n.*

Mid·east (mid′ēst′), MIDDLE EAST (def. 1). **—Mid′east′ern,** *adj.*

Mid′ Gla·mor′gan (glə môr′gən), a county in S Wales. 534,700. 393 sq. mi. (1019 sq. km).

Mi·di (mē dē′), the south of France.

Mi·di-Py·ré·nées′ (-pē rā nā′), a metropolitan region in SW France. 2,355,100; 17,509 sq. mi. (45,348 sq. km).

Mid·land (mid′lənd), a city in W Texas. 89,443.

Mid·lands (mid′ləndz), the central part of England; the midland counties.

Mid·lo·thi·an (mid lō′thē ən), a historic county in SE Scotland.

Mid·way (mid′wā′), several U.S. islets in the N Pacific, about 1300 mi. (2095 km) NW of Hawaii. 2 sq. mi. (5 sq. km).

Mid·west (mid′west′), a region in the N central United States, including the states of Illinois, Indiana, Iowa, Kansas, Michigan, Minnesota, Missouri, Nebraska, North Dakota, Ohio, South Dakota, and Wisconsin. Also called **Middle West. —Mid′west′ern,** *adj.* **—Mid′west′ern·er,** *n.*

Mid′west Cit′y, a city in central Oklahoma, near Oklahoma City. 52,267.

Mi·ko·nos (mē′kə nōs′), MYKONOS.

Mi·lan (mi lan′, -län′), an industrial city in central Lombardy, in N Italy. 1,478,505. Italian, **Mi·la·no** (mē lä′nô). **—Mil·an·ese** (mil′ə nēz′, -nēs′), *n., pl.* **-ese,** *adj.*

Mi·le·tus (mi lē′təs), an ancient city in Asia Minor, on the Aegean. **—Mi·le·sian** (mi lē′zhən, -shən, mi-), *adj., n.*

Mi·los (mē′los, -lōs, mi′-), MELOS.

Mil·pi·tas (mil pē′təs), a town in W California. 50,686.

Mil·wau·kee (mil wô′kē), a port in SE Wisconsin, on Lake Michigan. 628,088. **—Mil·wau·kee·an,** *n., adj.*

Mi′nas Ba′sin (mi′nəs), a bay in E Canada, the easternmost arm of the Bay of Fundy, in N Nova Scotia.

Mi·nas Ge·rais (mē′nəs zhi ris′), a state in E Brazil. 15,099,700; 226,708 sq. mi. (587,172 sq. km). *Cap.:* Belo Horizonte.

Min·da·na·o (min′də nä′ō, -nou′), the second

largest island of the Philippines, in the S part. 36,537 sq. mi. (94,631 sq. km).

Min′dana′o Deep′, an area in the Pacific Ocean W of the Philippines: one of the deepest points in any ocean. 34,440 ft. (10,497 m).

Min·do·ro (min dôr′ō, -dōr′ō), a central island of the Philippines. 3759 sq. mi. (9735 sq. km).

Minn., Minnesota.

Min·ne·ap·o·lis (min′ē ap′ə lis), a city in SE Minnesota, on the Mississippi. 368,383. —**Min′- ne·a·pol′i·tan** (-ə pol′i tn), *n.*

Min·ne·so·ta (min′ə sō′tə), **1.** a state in the N central United States. 4,375,099; 84,068 sq. mi. (217,735 sq. km). *Cap.:* St. Paul; *Abbr.:* MN, Minn. **2.** a river flowing SE from the W border of Minne- sota into the Mississippi near St. Paul. 332 mi. (535 km) long. —**Min′ne·so′tan,** *adj., n.*

Mi·nor·ca (mi nôr′kə), one of the Balearic Is- lands, in the W Mediterranean. 57,000; 271 sq. mi. (700 sq. km). Spanish, **Menorca.** —**Mi·nor′- can,** *adj., n.*

Minsk (minsk), the capital of Belarus, in the cen- tral part, on a tributary of the Berezina. 1,589,000.

Miq·ue·lon (mik′ə lon′; *Fr.* mēk° lôn′), Sᴛ. Pɪᴇʀʀᴇ ᴀɴᴅ Mɪǫᴜᴇʟᴏɴ.

Mi·rim (mi rim′), **Lake,** a lake on the E Uru- guay–S Brazil border. ab. 108 mi. (174 km) long. Spanish, **Merín.**

Mi·ri·na (mē′rē nä), the capital of the Greek is- land Lemnos.

Mis·kolc (mish′kôlts), a city in N Hungary. 210,000.

Misr (mis′rə), Arabic name of Eɢʏᴘᴛ.

Miss., Mississippi.

Mis′sionary Ridge′, a ridge in NW Georgia and SE Tennessee: Civil War battle 1863.

Mis′sion Vi·e′jo (vē ā′hō), a city in SW Califor- nia. 72,820.

Mis·sis·sau·ga (mis′ə sô′gə), a city in SE On- tario, in S Canada, on the SW shore of Lake On- tario: suburb of Toronto. 374,005.

Mis·sis·sip·pi (mis′ə sip′ē), **1.** a state in the S United States. 2,573,216; 47,716 sq. mi. (123,585 sq. km). *Cap.:* Jackson. *Abbr.:* MS, Miss. **2.** a river flowing S from N Minnesota to the Gulf of Mexico: the principal river of the U.S. 2470 mi.

(3975 km) long; from the headwaters of the Missouri to the Gulf of Mexico 3988 mi. (6418 km) long. —**Mis·sis·sip′pi·an**, *adj.*, *n.*

Mis·so·lon·ghi (mis′ə lông′gē), a town in W Greece, on the Gulf of Patras: Byron died here 1824. 10,164. Greek, **Mesolongion**.

Mis·sou·la (mi zōō′lə), a city in W Montana. 33,388.

Mis·sour·i (mi zŏŏr′ē, -zŏŏr′ə), **1.** a state in the central United States. 5,117,073; 69,674 sq. mi. (180,455 sq. km). *Cap.:* Jefferson City. *Abbr.:* MO, Mo. **2.** a river flowing from SW Montana into the Mississippi N of St. Louis, Mo. 2723 mi. (4382 km) long. —**Mis·sour′i·an**, *adj.*, *n.*

Mis·tas·si·ni (mis′tə sē′nē), a lake in E Canada, in Quebec province. 840 sq. mi. (2176 sq. km).

Mis·ti (mēs′tē), EL MISTI.

Mitch·ell (mich′əl), **Mount**, a mountain in W North Carolina: highest peak in the eastern U.S., 6684 ft. (2037 m).

Mit·i·li·ni (mit′l ē′nē, mēt′-), MYTILENE.

Mi·ya·za·ki (mē′yä zä′kē), a city on SE Kyushu, in Japan. 280,000.

Miz·o·ram (miz′ə ram′), a state in NE India. 493,757; 8140 sq. mi. (21,081 sq. km). *Cap.:* Aizawl.

Mma·ba·tho (mä bä′tō), the capital of Bophuthatswana, W of Johannesburg.

MN, Minnesota.

MO or **Mo.,** Missouri.

Mo·ab (mō′ab), an ancient kingdom E of the Dead Sea, in what is now Jordan. —**Mo′ab·ite′** (-ə bīt′), *n.*, *adj.*

Mo·bile (mō bēl′, mō′bēl), **1.** a seaport in SW Alabama at the mouth of the Mobile River. 196,278. **2.** a river in SW Alabama, formed by the confluence of the Alabama and Tombigbee rivers. 38 mi. (61 km) long.

Mo′bile Bay′, a bay of the Gulf of Mexico, in SW Alabama: Civil War naval battle 1864. 36 mi. (58 km) long.

Mo·çam·bi·que (*Port.* mōō′səm bē′kə), MOZAMBIQUE.

Mo·cha (mō′kə) also **Mukha,** a seaport in the Republic of Yemen, on the Red Sea. 25,000.

Mo·de·na (mōd′n ə), a city in N Italy, NW of Bologna. 176,556.

Mo·des·to (mə des′tō), a city in central California. 164,730.

Moe·si·a (mē′shē ə), an ancient country in S Europe, S of the Danube and N of ancient Thrace and Macedonia: later a Roman province.

Mo·ga·di·shu (mō′gə dē′shoo), the capital of Somalia, in the S part. 444,882. Italian, **Mo·ga·di·scio** (mō′gä dē′shô).

Mo·gi·lev (mō′gi lef′, -lôf′, -lof′), a city in E Belarus, on the Dnieper. 359,000.

Mo·gol·lon (mō′gə yōn′), **1.** an extensive plateau or mesa in central Arizona: the SW margin of the Colorado Plateau. **2.** a mountain range in W New Mexico.

Mo·hawk (mō′hôk), a river flowing E from central New York to the Hudson. 148 mi. (240 km) long.

Mo·hen·jo-Da·ro (mō hen′jō där′ō), an archaeological site in Pakistan, near the Indus River: six successive ancient cities were built here.

Mo·ja·ve (or **Mo·ha·ve**) **Des′ert** (mō hä′vē), a desert in S California: part of the Great Basin. ab. 15,000 sq. mi. (38,850 sq. km).

Mok·po (môk′pō), a seaport in SW South Korea. 221,816.

Mol·dau (môl′dou, mōl′-), German name of the **Vltava**.

Mol·da·vi·a (mol dā′vē ə, -vyə), **1.** a region in NE Romania: formerly a principality that united with Wallachia to form Romania. *Cap.:* Jassy. **2.** former name of **Moldova**. —**Mol·da′vi·an,** *adj.,* *n.*

Mol·do·va (môl dō′və), a republic in SE Europe. 4,341,000; 13,100 sq. mi. (33,929 sq. km). *Cap.:* Kishinev. Formerly, **Moldavia, Molda′vian So′viet So′cialist Repub′lic.** —**Mol·do′van,** *adj., n.*

Mo·li·se (mə lē′zā), a region in S central Italy. 334,680; 1713 sq. mi. (4438 sq. km).

Mo·lo·ka·i (mō′lə ki′, -kä′ē, mol′ə-), an island in central Hawaii. 6700; 264 sq. mi. (684 sq. km).

Mo·lo·po (mə lō′pō), a river in S Africa, flowing SW along the S Botswana-N South Africa border to the Orange River. ab. 600 mi. (965 km) long.

Mo·lo·tov (mol′ə tôf′, -tof′, mō′lə-), former name of **Perm**.

Mo·luc·cas (mə luk′əz), a group of islands in Indonesia, between Sulawesi and New Guinea.

1,411,006; ab. 30,000 sq. mi. (78,000 sq. km). Also called **Moluc′ca Is′lands.** Formerly, **Spice Islands. —Mo·luc′can,** *adj., n.*

Mom·ba·sa (mom bä′sä, -bas′ə), **1.** an island in S Kenya. **2.** a seaport on this island. 341,148.

Mon·a·co (mon′ə kō′, mə nä′kō), **1.** a principality on the Mediterranean coast, bordering SE France. 27,063; ½ sq. mi. (1.3 sq. km). **2.** the capital of this principality. 1234. **—Mon′a·can** (-kən), *n., adj.* **—Mon′e·gasque′** (-i gask′), *n., adj.*

Mo·nad·nock (mə nad′nok), **Mount,** a mountain peak in SW New Hampshire. 3186 ft. (971 m).

Mon·a·ghan (mon′ə gən, -han′), a county in the NE Republic of Ireland. 51,174; 498 sq. mi. (1290 sq. km).

Mo′na Pas′sage (mō′nə), a strait between Hispaniola and Puerto Rico. 80 mi. (129 km) wide.

Mo·na·stir (mô′nä stēr′), Turkish name of BITOLA.

Mön·chen·glad·bach (mœn′kнən glät′bäkн), a city in W North Rhine-Westphalia, in W Germany. 249,600. Formerly, **München-Gladbach.**

Monc·ton (mungk′tən), a city in SE New Brunswick, in E Canada. 55,468.

Mon·go·li·a (mong gō′lē ə, mon-), **1.** a region in Asia including Inner Mongolia in China and the Mongolian People's Republic. **2.** MONGOLIAN PEOPLE'S REPUBLIC. **—Mon·go′li·an,** *n., adj.*

Mongo′lian Peo′ple's Repub′lic, a republic in E central Asia, in N Mongolia. 2,001,000; ab. 600,000 sq. mi. (1,500,000 sq. km). *Cap.:* Ulan Bator. Formerly, **Outer Mongolia.** Also called **Mongolia.**

Mon·mouth·shire (mon′məth shēr′, -shər), a historic county in E Wales, now part of Gwent, Mid Glamorgan, and South Glamorgan. Also called **Mon′mouth.**

Mo·non·ga·he·la (mə nong′gə hē′lə), a river flowing from N West Virginia through SW Pennsylvania into the Ohio River. 128 mi. (205 km) long.

Mon·roe (mən rō′), a city in N Louisiana. 54,909.

Mon·ro·vi·a (mən rō′vē ə), the capital of Liberia, in W Africa. 425,000.

Mons (môɴs), a city in SW Belgium. 89,515.

Mont., Montana.

Mon·tan·a (mon tan′ə), a state in the NW United

States. 799,065; 147,138 sq. mi. (381,085 sq. km). *Cap.:* Helena. *Abbr.:* MT, Mont. —**Mon·tan'an,** *adj., n.*

Mon'tauk Point' (mon'tôk), the SE end of Long Island, in SE New York.

Mont Blanc (môn bläN'), a mountain in SE France, near the Italian border: highest peak of the Alps, 15,781 ft. (4810 m).

Mont Cer·vin (môn ser vaN'), French name of the **MATTERHORN.**

Mon·te Al·bán (môn'te äl bän'), a major ceremonial center of the Zapotec culture, near the city of Oaxaca, Mexico, occupied 600 B.C.–A.D. 700.

Mon·te·bel·lo (mon'tə bel'ō), a city in SW California, SE of Los Angeles. 59,564.

Mon·te Car·lo (mon'tē kär'lō, -ti), a town in Monaco principality, in SE France: gambling resort. 13,154.

Mon·te'go Bay' (mon tē'gō), a city in NW Jamaica: seaside resort. 70,265.

Mon·te·ne·gro (mon'tə nē'grō, -neg'rō), a constituent republic of Yugoslavia, in the SW central part. 620,000; 5333 sq. mi. (13,812 sq. km). *Cap.:* Podgorica. —**Mon'te·ne'grin** (-nē'grin, -neg'rin), *adj., n.*

Mon·te·rey (mon'tə rā'), a city in W California, on Monterey Bay: the capital of California until 1847. 31,954.

Mon'terey Bay', an inlet of the Pacific in W California. 26 mi. (42 km) long.

Mon'terey Park', a city in SW California, E of Los Angeles. 60,738.

Mon·ter·rey (mon'tə rā'), the capital of Nuevo León, in NE Mexico. 1,916,472.

Mon·te·vi·de·o (mon'tə vi dā'ō), the capital of Uruguay. 1,309,100.

Mont·gom·er·y (mont gum'ə rē, -gum'rē), **1.** the capital of Alabama, in the central part, on the Alabama River. 187,106. **2.** MONTGOMERYSHIRE.

Mont·gom·er·y·shire (mont gum'ə rē shēr', -shər, -gum'rē-), a historic county in Powys, in central Wales. Also called **Montgomery.**

Mon·ti·cel·lo (mon'ti chel'ō, -sel'ō), the estate and residence of Thomas Jefferson, in central Virginia, near Charlottesville.

Mont·mar·tre (môn mar'trª), a hilly section in

the N part of Paris, France: noted for the artists who have frequented and lived in the area.

Mont·pel·ier (mont pēl′yər), the capital of Vermont, in the central part. 8247.

Mont·pel·lier (môn pe lyā′), a city in S France. 221,307.

Mont·re·al (mon′trē ôl′, mun′-), a port in S Quebec, in E Canada, on an island (**Mon′treal Is′land**) in the St. Lawrence. 1,015,420. French, **Mont·ré·al** (môn Rā Al′). —**Mont′re·al′er**, n.

Mon′treal North′, a city in S Quebec, in E Canada, N of Montreal. 94,914. French, **Mont·ré·al–Nord** (môn Rā Al nôR′).

Mon·treuil (môn trœ′yə), a suburb of Paris, in N France. 93,394.

Mont-Saint-Mi·chel or **Mont Saint Mi·chel** (môn saN mē shel′), a rocky islet near the coast of NW France, in an inlet of the Gulf of St. Malo: famous abbey and fortress.

Mont·ser·rat (mont′sə rat′), an island in the Leeward Islands, in the SE West Indies: a British crown colony. 11,852; 39 sq.mi. (102 sq. km).

Mon·za (mon′zə), a city in N Italy, NNE of Milan. 122,103.

Moose′head Lake′ (mōōs′hed′), a lake in central Maine. 42 mi. (68 km) long; 300 sq. mi. (780 sq. km).

Moose′ Jaw′, a city in S Saskatchewan, in SW Canada. 32,581.

Mo·ra·da·bad (môr′ə də bad′, mōr′-, mə rä′dəbäd′), a city in N Uttar Pradesh, in N India. 348,000.

Mo·ra·tu·wa (mô rä′tŏō wə), a city in W Sri Lanka. 134,826.

Mo·ra·va (môr′ə və; Czech. mô′Rä vä), **1.** a river flowing S from N Czechoslovakia to the Danube. 240 mi. (385 km) long. **2.** a river in E Yugoslavia, flowing N to the Danube. 134 mi. (216 km) long. **3.** Czech name of Moravia.

Mo·ra·vi·a (mô rä′vē ə, -rä′-, mō-), a region in central Czechoslovakia, between Bohemia and Slovakia. Czech, **Morava**. —**Mo·ra′vi·an**, adj., n.

Mora′vian Gate′, a mountain pass between the E Sudeten and W Carpathian mountains, in N central Czechoslovakia.

Mo·rav·ská Os·tra·va (mô′Räf skä ôs′tRä vä), former name of Ostrava.

Mor·ay (mûr′ē), a historic county in NE Scotland, on Moray Firth.

Mor·ay Firth′ (fûrth), an arm of the North Sea projecting into the NE coast of Scotland. Inland portion ab. 30 mi. (48 km) long.

Mor·do·vi·an (or **Mord·vin/i·an**) **Auton/-omous Repub/lic** (môr dō′vē ən or môrd vin′-ē ən), an autonomous republic in the Russian Federation in Europe. 964,000; 9843 sq. mi. (25,493 sq. km). *Cap.*: Saransk.

Mo·re·a (mô rē′ə, mō-), PELOPONNESUS.

Mo·re·lia (mô Re′lyä), the capital of Michoacán, in central Mexico. 353,055.

Mo·re·los (mô Re′lôs), a state in S central Mexico. 947,089; 1916 sq. mi. (4960 sq. km). *Cap.*: Cuernavaca.

Mo·re/no Val/ley (mə rē′nō), a city in SW California, E of Riverside. 118,779.

Mor·gan·town (môr′gən toun′), a city in N West Virginia. 25,879.

Mo·roc·co (mə rok′ō), a kingdom in NW Africa: formed from a sultanate that was divided into two protectorates (**French Morocco** and **Spanish Morocco**) and an international zone. 23,000,000; 172,104 sq. mi. (445,749 sq. km). *Cap.*: Rabat. French, **Maroc**. Compare TANGIER ZONE. —**Mo·roc/can**, *adj.*, *n.*

Mo·ro·ni (mô rō′nē), the capital of the Comoros. 20,112.

Mor·ris·town (môr′is toun′, mor′-), a city in N New Jersey: Washington's winter headquarters 1776–77, 1779–80. 16,189.

Mos·cow (mos′kō, -kou), the capital of the Russian Federation: capital of the former Soviet Union. 8,967,000. Russian, **Mo·skva** (mu skvä′). —**Mus·co·vite** (mus′kə vit′), *n.*, *adj.*

Mo·selle (mō zel′), a river in W central Europe, flowing from the Vosges Mountains in NE France into the Rhine at Coblenz, in W Germany. 320 mi. (515 km) long. German, **Mo·sel** (mō′zəl).

Mosqui/to Coast/, a coastal region in Central America bordering on the Caribbean Sea in E Honduras and Nicaragua.

Mo·sul (mō sōōl′), a city in N Iraq, on the Tigris, opposite the ruins of Nineveh. 570,926.

Mo·town (mō′toun′), Detroit, Michigan: a nickname.

Moul·mein (mōōl mān′, mōl-) a seaport in S Burma at the mouth of the Salween River. 219,991.

Moun′tain States′, Arizona, Colorado, Idaho, Montana, Nevada, New Mexico, Utah, and Wyoming.

Moun′tain View′, a city in central California, S of San Francisco. 67,460.

Mount′ Des′ert Is′land (dez′ərt, di zûrt′), an island off the coast of E central Maine: forms part of Acadia National Park. 14 mi. (23 km) long; 8 mi. (13 km) wide.

Mount′ McKin′ley Na′tional Park′, former name of DENALI NATIONAL PARK.

Mount′ Pros′pect, a city in NE Illinois, near Chicago. 53,170.

Mount′ Rainier′ Na′tional Park′, a national park in W Washington, including Mount Rainier. 378 sq. mi. (980 sq. km).

Mount′ Rush′more Na′tional Memo′rial, See under RUSHMORE.

Mount′ Ver′non, 1. the home and tomb of George Washington in NE Virginia, on the Potomac, near Washington, D.C. **2.** a city in SE New York, near New York City. 67,153.

Mo·zam·bique (mō′zam bēk′, -zəm-), a republic in SE Africa: formerly an overseas province of Portugal; gained independence in 1975. 14,900,000; 297,731 sq. mi. (771,123 sq. km). *Cap.:* Maputo. Portuguese, **Moçambique.** Formerly, **Portuguese East Africa.** —**Mo′zam·bi′-can,** *n., adj.*

Mo′zambique Chan′nel, a channel in SE Africa, between Mozambique and Madagascar. 950 mi. (1530 km) long; 250–550 mi. (400–885 km) wide.

MS, Mississippi.

MT, Montana.

Mu·kal·la (mōō kal′ə), a seaport in SE Yemen, on the Gulf of Aden. 158,000.

Muk·den (mōōk′den′, mōōk′-), a former name of SHENYANG.

Mu·kha (mōō kä′), MOCHA.

Mul·ha·cén (mōō′lä then′, -sen′), a mountain in S Spain: the highest peak in Spain. 11,411 ft. (3478 m).

Mül·heim an der Ruhr (mүl′him än deR RōōR′),

a city in North Rhine-Westphalia, W Germany, near Essen. 176,100.

Mul·house (mʏ lōōz′), a city in E France, near the Rhine. 113,794. German, **Mül·hau·sen** (mʏl hou′zən).

Mull (mul), an island in the Hebrides, in W Scotland. ab. 351 sq. mi. (910 sq. km).

Mul·tan (mŏŏl tän′), a city in E central Pakistan. 742,000.

Mün′chen-Glad′bach (mʏn′kʜən glät′bäkʜ), former name of Mönchengladbach.

Mun·cie (mun′sē), a city in E Indiana. 71,035.

Mu·nich (myōō′nik), the capital of Bavaria, in SW Germany. 1,188,800. German, **Mün′chen.**

Mun·ster (mun′stər), a province in SW Republic of Ireland. 1,019,694; 9316 sq. mi. (24,130 sq. km).

Mün·ster (mʏn′stər), a city in NW Germany. 246,300.

Mu·ra·no (mŏŏ rä′nō), an island suburb of Venice, noted for Venetian glass manufacture.

Mu·rat (mŏŏ rät′), a river in E Turkey, flowing W to the Euphrates. 425 mi. (685 km) long. Also called **Mu·rad Su** (mŏŏ räd′ sōō′).

Mur·cia (mŏŏr′shə), **1.** a city in SE Spain. 309,504. **2.** a region in SE Spain: formerly a kingdom.

Mu·reş (mŏŏr′esh), a river in SE central Europe, flowing W from the Carpathian Mountains in central Romania to the Tisza River in S Hungary. 400 mi. (645 km) long. Hungarian, **Maros.**

Mur·frees·bor·o (mûr′frēz bûr′ō, -bur′ō), a city in central Tennessee: Civil War battle 1862. 44,922.

Mur·mansk (mŏŏr mänsk′), a seaport and railroad terminus in the NW Russian Federation, on the Kola Inlet, an arm of the Arctic Ocean. 432,000.

Mu·rom (mŏŏr′əm), a city in the W Russian Federation in Europe, SW of Nizhni Novgorod. 120,000.

Mur·ray (mûr′ē, mur′ē), a river in SE Australia, flowing W along the border between Victoria and New South Wales, through SE South Australia into the Indian Ocean. 1200 mi. (1930 km) long.

Mur·rum·bidg·ee (mûr′əm bij′ē), a river in SE

Australia, flowing W through New South Wales to the Murray River. 1050 mi. (1690 km) long.

Mus·cat or **Mas·qat** (mus kat′), the capital of Oman. 250,000.

Muscat′ and Oman′, former name of OMAN.

Mus′cle Shoals′, former rapids of the Tennessee River in SW Alabama, changed into a lake by Wilson Dam: part of the Tennessee Valley Authority.

Mus·co·vy (mus′kə vē), **1.** Also called **Grand Duchy of Muscovy.** a principality founded c1271 and centered on Moscow: gained control over the neighboring Great Russian principalities and established the Russian Empire under the czars. **2.** *Archaic.* Moscow. **3.** *Archaic.* RUSSIA.

Mu·shin (mōō′shin), a city in SW Nigeria, NW of Lagos. 240,700.

Mut·tra (mu′trə), former name of MATHURA.

Mwe·ru (mwä′rōō), a lake in S central Africa, between Zaire and Zambia. 68 mi. (109 km) long.

My·an·mar (mi än′mär), **Union of,** official name of BURMA.

My·ce·nae (mi sē′nē), an ancient city in S Greece, in Argolis: important ruins. —**My′ce·nae′an, My′ce·ne′an** (-sə nē′ən) *adj., n.*

My·ko·nos or **Mi·ko·nos** (mē′kə nōs′), a mountainous island in SE Greece, in the S Aegean: resort. 3823; 35 sq. mi. (90 sq. km).

My·ra (mi′rə), an ancient city in SW Asia Minor, in Lycia.

My·si·a (mish′ē ə), an ancient country in NW Asia Minor. —**My′si·an,** *adj., n.*

My·sore (mi sôr′, -sōr′), **1.** a city in S Karnataka, in S India. 476,000. **2.** former name of KARNATAKA.

Mys·tic (mis′tik), a historic seaport in SE Connecticut: maritime museum. 2618.

Myt·i·le·ne or **Mit·i·li·ni** (mit/l ē′nē, mēt′-), **1.** the capital of the Greek island Lesbos. 24,115. **2.** LESBOS.

My·ti·shchi (mi tē′shē), a city in the W Russian Federation in Europe, NE of Moscow. 150,000.

N

Na·be·re·zhny·e Chel·ny (nä′bə rezh′nē ə chel nē′), a port in the Tatar Autonomous Republic, in the Russian Federation in Asia, E of Kazan, on the Kama River. 501,000.

Nab·lus (nab′ləs, nä′bləs), a city in Samaria, formerly in W Jordan, occupied by Israel since 1967: near site of ancient town of Shechem. 50,000. Hebrew, **Shechem**.

Na·fud (na fōōd′), NEFUD DESERT.

Na·ga (nä′gä), a city on E Cebú, in the S central Philippines. 90,712.

Na·ga·land (nä′gə land′), a state in NE India. 773,281; 6366 sq. mi. (16,488 sq. km). *Cap.:* Kohima.

Na·ga·no (nä gä′nō), a city on central Honshu, in central Japan. 339,000.

Na·ga·sa·ki (nä′gə sä′kē, nag′ə sak′ē), a seaport on W Kyushu, in SW Japan: second military use of the atomic bomb August 9, 1945. 447,000.

Na·gor′no-Ka·ra·bakh′ Auton′omous Region (nə gôr′nō kär′ə bäk′), an autonomous region in SW Azerbaijan. 188,000; 1700 sq. mi. (4400 sq. km). *Cap.:* Stepanakert.

Na·go·ya (nə goi′ə), a city on S Honshu, in central Japan. 2,091,884.

Nag·pur (näg′pŏŏr), a city in NE Maharashtra, in central India. 1,298,000.

Nagy·vá·rad (nod/y° vä′rod, noj′-), Hungarian name of ORADEA.

Na·ha (nä′hä), a port on SW Okinawa, in S Japan. 295,801.

Nai·ro·bi (ni rō′bē), the capital of Kenya, in the SW part. 827,775.

Na·jaf (naj′af) also **An-Najaf**, a city in central Iraq: holy city of the Shi'ites; shrine contains tomb of Ali (A.D. c600–661). 472,103.

Na·khi·che·van′ Auton′omous Repub′lic (nə кнē′che vän′), an autonomous republic of Azerbaijan, surrounded by Armenia, Iran, and Turkey. 295,000; 2277 sq. mi. (5500 sq. km). *Cap.:* Nakhichevan.

Na·khod·ka (nə кнôt′kə), a port in the SE Russian Federation in Asia, SE of Vladivostok, on the Sea of Japan. 148,000.

Nal·chik (näl′chik), the capital of the Kabardino-Balkar Autonomous Republic in the S Russian Federation in Europe. 235,000.

Na·man·gan (nä′mən gän′), a city in E Uzbekistan, NW of Andizhan. 291,000.

Na·ma·qua·land (nə mä′kwə land′) also **Na·ma·land′** (nä′mə-), a region in the S part of Namibia, extending into the Republic of South Africa.

Na·mib Des·ert (nä′mib), a desert region in SW Africa, along the entire coast of Namibia.

Na·mib·i·a (nə mib′ē ə), a republic in SW Africa: a former German protectorate; a mandate of South Africa (1919–66); gained independence 1990. 1,400,000; 318,261 sq. mi. (824,296 sq. km). *Cap.:* Windhoek. Formerly, **German Southwest Africa** (1884–1919), **South West Africa** (1920–68). —**Na·mib′i·an,** *adj., n.*

Na·mur (nä mŏŏr′; *Fr.* nA myR′), **1.** a province in S Belgium. 415,326; 1413 sq. mi. (3660 sq. km). **2.** the capital of this province. 103,104.

Na·nai·mo (nə ni′mō), a port in SW British Columbia, in SW Canada, on the SE part of Vancouver Island. 50,890.

Nan·chang (nän′chäng′), the capital of Jiangxi province, in SE China. 1,120,000.

Nan·chong (nän′chông′) also **Nan·chung** (-chŏŏng′), a city in E central Sichuan province, in central China. 220,500.

Nan·cy (nan′sē; *Fr.* näN sē′), a city in NE France. 306,982.

Nan·da De·vi (nun′dä dā′vē), a mountain in N India, in Uttar Pradesh: a peak of the Himalayas. 25,645 ft. (7817 m).

Nan·di (nän′dē), a town on W Viti Levu, in the Fiji Islands.

Nan·ga Par·bat (nung′gə pur′but), a mountain in NW Kashmir, in the Himalayas. 26,660 ft. (8125 m).

Nan·jing (nän′jing′, nan′-) also **Nan·king′** (-king′), the capital of Jiangsu province, in E China: a former capital of China. 2,250,000.

Nan Ling (nän′ ling′), a mountain range in S China.

Nan·ning (nän′ning′, nan′-), the capital of Guangxi Zhuang region, in S China. 862,732.

Nan Shan (nän′ shän′), former name of QILIAN SHAN.

Nan·terre (nän teR′), a city in N France: W suburb of Paris. 90,371.

Nantes (nants; *Fr.* nänt), a seaport in W France, on the Loire River. 263,689.

Nan·tong (nän′tông′) also **Nan·tung** (-tŏŏng′), a city in SE Jiangsu province, in E China, on the Chang Jiang. 389,988.

Nan·tuck·et (nan tuk′it), an island off SE Massachusetts: summer resort. 6012; 15 mi. (24 km) long.

Nap·a (nap′ə), a city in W California: center of wine-producing region. 61,842.

Na·per·ville (nā′pər vil′), a city in NE Illinois. 85,351.

Na·ples (nā′pəlz), **1.** Italian, **Na·po·li** (nä′pô lē), a seaport in SW Italy, on the Bay of Naples. 1,200,958. **2. Bay of,** an inlet of the Tyrrhenian Sea. 22 mi. (35 km) long. —**Ne·a·pol·i·tan** (nē′ə·pol′i tn), *adj., n.*

Na·po (nä′pō), a South American river flowing from central Ecuador through NE Peru to the Amazon River. ab. 700 mi. (1125 km) long.

Na·ra (när′ə), a city on S Honshu, in central Japan: chief Buddhist center of ancient Japan; first capital of Japan A.D. 710–84. 334,000.

Nar·ba·da (nər bud′ə), a river flowing W from central India to the Arabian Sea. 800 mi. (1290 km) long.

Nar·bonne (naR bôn′), a city in S France. 41,565.

Na·rew (nä′Ref), a river in NE Poland, flowing S and SW into the Bug River: battle 1915. 290 mi. (465 km) long.

Nar·ra·gan′sett Bay′ (nar′ə gan′sit, nar′-), an inlet of the Atlantic in E Rhode Island. 28 mi. (45 km) long.

Nar·rows (nar′ōz), **the,** a strait between Staten Island and Long Island in New York Bay. 2 mi. (3.2 km) long.

Nar·vik (när′vik), a seaport in N Norway. 19,308.

Nase·by (nāz′bē), a village in W Northampton-shire, in central England: Royalist defeat 1645.

Nash·u·a (nash′o͞o ə), a city in S New Hampshire, on the Merrimack River. 79,662.

Nash·ville (nash′vil), the capital of Tennessee, in the central part. 488,374.

Nas·sau (nas′ô; *for 2 also* nä′sou), **1.** a seaport on New Providence Island: capital of the Bahamas. 132,000. **2.** a former duchy in central Germany: now a part of Hesse.

Nas·ser (nä′sər, nas′ər), **Lake,** a reservoir in SE Egypt, formed in the Nile River S of the Aswan High Dam. ab. 300 mi. (500 km) long; 6 mi. (10 km) wide.

Na·tal (nə tal′, -täl′; *for 2 also* nə tôl′), **1.** a province in the E part of the Republic of South Africa. 2,145,018; 35,284 sq. mi. (91,886 sq. km). *Cap.:* Pietermaritzburg. **2.** the capital of Rio Grande do Norte, in NE Brazil. 376,446. —**Na·tal′i·an,** *adj., n.*

Natch·ez (nach′iz), a port in SW Mississippi, on the Mississippi River. 19,460.

Na′tional Cit′y, a city in SW California, near San Diego. 54,249.

Na′tive States′, Indian States and Agencies.

Nat′ural Bridg′es, a national monument in SE Utah containing three natural bridges. Largest, 222 ft. (68 m) high; 261 ft. (80 m) span.

Nau·cra·tis (nô′krə tis), an ancient Greek city in N Egypt, on the Nile delta.

Na·u·ru (nä o͞o′ro͞o), **Republic of,** an island republic in the Pacific, near the equator, W of Kiribati: a UN trusteeship until 1968. 8042; 8 sq. mi. (21 sq. km). Formerly, **Pleasant Island.** —**Na·u′ru·an,** *n., adj.*

Na·varre (nə vär′), a former kingdom in SW France and N Spain. Spanish, **Na·var·ra** (nä vär′rä). —**Na·var′ri·an,** *adj.*

Nav·pak·tos (näf′päk tôs), Greek name of Lepanto.

Nax·os (nak′sos, näk′sōs), a Greek island in the S Aegean: the largest of the Cyclades group. 14,201; 169 sq. mi. (438 sq. km).

Na·ya·rit (nä′yä rēt′), a state in W Mexico. 846,278; 10,442 sq. mi. (27,045 sq. km). *Cap.:* Tepic.

Naz·a·reth (naz′ər əth), a town in N Israel: the childhood home of Jesus. 45,600. —**Naz·a·rene** (naz′ə rēn′, naz′ə rēn′), *n., adj.*

NC or **N.C.,** North Carolina.

ND or **N.D.,** North Dakota.

N.Dak., North Dakota.

N'Dja•me•na (ən jä mä′nä), the capital of Chad, in the SW part. 511,700. Formerly, **Fort-Lamy.**

Ndo•la (ən dō′lə), a city in N Zambia. 418,142.

NE, Nebraska.

Neagh (nā), **Lough,** a lake in E central Northern Ireland: largest freshwater lake in the British Isles. ab. 18 mi. (29 km) long and 11 mi. (18 km) wide.

Near′ East′, an indefinite geographical region usu. considered to encompass the countries of SW Asia, including Turkey, Lebanon, Syria, Iraq, Israel, Jordan, Saudi Arabia, and the other nations of the Arabian Peninsula. Compare MIDDLE EAST (def. 1). —**Near′ East′ern,** adj.

Neb. or **Nebr.,** Nebraska.

Ne•bo (nē′bō), **Mount.** See under PISGAH.

Ne•bras•ka (nə bras′kə), a state in the central United States. 1,578,385; 77,237 sq. mi. (200,044 sq. km). Cap.: Lincoln. Abbr.: NE, Nebr., Neb. —**Ne•bras′kan,** n., adj.

Neck•ar (nek′ər, -är), a river in SW Germany, flowing from the Black Forest, N and W to the Rhine River. 246 mi. (395 km) long.

Ne•der•land (nā′dər länt′), Dutch name of the NETHERLANDS.

Ne•fud′ Des′ert (nə fōōd′), a desert in N Saudi Arabia. ab. 50,000 sq. mi. (129,500 sq. km). Also called **An Nafud, Nafud.**

Neg•ev (neg′ev) also **Neg′eb** (-eb), a partly desert region in S Israel, bordering on the Sinai Peninsula. 4700 sq. mi. (12,173 sq. km).

Ne•gri Sem•bi•lan (nā′grē sem bē′län, sem′bē-län′, nə grē′), a state in Malaysia, on the SW Malay Peninsula. 573,578; 2580 sq. mi. (6682 sq. km).

Ne•gro (nā′grō; Sp. ne′grô; Port. ne′grŏŏ), **1.** a river in NW South America, flowing SE from Colombia into the Amazon. 1400 mi. (2255 km) long. **2.** a river in S Argentina, flowing E from the Andes to the Atlantic. 700 mi. (1125 km) long. **3.** a river in SE South America, flowing SW from Brazil into the Uruguay River. ab. 500 mi. (800 km) long. Portuguese, **Rio Negro.** Spanish, **Río Negro.**

Ne·gros (nā′grōs), an island of the central Philippines. 4906 sq. mi. (12,706 sq. km).

Neis·se (nī′sə), a river in N Europe, flowing N from NW Czechoslovakia into the Oder River. 145 mi. (233 km) long.

Nejd (nejd, nād), a region of Saudi Arabia in the E central part: formerly a sultanate. ab. 400,000 sq. mi. (1,000,000 sq. km).

Nel·son (nel′sən), **1.** a river in central Canada, flowing NE from Lake Winnipeg to Hudson Bay. 400 mi. (645 km) long. **2.** a seaport on N South Island, in New Zealand. 45,200.

Ne·man (nem′ən, nyem′-, nē′mən), a river rising in central Belarus, flowing W through Lithuania into the Baltic. 582 mi. (937 km) long. Lithuanian, **Ne·mu·nas** (nye′mŏŏ näs′).

Ne·me·a (nē′mē ə), a valley in SE Greece, in ancient Argolis. —**Ne·me·an** (ni mē′ən, nē′mē-), adj.

Ne·pal (nə pôl′, -päl′, -pal′, nā-), a constitutional monarchy in the Himalayas between N India and Tibet. 16,630,000; ab. 54,000 sq. mi. (140,000 sq. km). Cap.: Katmandu. —**Nep·a·lese** (nep′ə-lēz′, -lēs′), n., pl. **-lese**, adj.

Ness (nes), **Loch,** a lake in NW Scotland, near Inverness. 23 mi. (37 km) long.

Neth·er·lands (neth′ər ləndz), **the,** a kingdom in W Europe, on the North Sea. 14,715,000; 16,163 sq. mi. (41,863 sq. km). Capitals: Amsterdam and The Hague. Also called **Holland.** Dutch, **Nederland.** —**Neth′er·land·er** (-lan′dər, -lən-), n. —**Neth′er·land′i·an, Neth′er·land′ic,** adj.

Neth′erlands Antil′les, a Netherlands overseas territory in the Caribbean Sea, N and NE of Venezuela: includes the islands of Bonaire, Curaçao, Saba, and St. Eustatius, and the S part of St. Martin. 188,501; 308 sq. mi. (800 sq. km). Cap.: Willemstad. Formerly, **Curaçao.** —**Neth′erlands Antil′lean,** adj., n.

Neth′erlands East′ In′dies, a former name of the Republic of INDONESIA.

Neth′erlands Guian′a, a former name of SURINAME.

Neth′erlands New′ Guin′ea, a former name of IRIAN JAYA.

Né·thou (nā tŏŏ′), **Pic de** (pēk də), French name of Pico de ANETO.

Neu·châ·tel (nōō′shə tel′, nyōō′-, nœ′-), **1.** a canton in W Switzerland. 157,400; 309 sq. mi. (800 sq. km). **2.** the capital of this canton, on the Lake of Neuchâtel. 32,670. **3. Lake of,** a lake in W Switzerland. 85 sq. mi. (220 sq. km). German, **Neu·en·burg** (noi′ən bŏŏrk′).

Neuss (nois), a city in W Germany, near Düsseldorf. 142,200.

Neus·tri·a (nōō′strē ə, nyōō′-), the W part of the Frankish kingdom, corresponding roughly to N and NW France. **—Neus′tri·an,** adj.

Nev., Nevada.

Ne·va (nē′və; Russ. nyi vä′), a river in the NW Russian Federation in Europe, flowing through St. Petersburg into the Gulf of Finland. 40 mi. (65 km) long.

Ne·vad·a (nə vad′ə, -vä′də), a state in the W United States. 1,201,833; 110,540 sq. mi. (286,300 sq. km). Cap.: Carson City. Abbr.: NV, Nev. **—Ne·vad′an, Ne·vad′i·an,** adj., n.

Ne·va·do del Ruiz (ne vä′thô ṯhel Rwēs′), a volcano in W central Colombia, in the Andes. 17,720 ft. (5401 m).

Ne·vis (nē′vis, nev′is), one of the Leeward Islands, in the E West Indies: part of St. Kitts-Nevis; formerly a British colony. 9,428; 50 sq. mi. (130 sq. km). Compare ST. KITTS-NEVIS-ANGUILLA.

New′ Am′ster·dam (am′stər dam′), a former Dutch town on Manhattan Island: renamed New York by the British in 1664.

New·ark (nōō′ərk, nyōō′- for 1; -ärk′ for 2), **1.** a city in NE New Jersey. 275,221. **2.** a city in N Delaware. 25,098.

New′ark Bay′, a bay in NE New Jersey. 6 mi. (10 km) long; 1 mi. (1.6 km) wide.

New′ Bed′ford (bed′fərd), a seaport in SE Massachusetts. 99,922.

New′ Brit′ain, 1. the largest island in the Bismarck Archipelago, Papua New Guinea, in the W central Pacific Ocean. 268,400. ab. 14,600 sq. mi. (37,814 sq. km). Cap.: Rabaul. **2.** a city in central Connecticut. 75,491.

New′ Bruns′wick, a province in SE Canada. 709,442; 27,985 sq. mi. (72,480 sq. km). Cap.: Fredericton.

New′ Cale·do′nia, 1. an island in the S Pacific, ab. 800 mi. (1290 km) E of Australia. 127,885;

6224 sq. mi. (16,120 sq. km). **2.** an overseas territory of France comprising this island and other smaller islands: formerly a penal colony. 145,368; 7200 sq. mi. (18,650 sq. km). *Cap.*: Nouméa.

New′ Castile′, a region in central Spain: formerly a province. 27,933 sq. mi. (72,346 sq. km). Spanish, **Castilla la Nueva.**

New·cas·tle (nōō′kas/əl, -kä′səl, nyōō′-), **1.** Also called **New′cas·tle-up·on′-Tyne′.** a seaport in NE England, on the Tyne River: coal center. 282,700. **2.** a seaport in E New South Wales, in SE Australia. 429,300.

New′ Del′hi (del′ē), the capital of India, in the N part, adjacent to Delhi. 271,990. Compare DELHI (def. 2).

New′ Eng′land, an area in the NE United States including the states of Connecticut, Maine, Massachusetts, New Hampshire, Rhode Island, and Vermont. —**New′ Eng′land·er,** *n.*

New′ For′est, a forest region in S England, in Hampshire. 145 sq. mi. (376 sq. km).

New·found·land (nōō′fən lənd, -land′, -fənd-, nyōō′-; nōō found′lənd, nyōō-), **1.** a large island in E Canada. 42,734 sq. mi. (110,680 sq. km). **2.** a province in E Canada, composed of Newfoundland island and Labrador. 568,349; 155,364 sq. mi. (402,390 sq. km). *Cap.*: St. John's. —**New′found·land·er,** *n.*

New′ Geor′gia, 1. a group of islands in the Solomon Islands, in the SW Pacific. **2.** the chief island of this group. 50 mi. (80 km) long; 20 mi. (32 km) wide.

New′ Guin′ea, 1. a large island N of Australia, politically divided into the Indonesian province of Irian Jaya (West Irian) and the independent country of Papua New Guinea. ab. 316,000 sq. mi. (818,000 sq. km). **2. Trust Territory of,** a former United Nations trust territory that included NE New Guinea, the Bismarck Archipelago, Bougainville, and other islands, administered by Australia jointly with the Territory of Papua until 1975: now part of Papua New Guinea. —**New′ Guin′ean,** *n., adj.*

New·ham (nōō′əm, nyōō′-), a borough of Greater London, England. 206,500.

New′ Hamp′shire (hamp′shər, -shēr) a state in the NE United States. 1,109,252; 9304 sq. mi.

(24,100 sq. km). *Cap.:* Concord. *Abbr.:* NH, N.H.
—**New/ Hamp/shir·ite/,** *n.*

New/ Ha/ven, a seaport in S Connecticut, on Long Island Sound. 130,474.

New/ Heb/rides, former name of VANUATU.

New/ Ire/land, an island in the Bismarck Archipelago, in the W central Pacific Ocean NE of New Guinea: part of Papua New Guinea. 78,900; ab. 3800 sq. mi. (9800 sq. km).

New/ Jer/sey, a state in the E United States, on the Atlantic coast. 7,730,188; 7836 sq. mi. (20,295 sq. km). *Cap.:* Trenton. *Abbr.:* NJ, N.J.
—**New/ Jer/sey·an, New/ Jer/sey·ite/,** *n.*

New·mar·ket (noō/mär/kit, nyoō/-), a town in W Suffolk, in E England: horse races. 12,934.

New/ Mex/ico, a state in the SW United States. 1,515,069; 121,666 sq. mi. (315,115 sq. km). *Cap.:* Santa Fe. *Abbr.:* NM, N. Mex., N.M. —**New/ Mex/ican,** *n., adj.*

New/ Or/le·ans (ôr/lē ənz, ôr lēnz/, ôr/lənz), a seaport in SE Louisiana, on the Mississippi. 496,938. —**New/ Or·lea/ni·an,** *n., adj.*

New·port (noō/pôrt/, -pōrt/, nyoō/-), **1.** a seaport in Gwent, in SE Wales, near the Severn estuary. 129,500. **2.** a seaport and summer resort in SE Rhode Island: naval base. 28,227.

New/port Beach/, a city in SW California, SE of Los Angeles. 66,643.

New/port News/ (noō/pôrt/, -pōrt/, -pərt, nyoō/-), a seaport in SE Virginia: shipbuilding and ship-repair center. 170,045.

New/ Prov/idence, an island in the N Bahamas. 135,437; 58 sq. mi. (150 sq. km).

New/ Ro·chelle/ (rə shel/, rō-), a city in SE New York. 67,265.

New/ Sibe/rian Is/lands, a group of islands in the Arctic Ocean, N of the Russian Federation in Asia: part of the Yakut Autonomous Republic. 14,826 sq. mi. (38,400 sq. km).

New/ South/ Wales/, a state in SE Australia. 5,570,000; 309,433 sq. mi. (801,430 sq. km). *Cap.:* Sydney.

New/ Spain/, a former Spanish viceroyalty (1535–1821) including Central America N of Panama, Mexico, the West Indies, the SW United States, and the Philippines.

New·ton (nōōt′n, nyōōt′n), a city in E Massachusetts, near Boston. 82,585.

New′ World′, WESTERN HEMISPHERE.

New′ York′, 1. Also called **New′ York′ State′.** a state in the NE United States. 17,990,455; 49,576 sq. mi. (128,400 sq. km). *Cap.:* Albany. *Abbr.:* NY, N.Y. **2.** Also called **New′ York′ Cit′y.** a seaport in SE New York at the mouth of the Hudson: comprising the boroughs of Manhattan, Queens, Brooklyn, the Bronx, and Staten Island. 7,322,564. **3. Greater,** New York City, the counties of Nassau, Suffolk, Rockland, and Westchester in New York, and the counties of Bergen, Essex, Hudson, Middlesex, Morris, Passaic, Somerset, and Union in New Jersey: the metropolitan area as defined by the U.S. census. 15,509,093. **4.** the borough of Manhattan. —**New′ York′er,** *n.*

New′ York′ Bay′, a bay of the Atlantic at the mouth of the Hudson, W of Long Island and E of Staten Island and New Jersey.

New′ York′ State′ Barge′ Canal′, a New York State waterway system, connecting Lake Champlain and the Hudson River with Lakes Erie and Ontario: consists of the rebuilt Erie Canal and three shorter canals. 524 mi. (843 km) long.

New′ Zea′land (zē′lənd), a country in the S Pacific, SE of Australia, consisting of North Island, South Island, and adjacent small islands: a member of the Commonwealth of Nations. 3,307,084; 103,416 sq. mi. (267,845 sq. km). *Cap.:* Wellington. —**New′ Zea′land·er,** *n.*

Nga·li·e·ma (əng gä′lē ā′mə), **Mount,** a mountain with two summits, in central Africa, between Uganda and Zaire: highest peak in the Ruwenzori group. 16,763 ft. (5109 m). Formerly, **Mount Stanley.**

NH or **N.H.,** New Hampshire.

Ni·ag·a·ra (nī ag′rə, -ag′ər ə), a river on the boundary between W New York and Ontario, Canada, flowing from Lake Erie into Lake Ontario. 34 mi. (55 km) long.

Niag′ara Falls′, 1. the falls of the Niagara River: in Canada, the Horseshoe Falls, 158 ft. (48 m) high; 2600 ft. (792 m) wide; in the U.S., the American Falls, 167 ft. (51 m) high; 1000 ft. (305 m) wide. **2.** a city in W New York, on the falls. 61,840. **3.** a city in SE Ontario, on the falls. 72,107.

Nia·mey (nyä mā′), the capital of Niger, in the SW part, on the Niger River. 399,100.

Ni·cae·a (ni sē′ə), an ancient city in NW Asia Minor: Nicene Creed formulated here A.D. 325.

Nic·a·ra·gua (nik′ə rä′gwə), **1.** a republic in Central America. 3,500,000; 57,143 sq. mi. (148,000 sq. km). *Cap.:* Managua. **2. Lake,** a lake in SW Nicaragua. 92 mi. (148 km) long; 34 mi. (55 km) wide; 3060 sq. mi. (7925 sq. km). —**Nic′a·ra′guan,** *n., adj.*

Nice (nēs), a seaport in SE France, on the Mediterranean: resort. 338,486.

Nic′o·bar Is′lands (nik′ə bär′), a group of islands of India in the E part of the Bay of Bengal, forming the S part of the Andaman and Nicobar Islands. 30,433; 635 sq. mi. (1645 sq. km).

Nic·o·si·a (nik′ə sē′ə), the capital of Cyprus, in the central part. 164,500.

Nid·wal·den (nēt′väl′dən), a canton in central Switzerland. 31,500; 106 sq. mi. (275 sq. km).

Nie·der·sach·sen (nē′dər zäk′sən), German name of LOWER SAXONY.

Ni·ger (ni′jər, nē zhâr′), **1.** a republic in NW Africa: formerly part of French West Africa. 7,190,000; 458,976 sq. mi. (1,188,748 sq. km). *Cap.:* Niamey. **2.** a river in W Africa, rising in S Guinea and flowing into the Gulf of Guinea. 2600 mi. (4185 km) long. —**Ni·ge·ri·en** (ni jēr′ē en′), *adj., n.*

Ni·ge·ri·a (ni jēr′ē ə), a republic in W Africa: member of the Commonwealth of Nations. 88,500,000; 356,669 sq. mi. (923,773 sq. km). *Cap.:* Abuja. Official name, **Fed′eral Repub′lic of Nige′ria.** —**Ni·ge′ri·an,** *adj., n.*

Ni·hon (nē′hôn′), a Japanese name of JAPAN.

Ni·i·ga·ta (nē′ē gä′tä), a seaport on NW Honshu, in central Japan. 467,000

Ni·i·ha·u (nē′ē hä′ōō, nē′hou), an island in NW Hawaii, W of Kauai. 207; 72 sq. mi. (186 sq. km).

Nij·me·gen (ni′mā gən), a city in the E Netherlands, on the Waal River: peace treaty 1678. 145,816.

Nik·ko (nēk′kô; *Eng.* nik′ō, nē′kō), a city on central Honshu, in central Japan: famous for shrines and temples. 28,502.

Ni·ko·la·yev or **Ni·ko·la·ev** (nik′ə lä′yəf), a city in S Ukraine, on the Bug River. 503,000.

Ni·ko·pol (ni kô′pəl, nē′kə-), a city in SE Ukraine, on the Dnieper River. 154,000.

Nile (nīl), a river in E Africa, the longest in the world, flowing N from Lake Victoria in Uganda to the Mediterranean. 3473 mi. (5592 km) long; from the headwaters of the Kagera River, 4000 mi. (6440 km) long.

Nil·gi·ri Hills′ (nil′gə rē), a group of mountains in S India, in Tamil Nadu state. Highest peak, 8760 ft. (2670 m).

Nîmes (nēm), a city in S France: Roman ruins. 132,343.

Nin·e·veh (nin′ə və), the ancient capital of Assyria: its ruins are opposite Mosul, on the Tigris River, in N Iraq. **—Nin′e·vite′** (-vīt′), *n.*

Ning·bo or **Ning·po** (ning′bō′), a seaport in E Zhejiang province, E China. 1,020,000.

Ning·xia Hui (ning′shyä′ hwē′), an autonomous region in N China. 4,240,000; 25,640 sq. mi. (66,400 sq. km). *Cap.:* Yinchuan.

Niort (nyôr, nē ôr′), a city in W France. 58,203.

Ni·os or **Ny·os** (nē′ōs), **Lake,** a volcanic lake in Cameroon, at the NW border: poisonous eruption 1986.

Nip·i·gon (nip′i gon′), **Lake,** a lake in SW Ontario, in S central Canada. ab. 1870 sq. mi. (4845 sq. km).

Nip·is·sing (nip′ə sing), a lake in SE Canada, in Ontario. 330 sq. mi. (855 sq. km).

Nip·pon (ni pon′, nip′on), a Japanese name of JAPAN.

Nip·pur (ni pŏŏr′), an ancient Sumerian and Babylonian city in what is now SE Iraq.

Niš (nēsh), a city in SE Serbia, in E Yugoslavia. 230,711.

Ni·shi·no·mi·ya (nē′shē nô′mē yä′), a city on S Honshu, in S Japan. 412,000.

Ni·te·rói (nē′tə roi′), a seaport in SE Brazil opposite Rio de Janeiro. 382,736.

Nizh·ni Nov·go·rod (nizh′nē nov′gə rod′), a city in the Russian Federation in Europe, E of Moscow, on the Volga River. 1,438,000. Formerly (1932–91), **Gorki.**

NM or **N.M.,** New Mexico.

N. Mex., New Mexico.

Nome (nōm), **1.** a seaport in W Alaska. 3500. **2.**

Cape, a cape in W Alaska, on Seward Peninsula, W of Nome.

Nor·den·skjöld Sea/ (nôr/dn shōld/, -shəld), former name of LAPTEV SEA.

Nord/kyn Cape/ (nōr/kyn, nōōr/-), a cape in N Norway: the northernmost point of the European mainland.

Nord-Pas-de-Ca·lais (nôr pädᵃ ᴋᴀ lе/), a metropolitan region in N France. 3,923,200; 4793 sq. mi. (12,414 sq. km).

Nord·rhein-West·fal·en (nôrt/rin vest/fä/lən), German name of NORTH RHINE-WESTPHALIA.

Nor·folk (nôr/fək; *for 2 also* nôr/fôk), **1.** a county in E England. 736,200; 2068 sq. mi. (5355 sq. km). **2.** a seaport in SE Virginia: naval base. 261,229.

Nor/folk Is/land, an island in the S Pacific between New Caledonia and New Zealand: a territory of Australia. 1977; 13 sq. mi. (34 sq. km).

Nor·ge (nôr/gə), Norwegian name of NORWAY.

No·rilsk (nə rēlsk/), a city in the N Russian Federation in Asia, near the mouth of the Yenisei River. 181,000.

Nor·man (nôr/mən), a city in central Oklahoma. 80,071.

Nor·man·dy (nôr/mən dē), a historic region in NW France along the English Channel: Allied invasion in World War II began here June 6, 1944.

Norr·kö·ping (nôr/chœ/ping), a seaport in SE Sweden. 119,001.

North (nôrth), **the,** the northern area of the United States, esp. the states that fought to preserve the Union in the Civil War. —**North·ern** (nôr/thərn), *adj.* —**North/ern·er,** *n.*

North/ Af/rica, the northern part of Africa, esp. the region between the Mediterranean Sea and the Sahara Desert. —**North/ Af/rican,** *n., adj.*

North/ Amer/ica, the northern continent of the Western Hemisphere, extending from Central America to the Arctic Ocean. Highest point, Mt. McKinley, 20,320 ft. (6194 m); lowest, Death Valley, 280 ft. (85 m) below sea level; ab. 9,360,000 sq. mi. (24,242,400 sq. km). —**North/ Amer/ican,** *n., adj.*

North·amp·ton (nôr thamp/tən, nôrth hamp/-), **1.** a city in Northamptonshire, in central England. 177,200. **2.** NORTHAMPTONSHIRE.

North·amp·ton·shire (nôr thamp′tən shēr′, -shər, nôrth hamp′-), a county in central England. 561,800; 914 sq. mi. (2365 sq. km). Also called **Northampton.**

North′ Bay′, a city in SE Ontario, in S Canada. 50,623.

North′ Bor′neo, former name of SABAH.

North′ Brabant′, a province in the S Netherlands. 2,156,280; 1965 sq. mi. (5090 sq. km). *Cap.:* 's Hertogenbosch.

North′ Caroli′na, a state in the SE United States. 6,628,637; 52,586 sq. mi. (136,198 sq. km). *Cap.:* Raleigh. *Abbr.:* NC, N.C. **—North′ Carolin′ian,** *n., adj.*

North′ Cascades′ Na′tional Park′, a national park in NW Washington: site of glaciers and mountain lakes. 789 sq. mi. (2043 sq. km).

North′ Charles′ton, a city in SE South Carolina. 70,218.

North′ Dako′ta, a state in the N central United States. 638,800; 70,665 sq. mi. (183,020 sq. km). *Cap.:* Bismarck. *Abbr.:* ND, N. Dak. **—North′ Dako′tan,** *n., adj.*

North·east (nôrth′ēst′), **the,** the northeastern part of the United States. **—North′east′ern,** *adj.* **—North′east′ern·er,** *n.*

North′east Pas′sage, a ship route between the Atlantic and the Pacific along the N coast of Europe and Asia.

North′ern Dvi′na, DVINA (def. 2).

North′ern Hem′isphere, the half of the earth between the North Pole and the equator.

North′ern Ire′land, a political division of the United Kingdom, in the NE part of Ireland. 1,583,000; 5452 sq. mi. (14,121 sq. km). *Cap.:* Belfast.

North′ern Maria′na Is′lands, a group of islands in the W Pacific, N of Guam: formerly a part of the Trust Territory of the Pacific Islands; since 1986 a commonwealth associated with the U.S. 18,400; 184 sq. mi. (477 sq. km). *Cap.:* Saipan.

North′ern Rhode′sia, former name of ZAMBIA. **—North′ern Rhode′sian,** *n., adj.*

North′ern Spor′ades, See under SPORADES.

North′ern Ter′ritories, a former British protectorate in W Africa; now a part of N Ghana.

North′ern Ter′ritory, a territory in N Australia.

155,000; 523,620 sq. mi. (1,356,175 sq. km). *Cap.:* Darwin.

North′ Fri′sians, See under FRISIAN ISLANDS.

North′ Hol′land, a province in the W Netherlands. 2,352,888; 1163 sq. mi. (3010 sq. km). *Cap.:* Haarlem.

North′ Is′land, the northernmost principal island of New Zealand. 2,438,249; 44,281 sq. mi. (114,690 sq. km).

North′ Kore′a, a country in E Asia: formed 1948 after the division of the former country of Korea at 38° N. 21,890,000; 50,000 sq. mi. (129,500 sq. km). *Cap.:* Pyongyang. Compare KOREA. Official name, **Democratic People's Republic of Korea.** —**North′ Kore′an,** *n., adj.*

North′ Las′ Ve′gas, a city in S Nevada. 51,450.

North′ Lit′tle Rock′, a city in central Arkansas. 61,741.

North′ Miam′i, a city in SE Florida. 49,998.

North′ Osse′tian Auton′omous Repub′lic, an autonomous republic in the Russian Federation in SE Europe. 634,000; 3088 sq. mi. (8000 sq. km). *Cap.:* Vladikavkaz.

North′ Platte′, a river flowing from N Colorado through SE Wyoming and W Nebraska into the Platte. 618 mi. (995 sq. km) long.

North′ Pole′, the northern end of the earth's axis of rotation, the northernmost point on earth.

North′ Rhine′-Westpha′lia, a state in W Germany. 16,874,000; 13,154 sq. mi. (34,070 sq. km). *Cap.:* Düsseldorf. German, **Nordrhein-Westfalen.**

North′ Ri′ding (ri′ding), a former administrative division of Yorkshire, in N England.

North′ Riv′er, a part of the Hudson River between NE New Jersey and SE New York.

North′ Sea′, an arm of the Atlantic between Great Britain and the European mainland. ab. 201,000 sq. mi. (520,600 sq. km); greatest depth, 1998 ft. (610 m). Formerly, **German Ocean.**

North′ Slope′, the northern coastal area of Alaska, between the Brooks Range and the Arctic Ocean.

North·um·ber·land (nôr thum′bər lənd), a county in NE England. 300,900; 1943 sq. mi. (5030 sq. km).

North·um/berland Strait/, the part of the Gulf of St. Lawrence that separates Prince Edward Island from New Brunswick and Nova Scotia, in SE Canada. ab. 200 mi. (320 km) long; 9–30 mi. (15–48 km) wide.

North·um·bri·a (nôr thum/brē ə), an early English kingdom extending N from the Humber to the Firth of Forth. **—North·um/bri·an,** *adj., n.*

North/ Vancou/ver, a city in SW British Columbia, in SW Canada. 68,241.

North/ Vietnam/, the part of Vietnam north of the 17th parallel: a separate state 1954–75; now part of reunified Vietnam. Compare SOUTH VIETNAM, VIETNAM.

North·west (nôrth/west/), **the, 1.** the northwestern part of the United States, esp. Washington, Oregon, and Idaho. **2.** the northwestern part of the United States when its western boundary was the Mississippi River. **3.** the northwestern part of Canada. **—North/west/ern,** *adj.* **—North/west/ern·er,** *n.*

North/-West/ Frontier/ Prov/ince, a province in NW Pakistan. 12,287,000; 28,773 sq. mi. (74,522 sq. km). *Cap.:* Peshawar.

North/west Pas/sage, a ship route along the Arctic coast of Canada and Alaska, joining the Atlantic and Pacific oceans.

North/west Ter/ritories, a territory of Canada lying N of the provinces and extending E from Yukon territory to Davis Strait. 52,238; 1,304,903 sq. mi. (3,379,700 sq. km). *Cap.:* Yellowknife.

North/west Ter/ritory, the region north of the Ohio River and east of the Mississippi, organized by Congress in 1787, comprising present-day Ohio, Indiana, Illinois, Michigan, Wisconsin, and the E part of Minnesota.

North/ Yem/en, YEMEN. (def. 2).

North/ York/shire, a county in NE England. 705,700; 3208 sq. mi. (8309 sq. km).

Nor·walk (nôr/wôk), **1.** a city in SW California. 94,279. **2.** a city in SW Connecticut. 78,331.

Nor·way (nôr/wā), a kingdom in N Europe, in the W part of the Scandinavian Peninsula. 4,200,000; 124,555 sq. mi. (322,597 sq. km). *Cap.:* Oslo. Norwegian, **Norge. —Nor·we/gian** (-wē/jən), *adj., n.*

Norwe/gian Sea/, part of the Arctic Ocean, N

and E of Iceland and between Greenland and
Norway.

Nor·wich (nôr′ich, -ij, nor′-, nôr′wich), a city in
E Norfolk, in E England: cathedral. 118,600.

Not·ting·ham (not′ing əm; *U.S. often* -ham′), **1.**
a city in SW Nottinghamshire, in central England.
276,300. **2.** NOTTINGHAMSHIRE.

Not′ting·ham·shire′ (-shēr′, -shər), a county in
central England. 1,007,800; 854 sq. mi. (2210 sq.
km). Also called **Nottingham, Notts** (nots).

Nouak·chott (nwäk shot′), the capital of Mauritania, on the W coast. 500,000.

Nou·mé·a (nōō mā′ə), the capital of New Caledonia, on the SW coast. 60,112.

No′va Iguaçu′ (nō′və), a city in SE Brazil, NW
of Rio de Janeiro. 491,766.

No′va Lis·bo′a (lēzh bō′ə), former name of
HUAMBO.

No·va·ra (nō vär′ə), a city in NE Piedmont, in NW
Italy. 102,086.

No′va Sco′tia, a peninsula and province in SE
Canada: once a part of the French province of
Acadia. 873,176; 21,068 sq. mi. (54,565 sq. km).
Cap.: Halifax. —**No′va Sco′tian,** *n., adj.*

No·va·ya Zem·lya (nō′və yə zem′lē ä′), two
large islands in the Arctic Ocean, N of and belonging to the Russian Federation. 35,000 sq. mi.
(90,650 sq. km).

Nov·go·rod (nov′gə rod′), a city in the W Russian Federation in Europe, SE of St. Petersburg.
228,000.

No·vi Sad (nô′vē säd′), the capital of Vojvodina,
in N Yugoslavia, on the Danube. 257,685.

No·vo·cher·kassk (nō′və chər käsk′), a city in
the SW Russian Federation in Europe, NE of Rostov. 188,000.

No·vo·kuz·netsk (nō′və kŏŏz netsk′), a city in
the S Russian Federation in central Asia, SE of
Novosibirsk. 600,000. Formerly, **Stalinsk.**

No·vo·ros·siysk or **No·vo·ros·siisk** (nō′və rə-
sēsk′), a seaport in the SW Russian Federation in
Europe, on the Black Sea. 179,000.

No·vo·si·birsk (nō′və sə bērsk′), a city in the
Russian Federation in Asia, on the Ob. 1,436,000.

Nu·bi·a (nōō′bē ə, nyōō′-), **1.** a region in S Egypt
and N Sudan, extending from the Nile to the Red

Sea. **2.** an ancient kingdom in this region. —**Nu′-bi·an,** *adj., n.*

Nu′bian Des′ert, an arid region in the NE Sudan.

Nu·e·ces (nō̄ ā′səs, yō̄-), a river in S Texas, flowing SE to Corpus Christi Bay, on the Gulf of Mexico. 338 mi. (545 km) long.

Nue·vo La·re·do (nwā′vō lə rā′dō, nō̄ ā′-), a city in NE Mexico, on the Rio Grande opposite Laredo, Texas. 203,286.

Nue·vo Le·ón (nwā′vō lā ōn′, nō̄ ā′-), a state in NE Mexico. 2,344,000; 25,136 sq. mi. (65,102 sq. km). *Cap.:* Monterrey.

Nu·ku·a·lo·fa (nō̄′kō̄ ə lô′fə), the capital of Tonga, in the S Pacific Ocean. 28,899.

Nu·mid·i·a (nō̄ mid′ē ə, nyō̄-), an ancient country in N Africa, corresponding roughly to modern Algeria. —**Nu·mid′i·an,** *adj., n.*

Nu·rem·berg (nō̄r′əm bûrg′, nyō̄r′-), a city in central Bavaria, in SE Germany: site of international trials (1945–46) of Nazis accused of war crimes. 471,800. German, **Nürn·berg** (nʏʀn′-berk′).

Nu·ri·stan (nōōr′ə stan′, -stän′), a mountainous region in NE Afghanistan. 5000 sq. mi. (12,950 sq. km). Formerly, **Kafiristan.**

Nuuk (nōōk), Godthåb.

NV, Nevada.

NY or **N.Y.,** New York.

Nya·sa or **Nyas·sa** (nyä′sä, ni as′ə), former name of Lake Malawi.

Nya·sa·land (nyä′sä land′, ni as′ə-), former name of Malawi (def. 1).

NYC or **N.Y.C.,** New York City.

O

O·a·hu (ō ä′hōō), an island in central Hawaii: third largest island of the state. *Chief city:* Honolulu. 838,500; 589 sq. mi. (1525 sq. km).

Oak·land (ōk′lənd), a seaport in W California, on San Francisco Bay. 372,242.

Oak′ Lawn′, a city in NE Illinois, near Chicago. 56,182.

Oak′ Park′, a city in NE Illinois, near Chicago. 53,648.

Oak′ Ridge′, a city in E Tennessee, near Knoxville: atomic research center. 27,310.

Oak·ville (ōk′vil), a town in SE Ontario, in S Canada, SW of Toronto, on Lake Ontario. 87,107.

Oa·xa·ca (wə hä′kə, wä-), **1.** a state in S Mexico. 2,650,232; 36,375 sq. mi. (94,210 sq. km). **2.** the capital of this state, in the central part. 157,284. —**Oa·xa′can,** *adj.*

Ob (ôb, ob), **1.** a river in the W Russian Federation in Asia, flowing NW to the Gulf of Ob. 2500 mi. (4025 km) long. **2. Gulf of,** an inlet of the Arctic Ocean. ab. 500 mi. (800 km) long.

O·ber·am·mer·gau (ō′bər ä′mər gou′), a village in S Germany, SW of Munich: passion play performed every ten years. 4664.

O·ber·hau·sen (ō′bər hou′zən), a city in W Germany, in the lower Ruhr valley. 220,400.

O·ber·land (ō′bər land′, -länt′), BERNESE ALPS.

Ob·wal·den (*Ger.* ôp′väl′dən), a canton in central Switzerland. 27,900; 189 sq. mi. (490 sq. km).

O·ce·an·i·a (ō′shē an′ē ə, -ä′nē ə) also **O′ce·an′i·ca** (-an′i kə), the islands of the central and S Pacific, including Micronesia, Melanesia, Polynesia, and usu. Australasia. —**O′ce·an′i·an,** *adj., n.*

O·cean·side (ō′shən sīd′), a city in SW California. 128,398.

O·den·se (ō′тнən sə, ō′dən-), a seaport on Fyn island, in S Denmark. 174,016.

O·der (ō′dər), a river in central Europe, flowing from the Carpathians in N Czechoslovakia N through Poland and along the border between Germany and Poland into the Baltic. 562 mi. (905 km) long.

O·des·sa (ō des′ə), **1.** a seaport in S Ukraine, on

the Black Sea. 1,115,000. **2.** a city in W Texas. 89,699.

Of·fa·ly (ô′fə lē, of′ə-), a county in Leinster, in the central Republic of Ireland. 59,806; 760 sq. mi. (1970 sq. km).

Of·fen·bach (ô′fən bäk′, of′ən-), a city in S Hesse, in central Germany, on the Main River, near Frankfurt. 111,300.

O·ga·den (ō gä′dän), an arid region in SE Ethiopia.

Og·bo·mo·sho (og′bə mō′shō), a city in SW Nigeria. 527,400.

Og·den (ôg′dən, og′-), a city in N Utah. 63,909.

OH, Ohio.

O·hi·o (ō hi′ō), **1.** a state in the NE central United States. 10,847,115; 41,222 sq. mi. (106,765 sq. km). *Cap.*: Columbus. *Abbr.*: OH **2.** a river formed by the confluence of the Allegheny and Monongahela rivers, flowing SW from Pittsburgh, Pa., to the Mississippi in S Illinois. 981 mi. (1580 km) long. —**O·hi′o·an**, *adj.*, *n.*

O·hře (ôr′zhə), a river in central Europe, flowing from Germany through W Czechoslovakia NE to the Elbe. 193 mi. (310 km) long. German, **Eger.**

Oise (waz), a river in W Europe, flowing SW from S Belgium through N France to the Seine, near Paris. 186 mi. (300 km) long.

O·i·ta (ô′ē tä′), a seaport on NE Kyushu, in S Japan. 389,000.

OK, Oklahoma.

O·ka (ō kä′), a river in the central Russian Federation in Europe, flowing NE to the Volga at Nizhni Novgorod. 950 mi. (1530 km) long.

O·ka·van·go (ō′kə vang′gō, -väng′-), a river in central Africa, flowing SE from Angola to Botswana. ab. 1000 mi. (1610 km) long. Portuguese, **Cubango.**

O·ka·ya·ma (ô′kä yä′mä), a city on SW Honshu, in SW Japan. 570,000.

O·ka·za·ki (ô′kä zä′kē), a city on S central Honshu, in central Japan. 287,000.

O·kee·cho·bee (ō′ki chō′bē), **Lake,** a lake in S Florida, in the N part of the Everglades. 700 sq. mi. (1813 sq. km).

O′ke·fe·no′kee Swamp′ (ō′kə fə nō′kē, ō′kē-), a large wooded swamp area in SE Georgia. 660 sq. mi. (1709 sq. km).

O·khotsk (ō kotsk′), **Sea of,** an arm of the N Pacific enclosed by the Kamchatka Peninsula, the Kurile Islands, Sakhalin, and the Russian Federation in Asia.

O·ki·na·wa (ō′kə nou′wə, -nä′wə), the largest of the Ryukyu Islands: occupied by U.S. 1945–72. 544 sq. mi. (1409 sq. km). *Cap.:* Naha. —**O′ki·na′wan,** *adj., n.*

Okla., Oklahoma.

O·kla·ho·ma (ō′klə hō′mə), a state in the S central United States. 3,145,585; 69,919 sq. mi. (181,090 sq. km). *Cap.:* Oklahoma City. *Abbr.:* OK, Okla. —**O′kla·ho′man,** *adj., n.*

O′klaho′ma Cit′y, the capital of Oklahoma, in the central part. 444,719.

Ö·land (œ′länd′), an island of Sweden, off the SE coast, in the Baltic. 519 sq. mi. (1345 sq. km).

O·la·the (ō lā′thə), a city in E Kansas. 63,352.

Old′ Cas·tile′, a region in N Spain: formerly a province. Spanish, **Castilla la Vieja.**

Ol·den·burg (ōl′dən bûrg′), **1.** a former state in NW Germany, now part of Lower Saxony. **2.** a city in Lower Saxony, in NW Germany. 140,200.

Old′ Faith′ful, the best known geyser in Yellowstone National Park.

Old·ham (ōl′dəm; *locally* ou′dəm), a city in Greater Manchester, in NW England. 219,500.

Ol′du·vai Gorge′ (ōl′dōō vī′), a gorge in Tanzania containing australopithecine and human skeletal and cultural remains.

Old′ World′, 1. a. Europe, Asia, and Africa. **b.** Europe. **2.** EASTERN HEMISPHERE.

O·lin·da (ō lin′də; *Port.* ōō lēn′dä), a city in NE Brazil, N suburb of Recife, on the Atlantic coast: beach resort. 266,751.

Ol·ives (ol′ivz), **Mount of,** a small ridge E of Jerusalem. Highest point, 2737 ft. (834 m). Also called **Ol′i·vet′** (-ə vet′, -vit).

O·lo·mouc (ô′lô mōts), a city in central Moravia, in central Czechoslovakia. 107,000.

Ol·sztyn (ôl′shtin), a city in NE Poland. 739,000.

O·lym·pi·a (ə lim′pē ə, ō lim′-), **1.** a plain in Elis, Greece, where the ancient Olympic Games were held. **2.** the capital of Washington, in the W part, on Puget Sound. 27,447. —**O·lym′pi·an,** *adj., n.*

O·lym′pic Moun′tains (ə lim′pik, ō lim′-), a

mountain system in NW Washington, part of the Coast Ranges. Highest peak, Mt. Olympus, 7954 ft. (2424 m).

Olym'pic Na'tional Park', a national park in NW Washington: rain forest, glaciers. 1429 sq. mi. (3702 sq. km).

Olym'pic Penin'sula, a peninsula in NW Washington, between the Pacific Ocean and Puget Sound.

O·lym·pus (ə lim'pəs, ō lim'-), **Mount, 1.** a mountain in NE Greece, on the boundary between Thessaly and Macedonia: mythical abode of the Greek gods. 9730 ft. (2966 m). **2.** a mountain in NW Washington: highest peak of the Olympic Mountains. 7954 ft. (2424 m).

O·lyn·thus (ō lin'thəs), an ancient city in NE Greece, on the Chalcidice Peninsula. —**O·lyn'thi·an**, adj., n.

O·ma·ha (ō'mə hô', -hä'), a city in E Nebraska, on the Missouri River. 335,795.

O·man (ō män'), **1. Sultanate of.** Formerly, **Muscat and Oman,** an independent sultanate in SE Arabia. 1,200,000; ab. 82,800 sq. mi. (212,380 sq. km). Cap.: Muscat. **2. Gulf of,** a NW arm of the Arabian Sea, at the entrance to the Persian Gulf. —**O·ma·ni** (ō mä'nē), n., pl. **-nis,** adj.

Om·dur·man (om'dŏŏr män'), a city in central Sudan, on the White Nile opposite Khartoum. 526,287.

O·mi·ya (ô'mē yä'), a city on E Honshu, in Japan, NW of Tokyo. 377,000.

Omsk (ômsk), a city in the SW Russian Federation in Asia, on the Irtysh River. 1,148,000.

On (on), Biblical name of Heliopolis.

O·ne·ga (ō neg'ə), **Lake,** a lake in the NW Russian Federation in Europe: second largest lake in Europe. 3764 sq. mi. (9750 sq. km).

O·nei·da Lake' (ō nī'də), a lake in central New York. 80 sq. mi. (207 sq. km).

On'on·da·ga Lake' (on'ən dô'gə, -dä'-, -dâ'-, on'-), a salt lake in central New York, W of Syracuse. 5 sq. mi. (13 sq. km).

On·tar·i·o (on târ'ē ō'), **1.** a province in S Canada, bordering on the Great Lakes. 9,101,694; 412,582 sq. mi. (1,068,585 sq. km). Cap.: Toronto. **2. Lake,** a lake between the northeastern

U.S. and S Canada, between New York and Ontario: the smallest of the Great Lakes. 7540 sq. mi. (19,530 sq. km). **3.** a city in SW California, E of Los Angeles. 133,179. —**On·tar/i·an,** *adj., n.*

O·por·to (ō pôr/tō, ō pōr/-), a port in NW Portugal, near the mouth of the Douro River. 327,368. Portuguese, **Pôrto.**

OR, Oregon.

O·ra·dea (ô Rä/dyä), a city in NW Romania. 208,507. Hungarian, **Nagyvárad.**

O·ran (ô ran/, ō ran/, ô rän/), a seaport in NW Algeria. 916,578.

Or·ange (ôr/inj, or/-; *Fr.* ô RäNzh/ *for 3*), **1.** a river in the Republic of South Africa, flowing W from Lesotho to the Atlantic. 1300 mi. (2095 km) long. **2.** a city in SW California, near Los Angeles. 110,658. **3.** a town in SE France, near Avignon: Roman ruins. 26,468.

Or/ange Free/ State/, a province in central Republic of South Africa: a Boer republic 1854–1900; a British colony (**Or/ange Riv/er Col/ony**) 1900–10. 1,863,327; 49,647 sq. mi. (128,586 sq. km). *Cap.:* Bloemfontein.

Or·dzho·ni·ki·dze (ôr/jon i kid/zə), former name (1944–91) of VLADIKAVKAZ.

Ore. or **Oreg.,** Oregon.

Or·e·gon (ôr/i gən, -gon/, or/-), a state in the NW United States, on the Pacific coast. 2,842,321; 96,981 sq. mi. (251,180 sq. km). *Cap.:* Salem. *Abbr.:* OR, Oreg., Ore. —**Or/e·go/ni·an** (-gō/nē-ən), *adj., n.*

Or/egon Trail/, a route used during the U.S. westward migrations, esp. in the period from 1840 to 1860, starting in Missouri and ending in Oregon. ab. 2000 mi. (3200 km) long.

O·rel (ô rel/, ō rel/; *Russ.* u ryôl/), a city in the W Russian Federation in Europe, S of Moscow. 335,000.

O·rem (ôr/əm, ōr/-), a city in N Utah. 67,561.

O·ren·burg (ôr/ən bûrg/, ōr/-), a city in the SW Russian Federation in Asia, on the Ural River. 547,000. Formerly, **Chkalov.**

Ø·re·sund (*Dan.* œ/Rə sōōn/) also **Ö·re·sund** (*Sw.* -sŏŏnd/), a strait between the Danish island of Zealand and SW Sweden, connecting the Kattegat with the Baltic. English, **The Sound.**

O·ri·ent (ôr/ē ənt, -ē ent/, ōr/-), **the, 1.** the

countries of Asia, esp. East Asia. **2.** (formerly) the
countries to the E of the Mediterranean. —**O′ri‧**
en′tal, *adj. n.*

O‧ri‧no‧co (ôr′ə nō′kō, ōr′-), a river in N South
America, flowing N from the border of Brazil,
along the E border of Colombia, and NE through
Venezuela into the Atlantic. 1600 mi. (2575 km)
long.

O‧ris‧sa (ô ris′ə, ō ris′ə), a state in E India.
26,272,054; 60,136 sq. mi. (155,752 sq. km).
Cap.: Bhubaneswar.

O‧ri‧za‧ba (ôr′ə zä′bə, -sä′-, ōr′-), **1.** Also called
Citlaltepetl. an inactive volcano in SE Mexico, in
Veracruz state. 18,546 ft. (5653 m). **2.** a city near
this peak. 114,848.

Ork′ney Is′lands (ôrk′nē), an island group off
the NE tip of Scotland. 19,338; 340 sq. mi. (880
sq. km).

Or‧lan‧do (ôr lan′dō), a city in central Florida.
164,693.

Or‧lé‧a‧nais (ôʀ lā A ne′), a former province in N
France. *Cap.:* Orléans.

Or‧lé‧ans (ôr′lē ənz; *Fr.* ôʀ lā än′), a city in central France, SSW of Paris. 105,589.

Or‧muz (ôr mōōz′, ôr′muz), **Strait of,** HORMUZ,
Strait of.

O‧ron‧tes (ô ron′tēz, ō ron′-), a river in W Asia,
flowing N from Lebanon through Syria and Turkey
to the Mediterranean. 250 mi. (405 km) long.

Orsk (ôrsk), a city in the S Russian Federation in
Europe, on the Ural River. 273,000.

Or‧te‧gal (ôr′ti gäl′), **Cape,** a cape in NW Spain,
on the Bay of Biscay.

Ort‧les (ôrt′lās), **1.** a range of the Alps in N Italy.
2. the highest peak of this range. 12,802 ft.
(3902 m). German, **Ort‧ler** (ôʀt′lər).

O‧ru‧mi‧yeh (ô rōō′mē ə), a city in NW Iran.
262,588. Formerly, **Rezaiyeh.**

O‧ru‧ro (ô rōōr′ō), a city in W Bolivia: a former
capital. 145,410; over 12,000 ft. (3660 m) high.

O‧sage (ō′sāj, ō sāj′), a river flowing E from E
Kansas to the Missouri River in central Missouri.
500 mi. (800 km) long.

O‧sa‧ka (ō sä′kə, ō′sä kä′), a city on S Honshu,
in S Japan. 2,546,000.

Osh‧a‧wa (osh′ə wə), a city in SE Ontario, in S

Canada, NE of Toronto, on Lake Ontario. 123,651.

Osh·kosh (osh′kosh), a city in E Wisconsin, on Lake Winnebago. 55,006.

O·shog·bo (ō shog′bō), a city in SW Nigeria. 344,500.

Os·lo (oz′lō, os′-), the capital of Norway, in the SE part, at the head of Oslo Fjord. 453,700. Formerly, **Christiania.**

Os′lo Fjord′, an inlet of the Skagerrak, in SE Norway.

Os·na·brück (oz′nə brook′), a city in Lower Saxony, in NW Germany. 150,900.

Os·sa (os′ə), a mountain in E Greece, in Thessaly. 6490 ft. (1978 m).

Os·se·tia (o sē′shə), a region in Caucasia: divided between the North Ossetian Autonomous Republic of the Russian Federation and the South Ossetian Autonomous Region of the Georgian Republic. —**Os·se′tian,** *n., adj.*

Os·si·ning (os′ə ning), a town in SE New York, on the Hudson: the site of a state prison. 34,124.

Ost·end (os tend′, os′tend), a seaport in NW Belgium. 68,397. French, **Os·tende** (ôs tänd′).

Ö·ster·reich (œ′stər RĪKH′), German name of AUSTRIA.

Os·ti·a (os′tē ə), a town in central Italy, SW of Rome: ruins from 4th century B.C.; site of ancient port of Rome.

O·stra·va (ôs′trä vä), a city in N Moravia, in N Czechoslovakia. 331,000. Formerly, **Moravská Ostrava.**

Oś·wię·cim (ôsh vyen′chēm), Polish name of AUSCHWITZ.

O·ta·ru (ō tär′oo, ō′tä roo′), a city in W Hokkaido, in N Japan. 183,000.

Ot·ta·wa (ot′ə wə), **1.** the capital of Canada, in SE Ontario. 300,763. **2.** a river in SE Canada, flowing SE into the St. Lawrence at Montreal. 685 mi. (1105 km) long.

Ot′to·man Em′pire (ot′ə mən), a Turkish state that was founded about 1300 by Osman and reached its greatest territorial extent under Suleiman in the 16th century: collapsed after World War I. *Cap.:* Constantinople.

Ouach·i·ta or **Wash·i·ta** (wosh′i tô′, wô′shi-), a

river flowing SE from Arkansas to the Red River in Louisiana. 605 mi. (975 km) long.

Ouach/ita Moun/tains, a range extending from SE Oklahoma to W Arkansas.

Oua·ga·dou·gou (wä′gə dōō′gōō), the capital of Burkina Faso, in the central part. 442,223.

Ouj·da (ōōj dä′), a city in NE Morocco. 780,762.

Ou·lu (ō′lōō, ou′-), a city in W Finland, on the Gulf of Bothnia. 98,582.

Ouse (ōōz), **1.** Also called **Great Ouse.** a river in E England flowing NE to the Wash. 160 mi. (260 km) long. **2.** a river in NE England, in Yorkshire, flowing SE to the Humber. 57 mi. (92 km) long.

Out/er Banks/, a chain of sandy barrier islands along the coast of North Carolina.

Out/er Heb/rides, See under HEBRIDES.

Out/er Mongo/lia, former name of MONGOLIAN PEOPLE'S REPUBLIC.

O·ver·ijs·sel (ō′vər ī′səl), a province in the E Netherlands. 1,009,997. *Cap.:* Zwolle.

O/ver·land Park/ (ō′vər lənd), a city in E Kansas, near Kansas City. 111,790.

O·vie·do (ô vye′₺hô), a city in NW Spain. 190,651.

Ow·ens·bor·o (ō′ənz bûr′ō, -bur′ō), a city in NW Kentucky, on the Ohio River. 53,549.

Ow/en Stan/ley Range/ (ō′ən), a mountain range on New Guinea in SE Papua New Guinea. Highest peak, Mt. Victoria, 13,240 ft. (4036 m).

Ox·ford (oks′fərd), **1.** a city in S Oxfordshire, in S England, NW of London: university, founded in 12th century. 115,800. **2.** OXFORDSHIRE. —**Ox·o·ni·an** (ok sō′nē ən), *adj., n.*

Ox/ford·shire/ (-shēr′, -shər), a county in S England. 578,000; 1008 sq. mi. (2610 sq. km). Also called **Oxford.**

Ox·nard (oks′närd), a city in SW California, NW of Los Angeles. 142,216.

Ox·us (ok′səs), AMU DARYA.

O/zark Moun/tains (ō′zärk), a group of low mountains in S Missouri, N Arkansas, and NE Oklahoma. Also called **Ozarks.**

O·zarks (ō′zärks), **1.** OZARK MOUNTAINS. **2. Lake of the,** a reservoir in central Missouri, formed by a dam on the Osage River. 130 mi. (209 km) long.

P

PA or **Pa.**, Pennsylvania.

Pa·chu·ca (pä chōō′kä), the capital of Hidalgo, in central Mexico: silver mines. 135,248.

Pa·cif′ic Is′lands, Trust′ Ter′ritory of the (pə sif′ik), a group of islands in the W Pacific, established in 1947 by the United Nations as a U.S. trusteeship: orig. composed of the Caroline, Marshall, and Mariana Islands (except Guam); the Republic of Palau is the only remaining trust territory. Compare MARSHALL ISLANDS, MICRONESIA (def. 2), NORTHERN MARIANA ISLANDS, PALAU.

Pacif′ic O′cean, an ocean bordered by the American continents, Asia, and Australia: largest ocean in the world; divided by the equator into the North Pacific and the South Pacific. 70,000,000 sq. mi. (181,300,000 sq. km); greatest known depth, 35,433 ft. (10,800 m). Also called **Pacific.**

Pacif′ic Rim′, the group of countries bordering on the Pacific Ocean, esp. the industrialized nations of Asia.

Pac·to·lus (pak tō′ləs), a small river in Asia Minor, in ancient Lydia: famous for the gold washed from its sands.

Pa·dang (pä däng′), a seaport in W central Sumatra, in W Indonesia. 480,922.

Pad·ding·ton (pad′ing tən), a former residential borough of Greater London, England, now part of Westminster.

Pad·u·a (paj′ōō ə), a city in NE Italy. 223,907. Italian, **Pa·do·va** (pä′dô vä). —**Pad′u·an**, *adj., n.*

Pa·du·cah (pə dōō′kə, -dyōō′-), a city in W Kentucky, at the junction of the Tennessee and Ohio rivers. 29,315.

Pa·dus (pā′dəs), ancient name of Po.

Paes·tum (pes′təm), an ancient coastal city of Lucania, in S Italy.

Pa·go Pa·go (päng′ō päng′ō, päng′gō päng′gō, pä′gō pä′gō), the chief harbor and town of American Samoa, on Tutuila Island. 2451.

Pa·hang (pä häng′, pə hang′), a state in Malaysia, on the SE Malay Peninsula. 798,782; 13,820 sq. mi. (35,794 sq. km).

Paint′ed Des′ert, a region in N central Arizona,

E of the Colorado River: many-colored rock surfaces.

Pais·ley (pāz′lē), a city in SW Scotland, W of Glasgow. 84,789.

Pak·i·stan (pak′ə stan′, pä′kə stän′), **Islamic Republic of,** a republic in S Asia, between India and Afghanistan: formerly part of British India; known as West Pakistan from 1947–71 to distinguish it from East Pakistan (now Bangladesh). 102,200,000; 310,403 sq. mi. (803,881 sq. km). *Cap.:* Islamabad. —**Pa′ki·sta′ni** (-stan′ē, -stä′nē), *n., pl.* **-nis, ni,** *adj.*

Pa·lat·i·nate (pə lat′n āt′, -it), **the,** either of two historic regions of Germany that constituted an electorate of the Holy Roman Empire: one **(Lower Palatinate** or **Rhine Palatinate)** is now part of Rhineland-Palatinate, and the other **(Upper Palatinate)** is now part of Bavaria. German, **Pfalz.** —**Pal·a·tine** (pal′ə tin′, -tin), *adj., n.*

Pal·a·tine (pal′ə tin′, -tin), one of the seven hills on which ancient Rome was built.

Pa·lau (pə lou′), **Republic of,** a group of islands in the W Pacific, part of the Caroline group: since 1947 part of the Trust Territory of the Pacific Islands. 14,106; 192 sq. mi. (497 sq. km).

Pa·la·wan (pä lä′wän), an island in the W Philippines. 4550 sq. mi. (11,784 sq. km).

Pa·lem·bang (pä′lem bäng′), a city in SE Sumatra, in W Indonesia. 787,187.

Pa·len·que (pä leng′kä), a village in SE Mexico, in Chiapas state: ruins of an ancient Mayan city.

Pa·ler·mo (pə lûr′mō, -lâr′-), the capital of Sicily, in the NW part. 701,782. —**Pa·ler′mi·tan** (-mi tn), *adj., n.*

Pal·es·tine (pal′ə stīn′), **1.** Also called **Holy Land.** Biblical name, **Canaan.** an ancient country in SW Asia, on the E coast of the Mediterranean. **2.** a former British mandate (1923–48) comprising part of this country, divided between Israel, Jordan, and Egypt in 1948: the Jordanian and Egyptian parts were occupied by Israel in 1967. —**Pal′es·tin′i·an** (-stin′ē ən), *n., adj.*

Pal·i·sades (pal′ə sādz′), a line of cliffs in NE New Jersey and SE New York extending along the W bank of the lower Hudson River. ab. 15 mi. (24 km) long; 300–500 ft. (91–152 m) high.

Pal·ma (päl′mä), the capital of the Balearic Islands, on W Majorca. 321,112. Also called **Pal′-ma de Mallor′ca.**

Palm′ Bay′, a town in E Florida. 62,632.

Palm′ Beach′, a town in SE Florida: seaside winter resort. 9814.

Palm·dale (päm′dāl′), a city in SW California, NE of Los Angeles. 68,842.

Palm′er Penin′sula (pä′mər), former name of ANTARCTIC PENINSULA.

Palm′ Springs′, a city in S California: resort. 40,181.

Pal·my·ra (pal mī′rə), an ancient city in central Syria, NE of Damascus: reputedly built by Solomon. Biblical name, **Tadmor.** —**Pal′my·rene′** (-mə rēn′), *adj., n.*

Pal·o Al·to (pal′ō al′tō), a city in W California, SE of San Francisco. 55,900.

Pal·o·mar (pal′ə mär′), **Mount,** a mountain in S California, NE of San Diego: observatory. 6126 ft. (1867 m) high.

Pa·los (pä′lōs), a seaport in SW Spain: starting point of Columbus's first voyage westward. 2540.

Pa·mirs (pä mērz′), **the,** a mountainous region in central Asia, largely in Tajikistan, where the Hindu Kush, Tien Shan, Kunlun, and Karakoram mountain ranges converge: highest peaks ab. 25,000 ft. (7600 m).

Pam·li·co Sound′ (pam′li kō′), a sound between the North Carolina mainland and coastal islands.

Pam·phyl·i·a (pam fil′ē ə), an ancient region in S Asia Minor: later a Roman province.

Pam·plo·na (pam plō′nə, päm-) also **Pam·pe·lu·na** (päm′pä lōō′nə), a city in N Spain. 183,703.

Pa·na·ji (pə nä′jē), the capital of Goa, in SW India. 76,839.

Pan·a·ma (pan′ə mä′, -mô′), **1.** a republic in S Central America. 2,320,000; 28,575 sq. mi. (74,000 sq. km). **2.** Also called **Pan′ama Cit′y.** the capital of Panama, at the Pacific end of the Panama Canal. 386,393. **3. Isthmus of,** an isthmus between North and South America. **4. Gulf of,** the portion of the Pacific in the bend of the Isthmus of Panama. Also, **Pa·na·má** (*Sp.* pä′nä-mä′) (for defs. 1, 2). —**Pan′a·ma′ni·an** (-mä′nē-ən), *adj., n.*

Pan′ama Canal′, a canal extending SE from the Atlantic to the Pacific across the Isthmus of Panama. 40 mi. (64 km) long.

Pan′ama Canal′ Zone′, CANAL ZONE.

Pan′a·mint Moun′tains (pan′ə mint), a mountain range in E California. Highest peak, 11,045 ft. (3365 m).

Pa·nay (pä nī′), an island in the central Philippines. 4446 sq. mi. (11,515 sq. km).

Pan·mun·jon (pän′mŏŏn′jon′), a village on the border of North Korea and South Korea: site of truce talks at the close of the Korean War.

Pan·no·ni·a (pə nō′nē ə), an ancient Roman province in central Europe, S and W of the Danube, whose territory is now mostly in Hungary and Yugoslavia. —**Pan·no′ni·an,** adj., n.

Pá·nu·co (pä′nə kō′, -nōō-), a river in E central Mexico, flowing E to the Gulf of Mexico. ab. 315 mi. (505 km) long.

Pao·chi (bou′jē′), BAOJI.

Pão de A·çú·car (poun′ di ä sōō′kär), Portuguese name of SUGARLOAF MOUNTAIN.

Pao·king (bou′king′), former name of SHAOYANG.

Pao·ting (bou′ding′), BAODING.

Pa′pal States′, the areas in central Italy ruled by the popes from A.D. 755 until the unification of Italy in 1870. Also called **States of the Church.**

Pa·pe·e·te (pä′pē ā′tā, pə pē′tē), a seaport on NW Tahiti, in the Society Islands: capital of the Society Islands and of French Polynesia. 22,967.

Paph·la·go·ni·a (paf′lə gō′nē ə, -gō′nyə), an ancient country and Roman province in N Asia Minor, on the S coast of the Black Sea.

Pa·phos (pā′fos), an ancient city in SW Cyprus.

Pap·u·a (pap′yŏŏ ə, pä′pŏŏ ä′), **1. Territory of,** a former Australian territory that included the SE part of New Guinea and adjacent islands: now part of Papua New Guinea. **2. Gulf of,** an inlet of the Coral Sea on the SE coast of New Guinea. —**Pap′u·an,** adj., n.

Pap′ua New′ Guin′ea, an independent country comprising the E part of the island of New Guinea and nearby islands: a former Australian Trusteeship Territory; independent since 1975; member of the Commonwealth of Nations. 3,400,000; 178,260 sq. mi. (461,693 sq. km). Cap.: Port Moresby. —**Pap′ua New′ Guin′ean,** n., adj.

Pa·rá (pə rä′), **1.** a state in N Brazil. 4,587,200; 481,869 sq. mi. (1,248,042 sq. km). *Cap.:* Belém. **2.** a river in N Brazil: the S estuary of the Amazon. 200 mi. (320 km) long.

Par·a·guay (par′ə gwī′, -gwā′), **1.** a republic in central South America between Bolivia, Brazil, and Argentina. 4,010,000; 157,047 sq. mi. (406,750 sq. km). *Cap.:* Asunción. **2.** a river in central South America, flowing S from W Brazil through Paraguay to the Paraná. 1500 mi. (2400 km) long. —**Par′a·guay′an,** *adj., n.*

Pa·ra·í·ba (par′ə ē′bə), a state in NE Brazil. 3,104,500; 21,760 sq. mi. (56,360 sq. km). *Cap.:* João Pessoa.

Par·a·mar·i·bo (par′ə mar′ə bō′), a seaport in and the capital of Suriname. 110,867.

Pa·ra·ná (par′ə nä′, pär′-), **1.** a state in SE Brazil. 8,530,000; 76,858 sq. mi. (199,060 sq. km). *Cap.:* Curitiba. **2.** a river flowing from S Brazil along the SE border of Paraguay into the Río de la Plata in E Argentina. 2050 mi. (3300 km) long. **3.** a city in E Argentina, on the Paraná River. 160,000.

Pa·ri·cu·tín (pə rē′kŏŏ tēn′), a volcano in W central Mexico: formed by eruptions 1943–52. 9213 ft. (2808 m).

Par·is (par′is; *Fr.* pA Rē′), the capital of France, in the N part, on the Seine. 2,188,918. —**Pa·ri·sian** (pə rē′zhən), *n., adj.*

Park′ Av′enue, an avenue in New York City traditionally associated with fashionable living.

Par·kers·burg (pär′kərz bûrg′), a city in NW West Virginia, on the Ohio River. 33,862.

Park′ Range′, a range of the Rocky Mountains in central Colorado. Highest peak, Mt. Lincoln, 14,287 ft. (4355 m).

Par·ma (pär′mə), **1.** a city in N Italy, SE of Milan. 179,019. **2.** a city in NE Ohio. 87,876.

Par·na·í·ba (pär′nä ē′bä), a river in NE Brazil, flowing NE to the Atlantic. 900 mi. (1450 km) long.

Par·nas·sus (pär nas′əs), **Mount,** a mountain in central Greece, N of the Gulf of Corinth and near Delphi. ab. 8000 ft. (2440 m).

Par·os (pâr′os, pär′ōs), a Greek island of the Cyclades, in the S Aegean: noted for its white marble. 6776; 77 sq. mi. (200 sq. km). —**Par·i·an** (pâr′ē ən, par′-), *adj.*

Par′ris Is′land (par′is), a U.S. Marine Corps training station in S South Carolina, NE of Savannah, Ga.

Par·sip·pa·ny–Troy′ Hills′ (pär sip′ə nē), a town in N New Jersey. 50,000.

Par·thi·a (pär′thē ə), an ancient country in W Asia, SE of the Caspian Sea, in what is now NE Iran. —**Par′thi·an,** *adj.*, *n.*

Pas·a·de·na (pas′ə dē′nə), **1.** a city in SW California, near Los Angeles. 131,591. **2.** a city in SE Texas, near Houston. 119,363.

Pa·sar·ga·dae (pə sär′gə dē′), an ancient ruined city in S Iran, NE of Persepolis: an early capital of ancient Persia; tomb of Cyrus the Great.

Pa·say (pä′si), a city in E Philippines, on Manila Bay, on E Luzon. 287,770.

Pas de Ca·lais (päd° ᴋᴀ le′), French name of the Strait of Dover.

Pas·sa·ic (pə sā′ik), a city in NE New Jersey. 58,041.

Pas′sa·ma·quod′dy Bay′ (pas′ə mə kwod′ē, pas′-), an inlet of the Bay of Fundy, between Maine and New Brunswick, at the mouth of the St. Croix River.

Pas·to (päs′tō), **1.** a city in SW Colombia. 244,700; ab. 8350 ft. (2545 m) above sea level. **2.** a volcanic peak in SW Colombia, near this city. 13,990 ft. (4265 m).

Pat·a·go·ni·a (pat′ə gō′nē ə, -gōn′yə), a region in S South America, in S Argentina and S Chile, extending from the Andes to the Atlantic. —**Pat′·a·go′ni·an,** *adj.*, *n.*

Pat·er·son (pat′ər sən), a city in NE New Jersey. 140,891.

Pat·mos (pat′mos, -məs, pät′mōs), one of the Dodecanese Islands, in the SE Aegean. 13 sq. mi. (34 sq. km).

Pat·na (put′nə, pat′-, put′nä′), the capital of Bihar state, in NE India, on the Ganges. 916,000.

Pa·tras (pə tras′, pa′trəs), **1.** Greek, **Pa·trai** (pä′ᴛʀɛ). a seaport in the Peloponnesus, in W Greece, on the Gulf of Patras. 112,000. **2. Gulf of,** an inlet of the Ionian Sea in the NW Peloponnesus. 10 mi. (16 km) long; 25 mi. (40 km) wide.

Pau (pō), a city in SW France: resort. 85,056.

Pa·vi·a (pä vē′ä), a city in N Italy, S of Milan. 85,056.

Pa·vlo·dar (pav′lə där′), a city in NE Kazakhstan. 331,000.

Paw·tuck·et (pô tuk′it), a city in NE Rhode Island. 72,644.

Pa·ys de la Loire (pä ē də lA lwAr′), a metropolitan region in NW France. 3,017,700; 12,387 sq. mi. (32,082 sq. km).

Peace′ Riv′er, a river in W Canada, flowing NE from E British Columbia through Alberta to the Slave River. 1050 mi. (1690 km) long.

Pearl′ Har′bor, a harbor near Honolulu, on S Oahu, in Hawaii: surprise attack by Japan on U.S. naval base Dec. 7, 1941.

Pearl′ Riv′er, 1. a river flowing from central Mississippi into the Gulf of Mexico. 485 mi. (780 km) long. **2.** ZHU JIANG.

Pe·cho·ra (pə chôr′ə, -chōr′ə), a river in the NE Russian Federation in Europe, flowing from the Ural Mountains to the Arctic Ocean. 1110 mi. (1785 km) long.

Pe·cos (pā′kəs, -kōs), a river flowing SE from N New Mexico through W Texas to the Rio Grande. 735 mi. (1183 km) long.

Pécs (pāch), a city in SW Hungary. 182,000.

Ped·er·nal·es (pûr′dn al′əs), a river in central Texas, flowing E to the Colorado River. ab. 105 mi. (169 km) long.

Pee·bles (pē′bəlz), a historic county in S Scotland. Also called **Pee′bles·shire′** (-bəlz shēr′, -shər, -bəl-), **Tweeddale.**

Pee Dee (pē′ dē′), a river flowing from North Carolina through South Carolina into the Atlantic. 435 mi. (700 km) long. Compare YADKIN.

PEI, Prince Edward Island.

Pei·ping (pā′ping′, bā′-), former name of BEIJING.

Pei·pus (pī′pəs), a lake in N Europe, on the border between Estonia and the W Russian Federation. 93 mi. (150 km) long; 356 sq. mi. (920 sq. km). Russian, **Chudskoye Ozero.** Estonian, **Peip·si** (pāp′sē).

Pei·rai·evs (pē′re efs′), Greek name of PIRAEUS.

Pe·king (pē′king′, pā′-), BEIJING.

Pe·lée (pə lā′), **Mount,** a volcano in the West Indies, on the island of Martinique: eruption 1902. 4428 ft. (1350 m).

Pel·la (pel′ə), a ruined city in N Greece, NW of Salonika: the capital of ancient Macedonia; birthplace of Alexander the Great.

Pel·o·pon·ne·sus (pel′ə pə nē′səs) also **Pel′o·pon′ne·sos** (-pō′nē sōs), a peninsula forming the S part of Greece: seat of the early Mycenaean civilization and the powerful city-states of Argos, Sparta, etc. 986,912; 8356 sq. mi. (21,640 sq. km). Also called **Morea.** —**Pel′o·pon·ne′sian** (-zhən, -shən), *adj., n.*

Pe·lo·tas (pə lō′təs), a city in S Brazil. 197,092.

Pem·ba (pem′bə), an island of Tanzania near the E coast of Africa. 207,919; 380 sq. mi. (984 sq. km).

Pem·broke (pem′brŏŏk, -brōk), PEMBROKESHIRE.

Pem′broke Pines′ (pem′brōk), a city in SE Florida, near Fort Lauderdale. 65,452.

Pem·broke·shire (pem′brŏŏk shēr′, -shər, -brōk-), a historic county in Dyfed, in SW Wales. Also called **Pembroke.**

Pe·nang (pi nang′, -näng′), **1.** an island in SE Asia, off the W coast of the Malay Peninsula. 110 sq. mi. (285 sq. km). **2.** a state in Malaysia including this island and parts of the adjacent mainland. 954,638; 400 sq. mi. (1036 sq. km). *Cap.:* Georgetown. Malay, **Pinang.**

Pen·chi (*Chin.* bun′chē′), BENXI.

Pen·del·i·kon (pen del′i kon′) also **Pentelikon,** a mountain in SE Greece, near Athens: noted for its fine marble. 3640 ft. (1110 m). Latin, **Pentelicus.**

Pe·ne·us (pə nē′əs), ancient name of PINIOS.

Peng·hu or **P′eng·hu** (pung′hōō′), a group of small islands off the coast of SE China, in the Taiwan Strait: controlled by Taiwan. 115,613; ab. 50 sq. mi. (130 sq. km). Also called **Pescadores.**

Peng·pu (pung′pōō′), BENGBU.

Pen·in·su·la (pə nin′sə lə, -nins′yə lə), **the, 1.** IBERIA (def. 1). **2.** a district in SE Virginia between the York and James rivers: Civil War battles.

Pen·ki (*Chin.* bun′jē′), BENXI.

Penn. or **Penna.,** Pennsylvania.

Pen′nine Alps′ (pen′īn), a mountain range on the border between Switzerland and Italy: part of the Alps. Highest peak, Monte Rosa, 15,217 ft. (4638 m).

Pen·nine Chain′, a range of hills in N England, extending from the S Midlands to the Cheviot Hills. Highest peak, 2930 ft. (893 m).

Penn·syl·va·nia (pen/səl vān′/yə, -vā′/nē ə), a state in the E United States. 11,881,643; 45,333 sq. mi. (117,410 sq. km). *Cap.:* Harrisburg. *Abbr.:* PA, Pa., Penn., Penna. **—Penn′syl·va′nian,** *n., adj.*

Pe·nob·scot (pə nob′/skot, -skət), a river flowing S from central Maine into Penobscot Bay. 350 mi. (565 km) long.

Penob′scot Bay′, an inlet of the Atlantic in S Maine. 30 mi. (48 km) long.

Pen·sa·co·la (pen/sə kō′lə), a seaport in NW Florida, on Pensacola Bay. 58,165.

Pen′saco′la Bay′, an inlet of the Gulf of Mexico, in NW Florida. ab. 30 mi. (48 km) long.

Pen·tel·i·cus (pen tel′i kəs), Latin name of PEN-DELIKON.

Pen·tel·i·kon (pen tel′i kon′; *Gk.* pen/de lē-kôn′), PENDELIKON.

Pen·za (pen′/zə), a city in the W Russian Federation in Europe, S of Nizhni Novgorod. 543,000.

Pen·zance (pen zans′), a seaport in SW Cornwall, in the SW extremity of England: resort. 19,521.

Peo/ple′s Democrat′ic Repub′lic of Yem′en, YEMEN (def. 3).

Pe·o·ri·a (pē ôr′ē ə, -ōr′-), **1.** a city in central Illinois, on the Illinois River. 113,504. **2.** a city in central Arizona, near Phoenix. 50,618.

Pe·ra (per′ə), former name of BEYOĞLU.

Pe·rae·a (pə rē′ə), a region in ancient Palestine, E of the Jordan and the Dead Sea.

Pe·rak (pā′rak, -räk, per′ə, pēr′ə), a state in Malaysia, on the SW Malay Peninsula. 1,805,198; 7980 sq. mi. (20,668 sq. km). *Cap.:* Ipoh.

Pe·rei·ra (pə rār′ə), a city in W Colombia. 287,999.

Per·ga·mum (pûr′/gə məm) also **Per′ga·mon** (-mən, -mon′), **1.** an ancient Greek kingdom on the coast of Asia Minor: later a Roman province. **2.** the ancient capital of this kingdom: now the site of Bergama, in W Turkey.

Pé·ri·gord (pā Rē gôR′), a division of the former province of Guienne, in SW France.

Perm (pûrm, pârm), a city in the E Russian Federation in Europe, on the Kama River. 1,091,000. Formerly, **Molotov.**

Per·nam·bu·co (pûr′nəm bōō′kō), **1.** a state in NE Brazil. 6,990,300; 38,000 sq. mi. (98,420 sq. km). *Cap.:* Recife. **2.** former name of Recife.

Per·nik (per′nik), a city in W Bulgaria, near Sofia. 97,225.

Per·pi·gnan (peR pē nyän′), a city in S France. 113,646.

Per·sep·o·lis (pər sep′ə lis), an ancient capital of Persia: its ruins are near Shiraz in SW Iran.

Per·sia (pûr′zhə, -shə), **1.** Also called **Per′sian Em′pire.** an ancient empire located in W and SW Asia: at its height it extended from Egypt and the Aegean to India; conquered by Alexander the Great 334–331 B.C. **2.** former official name (until 1935) of Iran. —**Per′sian,** *adj., n.*

Per′sian Gulf′, an arm of the Arabian Sea, between SW Iran and Arabia. 600 mi. (965 km) long. Also called **Arabian Gulf.**

Per′sian Gulf′ States′, Gulf States (def. 2).

Perth (pûrth), **1.** Also called **Perth′shire** (-shēr, -shər). a historic county in central Scotland. **2.** a city in this county: a port on the Tay River. 42,438. **3.** the capital of Western Australia, in SW Australia. 1,025,300.

Pe·ru (pə rōō′), a republic in W South America. 21,300,000; 496,222 sq. mi. (1,285,215 sq. km). *Cap.:* Lima. Spanish, **Pe·rú** (pe rōō′). —**Pe·ru·vi·an,** *adj., n.*

Pe·ru·gia (pə rōō′jə, -jē ə), a city in Umbria, in central Italy. 147,602. —**Pe·ru′gian,** *adj., n.*

Pe·sa·ro (pā′zə rō′, -sə-), a seaport in E Italy, on the Adriatic Sea. 90,147.

Pes·ca·do·res (pes′kə dôr′is, -ēz, -dōr′-), Penghu.

Pes·ca·ra (pe skär′ə), a city in E Italy, on the Adriatic Sea. 130,525.

Pe·sha·war (pə shä′wər), a city in N Pakistan, near the Khyber Pass: capital of the North-West Frontier Province. 555,000.

Pe·tach Tik·va or **Pe·tah Tiq·wa** (pe′täкн tik′vä), a city in W Israel, NE of Tel Aviv. 132,100.

Pe·ter·bor·ough (pē′tər bûr′ō, -bur′ō, -bər ə),

1. a city in Cambridgeshire, in central England. 151,200. **2.** a city in SE Ontario, in SE Canada. 61,049. **3. Soke of** (sōk), a former administrative division in Cambridgeshire, in central England.

Pe·ters·burg (pē'tərz bûrg'), a city in SE Virginia: besieged by Union forces 1864–65. 36,298.

Pe·tra (pē'trə, pe'-), an ancient city in what is now SW Jordan: capital of the Nabataeans.

Pet'ri·fied For'est Na'tional Park', a national park in E Arizona: buried tree trunks turned to stone by the action of mineral-laden water. 147 sq. mi. (381 sq. km).

Pet·ro·grad (pe'trə grad'), a former name (1914–24) of St. **Petersburg** (def. 1).

Pe·tro·pa·vlovsk (pe'trə pav'lôfsk, -lofsk), a city in N Kazakhstan. 233,000.

Petropa'vlovsk-Kam·chat'ski (kam chät'skē), a city in SE Kamchatka, in the E Russian Federation in Asia. 252,000.

Pe·tró·po·lis (pi trop'ə lis), a city in SE Brazil, NE of Rio de Janeiro. 149,427.

Pe·tro·za·vodsk (pe'trə zə votsk'), the capital of the Karelian Autonomous Republic, in the NW Russian Federation in Europe. 270,000.

Pfalz (pfälts), German name of the **Palatinate**.

Pforz·heim (fôrts'hīm', pfôrts'-), a city in W Baden-Württemberg, in SW Germany. 106,600.

Pha·ros (fâr'os), a small peninsula in N Egypt, near Alexandria: site of ancient lighthouse built during the reign of Ptolemy II.

Phar·sa·lus (fär sā'ləs), an ancient city in central Greece, in Thessaly: site of Caesar's victory over Pompey, 48 B.C.

Phe·ni·cia (fi nish'ə, -nē'shə), **Phoenicia**.

Phil·a·del·phi·a (fil'ə del'fē ə), a city in SE Pennsylvania, at the confluence of the Delaware and Schuylkill rivers. 1,585,577.

Phi·lae (fī'lē), an island in the Nile, in Upper Egypt: the site of ancient temples; now submerged by the waters of Lake Nasser.

Phi·lip·pi (fi lip'ī, fil'ə pī'), a ruined city in NE Greece, in Macedonia: Octavian and Mark Antony defeated Brutus and Cassius here, 42 B.C. —**Philip'pi·an** (-ē ən), adj., n.

Phil·ip·pines (fil'ə pēnz', fil'ə pēnz'), a republic comprising an archipelago of 7083 islands in the Pacific, SE of China: formerly (1898–1946) under the guardianship of the U.S. 60,477,000; 114,830 sq. mi. (297,410 sq. km). *Cap.:* Manila. Also called **Phil'ippine Is'lands.** Official name, **Republic of the Philippines.** —**Phil'ip·pine'**, *adj.*

Phil·ip·pop·o·lis (fil'ə pop'ə lis), Greek name of PLOVDIV.

Phi·lis·ti·a (fi lis'tē ə), an ancient country in SW Palestine on the Mediterranean coast: the land of the Philistines. —**Phil·is·tine** (fil'ə stēn', -stin'), *n., adj.*

Phnom (or **Pnom**) **Penh** (nom' pen', pə nôm'), the capital of Cambodia, in the S part. 500,000.

Pho·cae·a (fō sē'ə), an ancient seaport in Asia Minor: northernmost of the Ionian cities; later an important maritime state.

Pho·cis (fō'sis), an ancient district in central Greece, N of the Gulf of Corinth: site of Delphic oracle.

Phoe·ni·cia (fi nish'ə, -nē'shə), an ancient kingdom on the Mediterranean, in the region of modern Lebanon and Syria. —**Phoe·ni'cian,** *n., adj.*

Phoe·nix (fē'niks), the capital of Arizona, in the central part. 983,403.

Phoe'nix Is'lands, a group of eight coral islands in the central Pacific: part of Kiribati. 11 sq. mi. (28 sq. km).

Phryg·i·a (frij'ē ə), an ancient country in central and NW Asia Minor. —**Phryg'i·an,** *n., adj.*

Phu·ket (pōō'ket'), an island near the W coast of Thailand in the Andaman Sea: beach resorts. 146,400; 309 sq. mi. (801 sq. km).

P.I., Philippine Islands.

Pia·cen·za (pyä chen'tsä), a city in N Italy, on the Po River. 104,976. Ancient, **Placentia.**

Piau·í (pyou ē'), a state in NE Brazil. 2,517,900; 96,860 sq. mi. (250,870 sq. km). *Cap.:* Teresina.

Pic·ar·dy (pik'ər dē), **1.** a historic region and former province in N France. **2.** a metropolitan region in N France. 1,774,000; 7490 sq. mi. (19,399 sq. km). French, **Pi·car·die** (pē kàr dē').

Pic'ca·dil'ly Cir'cus (pik'ə dil'ē, -dil'ē), a traffic circle and open square in W London, England: theater and amusement center.

Pi·co Ri·ve·ra (pē'kō ri vâr'ə, -vēr'ə), a city in SW California, near Los Angeles. 59,177.

Pied·mont (pēd'mont), **1.** a plateau between the coastal plain and the Appalachian Mountains, including parts of Virginia, North Carolina, South Carolina, Georgia, and Alabama. **2.** Italian, **Pie·mon·te** (pye môn′te). a region in NW Italy. 4,377,229; 11,335 sq. mi. (29,360 sq. km). —**Pied·mon·tese** (pēd′mon tēz′, -tēs′), *n., pl.* **-tese,** *adj.*

Pi·e·ri·a (pi ēr′ē ə), a coastal region in NE Greece, in Macedonia, W of the Gulf of Salonika.

Pierre (pēr), the capital of South Dakota, in the central part, on the Missouri River. 12,906.

Pie·ter·mar·itz·burg (pē′tər mar′its bûrg′), the capital of Natal province, in the E Republic of South Africa. 192,417.

Pigs (pigz), **Bay of,** BAY OF PIGS.

Pikes′ Peak′ (piks), a mountain in central Colorado: a peak of the Rocky Mountains. 14,108 ft. (4300 m).

Pik Po·be·dy (pyēk′ pu bye′di), Russian name of POBEDA PEAK.

Pi·la·tus (pē lä′təs), a mountain in central Switzerland, near Lucerne: a peak of the Alps. 6998 ft. (2130 m).

Pil·co·ma·yo (pēl′kə mä′yō), a river in S central South America, flowing SE from Bolivia along the border between Paraguay and Argentina to the Paraguay River at Asunción. 1000 mi. (1610 km) long.

Pil′lars of Her′cules, the two promontories on either side of the eastern end of the Strait of Gibraltar, the Rock of Gibraltar in Europe and the Jebel Musa in Africa.

Pil·sen (pil′zən), German name of PLZE.

Pi·nang (pi nang′, -näng′), Malay name of PENANG.

Pi·nar del Rí·o (pi när′ del rē′ō), *n.* a city in W Cuba. 100,900.

Pin·dus (pin′dəs), a mountain range in central Greece: highest peak, 7665 ft. (2335 m).

Pine′ Bar′rens, the, an extensive coastal region in S and SE New Jersey, composed chiefly of pine stands, sandy soils, and swampy streams. ab. 2000 sq. mi. (5180 sq. km). Official name, **the Pine·lands** (pīn′landz′, -ləndz).

Pine′ Bluff′, a city in central Arkansas, on the Arkansas River. 57,140.

Pines (pīnz), **Isle of,** former name of the Isle of YOUTH.

Pi·ni·os (pē nyôs′), a river in N Greece, in Thessaly, flowing E to the Gulf of Salonika. 125 mi. (200 km) long. Ancient, **Peneus.** Formerly, **Salambria.**

Pinsk (pinsk), a city in SW Belarus, E of Brest. 106,000.

Pi·rae·us (pi rē′əs, pi rā′-), a seaport in SE Greece: the port of Athens. 196,389. Greek, **Peiraievs.**

Pi·sa (pē′zə, -zä), a city in NW Italy, on the Arno River: leaning tower. 103,527. —**Pi′san,** *adj., n.*

Pis·gah (piz′gə), **Mount,** a mountain ridge of ancient Moab, now in Jordan, NE of the Dead Sea: from its summit **(Mount Nebo)** Moses viewed the Promised Land.

Pish·pek (pish pek′), a former name (until 1926) of BISHKEK.

Pis·to·ia (pē stô′yä), a city in N Tuscany, in N Italy. 93,516.

Pit′cairn Is′land (pit′kârn), a small British island in the S Pacific, SE of Tuamotu Archipelago: settled 1790 by mutineers of H.M.S. *Bounty.* 59; 2 sq. mi. (5 sq. km).

Pitts·burgh (pits′bûrg), a port in SW Pennsylvania, at the confluence of the Allegheny and Monongahela rivers that forms the Ohio River. 369,879.

Pla·cen·tia (plə sen′shə, -shē ə), ancient name of PIACENZA.

Plains′ of A′bra·ham (ā′brə ham′, -həm), a high plain adjoining the city of Quebec, Canada: English victory over the French in 1759.

Pla·no (plā′nō), a city in N Texas. 128,713.

Plan·ta·tion (plan tā′shən), a city in S Florida. 66,692.

Pla·ta (plä′tä), **Rí·o de la** (rē′ō dä lä), an estuary on the SE coast of South America between Argentina and Uruguay, formed by the Uruguay and Paraná rivers. ab. 185 mi. (290 km) long. English, **River Plate.**

Pla·tae·a (plə tē′ə), an ancient city in Greece, in Boeotia: Greeks defeated Persians here 479 B.C. —**Pla·tae′an,** *adj., n.*

Plate (plāt), **River,** English name of Río de la PLATA.

Platte (plat), a river flowing E from the junction of the North and South Platte rivers in central Nebraska to the Missouri River S of Omaha. 310 mi. (500 km) long.

Platts·burgh (plats′bûrg), a city in NE New York, on Lake Champlain: battle 1814. 21,255.

Plau·en (plou′ən), a city in E Germany. 78,632.

Pleas′ant Is′land, former name of Nauru.

Pleas·an·ton (plez′ən tən), a city in W California. 50,553.

Plev·en (plev′en) also **Plev·na** (-nä), a city in N Bulgaria. 361,000.

Plo·eş·ti or **Plo·ieş·ti** (plô yesht′), a city in S Romania. 234,021.

Plov·div (plôv′dif), a city in S Bulgaria, on the Maritsa River. 356,596. Greek, **Philippopolis.**

Plym·outh (plim′əth), **1.** a seaport in SW Devonshire, in SW England, on the English Channel: the departing point of the *Mayflower* 1620. 257,900. **2.** a city in SE Massachusetts: the oldest town in New England, founded by the Pilgrims 1620. 45,608. **3.** a town in SE Minnesota. 50,889.

Pl·zeň (pōōl′zen/yə), a city in Bohemia, in W Czechoslovakia. 175,000. German, **Pilsen.**

Pnom Penh (nom′ pen′, pə nôm′), Phnom Penh.

Po (pō), a river in Italy, flowing E from the Alps in the NW to the Adriatic. 418 mi. (669 km) long. Ancient, **Padus.**

Po·be′da Peak′ (pə bed′ə), a mountain in central Asia: on the boundary between Kyrgyzstan and China: highest peak of the Tien Shan range. 24,406 ft. (7439 m). Russian, **Pik Pobedy.**

Po·ca·tel·lo (pō′kə tel′ō), a city in SE Idaho. 46,080.

Po′co·no Moun′tains (pō′kə nō′), a mountain range in NE Pennsylvania: resort area. ab. 2000 ft. (610 m) high. Also called **Po′co·nos′.**

Pod·go·ri·ca (pod′gə rēt′sə), the capital of Montenegro, in SW Yugoslavia. 132,290. Formerly (1945–92), **Titograd.**

Po·dolsk (pu dôlsk′), a city in the W Russian Federation in Europe, S of Moscow. 209,000.

Po·hai (*Chin.* bô′hī′), Bohai.

Pohn·pei (pōn′pā), an island in the W Pacific, in the Caroline group: part of the Federated States of Micronesia. 129 sq. mi. (334 sq. km). Formerly, **Ponape.**

Pointe-à-Pi·tre (*Fr.* pwan tA pē′tr°), a seaport

on central Guadeloupe, in the E West Indies. 25,310.

Pointe-Noire (*Fr.* pwANt NWAR⁄), a seaport in the S People's Republic of the Congo. 294,203.

Poi·tiers (pwä tyā⁄), a city in W France. 82,884.

Poi·tou (pwa tōō⁄), a region and former province in W France.

Poi·tou-Cha·rentes (pwa tōō shạ RäNt⁄), a metropolitan region in W France. 1,583,600; 9965 sq. mi. (25,810 sq. km).

Po·land (pō⁄lənd), a republic in E central Europe, on the Baltic Sea. 37,800,000; ab. 121,000 sq. mi. (313,400 sq. km). *Cap.:* Warsaw. Polish, **Pol·ska** (pôl⁄skä). **—Po⁄lish,** *adj.*

Po⁄lish Cor⁄ridor, a strip of land near the mouth of the Vistula River: formerly separated Germany from East Prussia; given to Poland in the Treaty of Versailles 1919 to provide it with access to the Baltic.

Pol·ta·va (pəl tä⁄və), a city in E Ukraine, SW of Kharkov: Russian defeat of Swedes 1709. 309,000.

Pol·to·ratsk (pol⁄tə rätsk⁄), former name of Ashkhabad.

Pol·y·ne·sia (pol⁄ə nē⁄zhə, -shə), one of the three principal divisions of Oceania, comprising those island groups in the Pacific lying E of Melanesia and Micronesia and extending from the Hawaiian Islands S to New Zealand. **—Pol⁄y·ne⁄sian,** *adj.*, *n.*

Pom·er·a·ni·a (pom⁄ə rā⁄nē ə, -rān⁄yə), a former province of NE Germany, now mostly in NW Poland. German, **Pom·mern** (pôm⁄ərn). **—Pom⁄er·a⁄ni·an,** *adj.*, *n.*

Po·mo·na (pə mō⁄nə), **1.** a city in SW California, E of Los Angeles. 131,723. **2.** Also called **Mainland.** the largest of the Orkney Islands, N of Scotland. 6502; 190 sq. mi. (490 sq. km).

Pom·pa·no Beach⁄ (pom⁄pə nō⁄), a city in SE Florida. 72,411.

Pom·peii (pom pā⁄, -pā⁄ē), an ancient city in SW Italy, on the Bay of Naples: buried along with Herculaneum by an eruption of nearby Mount Vesuvius in A.D. 79; much of the city has been excavated. **—Pom·pe⁄ian, Pom·pei⁄ian,** *n.*, *adj.*

Po·na·pe (pō⁄nə pā⁄, pon⁄ə-), former name of
Pohnpei.

Pon·ce (pon′sā), a seaport in S Puerto Rico. 190,679.

Pon·di·cher·ry (pon′di cher′ē, -sher′ē) also **Pon·di·ché·ry** (pôn dē shä rē′), **1.** a union territory of SE India, on the Coromandel Coast: formerly the chief settlement of French India. 604,136; 181 sq. mi. (469 sq. km). **2.** the capital of this territory. 251,471.

Pon·ta Del·ga·da (pon′tə del gä′də), a seaport on SW São Miguel island, in the E Azores. 69,930.

Pont·char·train (pon′chər trān′), **Lake,** a shallow extension of the Gulf of Mexico in SE Louisiana, N of New Orleans. 41 mi. (66 km) long; 25 mi. (40 km) wide.

Pon·ti·ac (pon′tē ak′), a city in SE Michigan. 71,166.

Pon·ti·a·nak (pon′tē ä′näk), a seaport on W Kalimantan (Borneo), in central Indonesia. 304,778.

Pon′tine Marsh′es (pon′tēn, -tin), an area in W Italy, SE of Rome: formerly marshy, now drained.

Pon·tus (pon′təs), an ancient country in NE Asia Minor, bordering on the Black Sea: later a Roman province. —**Pon′tic,** *adj.*

Pon′tus Eux·i′nus (yōōk si′nəs), ancient name of the BLACK SEA.

Poo·na (pōō′nə), a city in W Maharashtra, W India. 1,685,000.

Po·po·ca·té·petl (pô′pô kä te′pet′l, pō′pə kat′ə pet′l), a volcano in S central Mexico, SE of Mexico City. 17,887 ft. (5450 m).

Po·ri (pôr′ē), a seaport in W Finland, on the Gulf of Bothnia. 77,395.

Port′ Ar′thur (är′thər), **1.** a seaport in SE Texas, on Sabine Lake. 58,724. **2.** See under THUNDER BAY. **3.** LÜSHUN.

Port-au-Prince (pôrt′ō prins′, PRANS′, pōrt′-), the capital of Haiti, in the S part. 763,188.

Port′ Blair′ (blâr), the capital of the Andaman and Nicobar Islands, on S Andaman. 49,634.

Port′ Eliz′abeth, a seaport in the SE Cape of Good Hope province, in the S Republic of South Africa. 651,993.

Port′ Har′court (här′kərt, -kôrt, -kōrt), a seaport in S Nigeria. 296,200.

Port′ Jack′son, an inlet of the Pacific in SE Australia: the harbor of Sydney.

Port·land (pôrt′lənd, pōrt′-), **1.** a seaport in NW

Oregon, at the confluence of the Willamette and Columbia rivers. 437,319. **2.** a seaport in SW Maine, on Casco Bay. 64,358.

Port′ Lou′is (lōō′is, lōō′ē), a seaport in and the capital of Mauritius, on the NW coast. 139,038.

Port′ Mores′by (môrz′bē, môrz′-), a seaport in SE New Guinea: capital of Papua New Guinea. 152,100.

Pôr·to (pôr′tōō), Portuguese name of OPORTO.

Pôr·to A·le·gre (pôr′tōō ə leg′rə), the capital of Rio Grande do Sul, in S Brazil. 1,114,867.

Port′-of-Spain′, a seaport on NW Trinidad, in the SE West Indies: national capital of Trinidad and Tobago. 58,400.

Por·to No·vo (pôr′tō nō′vō, pōr′-), a seaport in and the capital of Benin, on the Gulf of Guinea. 208,258.

Por·to Ri·co (pôr′tə rē′kō, pōr′-), former name (until 1932) of PUERTO RICO. —**Por′to Ri′can,** adj., n.

Pôr·to Ve·lho (pôr′tōō vel′yōō), the capital of Rondônia, in W Brazil, on the Madeira River. 101,644.

Port′ Phil′lip Bay′ (fil′ip), a bay in SE Australia: the harbor of Melbourne. 31 mi. (50 km) long; 25 mi. (40 km) wide.

Port′ Roy′al, a historic town on SE Jamaica at the entrance to Kingston harbor: a former capital of Jamaica.

Port′ Sa·id′ (sä ēd′), a seaport in NE Egypt at the Mediterranean end of the Suez Canal. 382,000.

Ports·mouth (pôrts′məth, pōrts′-), **1.** a seaport in S Hampshire, in S England, on the English Channel. 186,800. **2.** a seaport in SE Virginia. 103,907. **3.** a seaport in SE New Hampshire: Russian-Japanese peace treaty 1905. 25,925.

Port St. Lu·cie (lōō′sē), a city in E Florida. 55,866.

Port′ Sudan′, a seaport in the NE Sudan, on the Red Sea. 206,727.

Por·tu·gal (pôr′chə gəl, pōr′-), a republic in SW Europe, on the Iberian Peninsula, W of Spain (including the Azores and the Madeira Islands). 10,290,000; 35,414 sq. mi. (91,720 sq. km). Cap.: Lisbon. —**Por′tu·guese′** (pôr′chə gēz′, -gēs′, pōr′-), n., pl. **-guese,** adj.

Por·tuguese East′ Af′rica, former name of MOZAMBIQUE.

Por·tuguese Guin′ea, former name of GUINEA-BISSAU.

Por·tuguese In′dia, a former Portuguese overseas territory on the W coast of India: annexed by India 1961. Compare DAMAN AND DIU, GOA.

Por·tuguese West′ Af′rica, former name of ANGOLA.

Po·to·mac (pə tō′mək), a river flowing SE from the Allegheny Mountains in West Virginia, along the boundary between Maryland and Virginia to Chesapeake Bay. 287 mi. (460 km) long.

Po·to·sí (pô′tô sē′), a city in S Bolivia. 113,000; 13,022 ft. (3970 m) above sea level.

Pots·dam (pots′dam), the capital of Brandenburg in NE Germany, SW of Berlin. 142,860.

Pow·ys (pō′is), a county in E Wales. 116,800; 1960 sq. mi. (5077 sq. km).

Po·yang (pô′yäng′), a lake in E China, in Jiangxi province. 90 mi. (145 km) long.

Poz·nań (pōz′nan, -nän; *Pol.* pôz′nän/yə), a city in W Poland, on the Warta River. 553,000.

Poz·zuo·li (pot swô′lē), a seaport in SW Italy, near Naples. 70,350.

PR or **P.R.,** Puerto Rico.

Prague (präg), the capital of Czechoslovakia and the Czech Republic, in the W part, on the river Vltava. 1,211,000. Czech, **Pra·ha** (prä′hä).

Prai·a (pri′ə), the capital of Cape Verde, in the S Atlantic Ocean, on S São Tiago Island. 39,000.

Prai′rie Prov′inces, the provinces of Manitoba, Saskatchewan, and Alberta, in W Canada.

Pra·to (prä′tō), a city in central Italy, near Florence. 160,220.

Pres·ton (pres′tən), a seaport in W Lancashire, in NW England. 126,700.

Pre·to·ri·a (pri tôr′ē ə, -tōr′-), the administrative capital of the Republic of South Africa, in the NE part: also the capital of Transvaal. 822,925.

Preus·sen (proi′sən), German name of PRUSSIA.

Prib/i·lof Is/lands (prib′ə lôf′, -lof′), a group of islands in the Bering Sea, SW of Alaska, belonging to the U.S.: breeding ground of fur seals.

Prince′ Al′bert, a city in central Saskatchewan, in S Canada. 28,631.

Prince′ Al′bert Na′tional Park′, a national

park in W Canada, in central Saskatchewan. 1496 sq. mi. (3875 sq. km).

Prince′ Ed′ward Is′land, an island in the Gulf of St. Lawrence, forming a province of Canada: 126,646; 2184 sq. mi. (5657 sq. km). *Cap.:* Charlottetown.

Prince′ George′, a city in central British Columbia, in W Canada. 67,621.

Prince′ of Wales′, Cape, a cape in W Alaska, on Bering Strait: the westernmost point of North America.

Prince′ of Wales′ Is′land, 1. the largest island in the Alexander Archipelago, in SE Alaska. 2231 sq. mi. (5778 sq. km). **2.** an island in N Canada, in the Northwest Territories. 12,830 sq. mi. (33,230 sq. km).

Prince′ Ru′pert (rōō′pərt), a seaport and railway terminus in W British Columbia, in W Canada. 14,754.

Prince·ton (prins′tən), a borough in central New Jersey. 12,016.

Prince′ Wil′liam Sound′, an inlet of the Gulf of Alaska, in S Alaska, E of the Kenai Peninsula.

Prín·ci·pe (prin′sə pə, -pā′), an island in the Gulf of Guinea, off the W coast of Africa. 5255; 54 sq. mi. (140 sq. km). Compare SO TOMÉ AND PRÍNCIPE.

Prip·et (prip′it, -et, prē′pet), a river in NW Ukraine and S Belarus, flowing E through the Pripet Marshes to the Dnieper River in NW Ukraine. 500 mi. (800 km) long. Russian, **Pri·pyat** (prȳē′pyit).

Prip′et Marsh′es, an extensive wooded marshland in S Belarus and NW Ukraine. 33,500 sq. mi. (86,765 sq. km).

Priš·ti·na (prish′ti nə), the capital of Kosovo, in S Yugoslavia. 210,040.

Pro·ko·pyevsk (prə kôp′yəfsk), a city in the S central Russian Federation in Asia, NW of Novokuznetsk. 278,000.

Pro·vence (prə väns′, -väns′), a region in SE France, bordering on the Mediterranean: formerly a province. —**Pro·ven·çal** (prō′vən säl′, prov′-ən-; *Fr.* prô vän sal′), *adj., n.*

Pro·vence-Côte d′A·zur (prô väns kōt dA zyr′), a metropolitan region in SE France. 4,058,800; 12,124 sq. mi. (31,400 sq. km).

Prov·i·dence (prov'i dəns), the capital of Rhode Island, in the NE part, at the head of Narragansett Bay. 160,728.

Prov·ince·town (prov'ins toun'), a town at the tip of Cape Cod, in SE Massachusetts: resort. 3536.

Pro·vo (prō'vō), a city in central Utah. 86,835.

Prud'hoe Bay' (prōō'dō), an inlet of the Beaufort Sea, N of Alaska: oil and gas fields.

Prus·sia (prush'ə), a former state in N Europe: became a military power in the 18th century and in 1871 led the formation of the German empire; formally abolished as an administrative unit in 1947. German, **Preussen.** Compare EAST PRUSSIA, WEST PRUSSIA. **—Prus'sian,** adj., n.

Prut (prōōt), a river in E Europe, flowing SE from the Carpathian Mountains in Ukraine along the boundary between Moldova and Romania into the Danube. 500 mi. (800 km) long.

Pskov (pskôf), **1.** a lake in N Europe, between Estonia and the W Russian Federation, forming the S part of Lake Peipus. **2.** a city near this lake, in the NW Russian Federation. 202,000.

Pue·bla (pweb'lä), **1.** a state in S central Mexico. 4,068,038; 13,124 sq. mi. (33,990 sq. km). **2.** the capital of this state, in the N part. 835,759.

Pueb·lo (pweb'lō), a city in central Colorado. 98,640.

Puer·to Ca·bel·lo (pwer'tō kä vä'yō, -bä'-), a seaport in N Venezuela. 94,000.

Puer'to Li·món' (lē mōn'), LIMÓN.

Puer'to Montt' (mônt), a city in S Chile. 113,488.

Puer·to Ri·co (pwer'tə rē'kō, pwer'tō, pôr'tə, pōr'-), an island in the central West Indies: a commonwealth associated with the U.S. 3,196,520; 3435 sq. mi. (8895 sq. km). Cap.: San Juan. Abbr: PR, P.R. Formerly (until 1932), **Porto Rico. —Puer'to Ri'can,** n., adj.

Puer·to Val·lar·ta (pwer'tō vä yär'tä), a city in W Mexico: resort. 70,000.

Pu'get Sound' (pyōō'jit), an arm of the Pacific, in NW Washington.

Pu·glia (pōō'lyä), Italian name of APULIA.

Pu·la (pōō'lä), a seaport in W Croatia, on the Istrian Peninsula, on the Adriatic. 77,057.

Pun·cak Ja·ya (pōōn'chäk jä'yä), a mountain in

Irian Jaya, Indonesia, on W New Guinea: highest island point in the world. 16,503 ft. (5030 m). Also called **Djaja Peak.** Formerly, **Mount Carstensz.**

Pun·jab (pun jäb′, pun′jäb), **1.** a former province in NW British India: now divided between India and Pakistan. **2.** a state in NW India. 16,788,915; 19,445 sq. mi. (50,362 sq. km). *Cap.:* Chandigarh. **3.** a province in NE Pakistan. 53,840,000; 79,284 sq. mi. (205,330 sq. km). *Cap.:* Lahore. —**Pun·ja′bi** (-jä′bē) *n., pl.* **-bis,** *adj.*

Pun·ta A·re·nas (pōōn′tä ä rā′näs), a seaport in S Chile, on the Strait of Magellan. 111,724.

Pu·ri (pŏōr′ē, pŏō rē′), a seaport in E Orissa, in E India, on the Bay of Bengal: temple of Krishna; Hindu pilgrimage center. 101,089.

Pu·rús (pŏō rōōs′), a river in NW central South America, flowing NE from E Peru through W Brazil to the Amazon. 2000 mi. (3200 km) long.

Pu·san (pōō′sän′), a seaport in SE South Korea. 3,516,768.

Pu·tu·ma·yo (pōō′tōō mä′yō), a river in NW South America, flowing SE from S Colombia into the Amazon in NW Brazil. 900 mi. (1450 km) long. Portuguese, **Içá.**

Pya·ti·gorsk (pyä′ti gôrsk′), a city in the SW Russian Federation in Europe, in Caucasia. 110,000.

Pyd·na (pid′nə), a town in ancient Macedonia, W of the Gulf of Salonika: decisive Roman victory over the Macedonians 186 B.C.

Pyong·yang (pyung′yäng′, -yang′, pyong′-), the capital of North Korea, in the SW part. 2,639,448.

Pyr·e·nees (pir′ə nēz′), a mountain range between Spain and France. Highest peak, Pico de Aneto, 11,165 ft. (3400 m). —**Pyr′e·ne′an,** *adj.*

Q

Qa·tar (kä′tär, kə tär′), an independent emirate on the Persian Gulf: under British protection until 1971. 371,863; 8500 sq. mi. (22,000 sq. km). *Cap.:* Doha. —**Qa·tar′i,** *adj., n.*

Qaz·vin or **Kaz·vin** (kaz vēn′), a city in NW Iran, NW of Teheran: capital of Persia in the 16th century. 248,591.

Qi·lian Shan (chē′lyän′ shän′), a mountain range in W China, on the border between Qinghai and Gansu provinces. Formerly, **Nan Shan.**

Qing·dao or **Tsing·tao** (ching′dou′), a seaport in E Shandong province, in E China, on the Yellow Sea. 1,250,000.

Qing·hai or **Ch'ing·hai** or **Tsing·hai** (ching′hī′), **1.** a province in NW China. 4,120,000; 278,400 sq. mi. (721,000 sq. km). *Cap.:* Xining. **2.** Also called **Koko Nor.** a salt lake in NE Qinghai province. 2300 sq. mi. (5950 sq. km).

Qin·huang·dao or **Chin·huang·tao** or **Chin·wang·tao** (chin′wäng′dou′), a seaport in NE Hebei province, in NE China, on the Bohai. 210,000.

Qiong′zhou′ Strait′ (chyông′jō′), a strait between Hainan island and Leizhou peninsula. 50 mi. (81 km) long; 15 mi. (24 km) wide. Also called **Hainan Strait.**

Qi·qi·har or **Chi·chi·haerh** or **Tsi·tsi·har** (chē′chē′här′), a city in W Heilongjiang province, in NE China. 1,260,000.

Qom or **Qum** (kōōm), a city in NW Iran, SW of Teheran. 543,139.

Quan·zhou or **Chuan·chow** (chwän′jō′), a seaport in SE Fujian province, in SE China, on Taiwan Strait. 410,229.

Quath·lam·ba (kwät läm′bə), DRAKENSBERG.

Que., Quebec.

Que·bec (kwi bek′, ki-), **1.** a province in E Canada. 6,532,461; 594,860 sq. mi. (1,540,685 sq. km). **2.** the capital of this province, on the St. Lawrence. 164,580. French, **Qué·bec′** (kā-). —**Que·bec′er, Que·beck′er,** *n.*

Queen′ Char′lotte Is′lands, a group of islands in British Columbia off the W coast of Canada. 4500; 3970 sq. mi. (10,280 sq. km).

Queen′ Eliz′abeth Is′lands, a group of islands, including Ellesmere Island, in the Arctic Ocean, in the N Northwest Territories, N Canada.

Queen′ Maud′ Land′ (môd), a coastal region of Antarctica, S of Africa: Norwegian explorations.

Queens (kwēnz), a borough of New York City, on Long Island. 1,951,598; 113.1 sq. mi. (295 sq. km).

Queens·land (kwēnz′land′, -lənd), a state in NE Australia. 2,649,600; 670,500 sq. mi. (1,736,595 sq. km). *Cap.:* Brisbane.

Queens·town (kwēnz′toun′), former name of Cóbh.

Quel·part (kwel′pärt), former name of Cheju.

Que·moy (ki moi′), an island off the SE coast of China, in the Taiwan Strait: administered by Taiwan. 61,305; 50 sq. mi. (130 sq. km).

Que·ré·ta·ro (kə ret′ə rō′, -rā′tə-), **1.** a state in central Mexico. 952,875; 4432 sq. mi. (11,480 sq. km). **2.** the capital of this state, in the SW part. 293,586.

Quet·ta (kwet′ə), a city in W central Pakistan: the capital of Baluchistan. 285,000.

Que·zal·te·nan·go (ke säl′tə näng′gō), a city in SW Guatemala. 65,733.

Que′zon Cit′y (kā′zon, -sōn), a city on W central Luzon Island, in the Philippines, NE of Manila: former national capital (1948–76). 1,632,000.

Quil·mes (kēl′mes), a city in E Argentina, near Buenos Aires. 445,662.

Quim·per (kaɴ peʀ′), a city in NW France: noted for pottery manufacture. 60,510.

Quin·cy (kwin′zē, -sē), a city in E Massachusetts, near Boston. 84,985.

Quin·ta·na Roo (kēn tä′nə rō′), a sparsely populated state in SE Mexico, on the E Yucatán peninsula. 393,398. *Cap.:* Chetumal. 19,435 sq. mi. (50,335 sq. km).

Quir·i·nal (kwir′ə nl), one of the seven hills on which ancient Rome was built.

Qui·to (kē′tō), the capital of Ecuador, in the N part. 1,110,248; 9348 ft. (2849 m) above sea level.

Qum (kŏŏm), Qom.

Qum·ran (kŏŏm′rän), Khirbet Qumran.

R

Ra·bat (rä bät′, rə-), the capital of Morocco, in the NW part on the Atlantic. 518,616.

Ra·baul (rə boul′), a seaport on NE New Britain island, in the Bismarck Archipelago, Papua New Guinea. 14,954.

Rab·bah (rab′ə) also **Rab·bath** (-əth), AMMAN.

Race (rās), **Cape,** a cape at the SE extremity of Newfoundland.

Ra·cine (rə sēn′, rā-), a city in SE Wisconsin. 84,298.

Rad·nor·shire (rad′nər shēr′, -shər), a historic county in Powys, in E Wales. Also called **Rad′nor.**

Ra·dom (rä′dôm), a city in E Poland. 214,000.

Ra·gu·sa (rä gōō′zä), **1.** a city in SE Sicily. 62,472. **2.** Italian name of DUBROVNIK.

Rain′bow Bridge′, a natural stone bridge in S Utah: a national monument. 290 ft. (88 m) high; 275-ft. (84-m) span.

Rai·nier (rə nēr′, rā-, rā′nēr), **Mount,** a volcanic peak in W Washington, in the Cascade Range. 14,408 ft. (4392 m).

Rai·pur (rī′pŏŏr), a city in SE Madhya Pradesh, in E central India. 339,000.

Ra·ja·sthan (rä′jə stän′), a state in NW India, bordering on Pakistan. 34,102,912; 132,078 sq. mi. (342,056 sq. km). *Cap.:* Jaipur.

Raj·kot (räj′kōt), a city in S Gujarat, in W India. 444,000.

Raj·pu·ta·na (räj′pŏŏ tä′nə), a region in NW India, largely coextensive with Rajasthan state.

Ra·leigh (rô′lē, rä′-), the capital of North Carolina, in the central part. 207,951.

Ra·mat Gan (rə mät′ gän′), a city in central Israel, near Tel Aviv. 115,600.

Ran·ca·gua (räng kä′gwä), a city in central Chile. 172,489.

Ran·chi (rän′chē), a city in S Bihar, in E India. 501,000.

Ran′cho Cu·ca·mon′ga (kōō′kə mung′gə, -mong′-), a city in SE California. 101,409.

Rand (rand), **The,** WITWATERSRAND.

Rand·ers (rä′nərz, -nərs), a seaport in E Jutland, in Denmark. 61,155.

Range/ley Lakes/ (rānj/lē), a group of lakes in W Maine.

Ran·goon (rang gōōn/), former name of YANGON.

Ra·pa Nu·i (rä/pə nōō/ē), EASTER ISLAND.

Rap·i·dan (rap/i dan/), a river in N Virginia, flowing E from the Blue Ridge Mountains into the Rappahannock River. 90 mi. (145 km) long.

Rap/id Cit/y, a city in SW South Dakota, in the Black Hills. 54,523.

Rap·pa·han·nock (rap/ə han/ək), a river flowing SE from N Virginia into the Chesapeake Bay. 185 mi. (300 km) long.

Ra·ro·tong·a (rär/ə tong/gə), one of the Cook Islands, in the S Pacific. 9281; 26 sq. mi. (67 sq. km). —**Ra/ro·tong/an,** *adj., n.*

Rasht (rasht) also **Resht,** a city in NW Iran, near the Caspian Sea. 290,897.

Ras Sham·ra (räs sham/rə), a locality in NW Syria, near the Mediterranean Sea: site of ancient Ugarit.

Ra·ven·na (rə ven/ə), a city in NE Italy. 136,324.

Ra·wal·pin·di (rä/wəl pin/dē), a city in N Pakistan. 928,000.

Read·ing (red/ing), **1.** a city in Berkshire, in S England. 132,900. **2.** a city in SE Pennsylvania. 78,380.

Re·ci·fe (rə sē/fə), the capital of Pernambuco state, in NE Brazil. 1,183,391. Formerly, **Pernambuco.**

Reck·ling·hau·sen (rek/ling hou/zən), a city in North Rhine-Westphalia, in NW Germany. 119,900.

Red·bridge (red/brij/), a borough of Greater London, England. 231,400.

Red/ Chi/na, CHINA, People's Republic of.

Red/ Deer/, a city in S central Alberta, in W Canada. 54,425.

Red·ding (red/ing), a city in N California. 66,462.

Red·lands (red/ləndz), a city in SW California, near Los Angeles. 60,394.

Re·don/do Beach/ (ri don/dō), a city in SW California. 60,167.

Red/ Riv/er, 1. a river flowing E from NW Texas along the S boundary of Oklahoma into the Mississippi River in Louisiana. ab. 1300 mi. (2095 km) long. **2.** Also called **Red/ Riv/er of the North/.** a river flowing N along the boundary be-

tween Minnesota and North Dakota to Lake Winnipeg in S Canada. 533 mi. (860 km) long. **3.** Vietnamese, **Song Hong.** Chinese, **Yuan Jiang.** a
river in SE Asia, flowing SE from Yunnan, China,
through N Vietnam to the Gulf of Tonkin. 500 mi.
(800 km) long.

Red/ Sea/, an arm of the Indian Ocean, extending NW between Africa and Arabia: connected to
the Mediterranean by the Suez Canal. 1450 mi.
(2335 km) long; 170,000 sq. mi. (440,300 sq.
km).

Red/wood Cit/y, a city in W California. 66,072.

Red/wood Na/tional Park/, a national park in
N California: redwood forest with some of the
world's tallest trees. 172 sq. mi. (445 sq. km).

Re·gens·burg (rā′gənz bûrg′, -bŏŏrg′), a city in
central Bavaria, in SE Germany, on the Danube.
118,600.

Reg·gio Ca·la·bri·a (rej′ō kə lä′brē ə, rej′ē ō′),
a seaport in S Italy, on the Strait of Messina.
178,094. Also called **Reg/gio di Cala/bria** (dē).

Reg/gio E·mi/lia (ə mēl/yə), a city in N Italy.
129,725. Also called **Reg/gio nel/l′E·mi/lia**
(nel/ə mēl/-).

Re·gi·na (ri ji/nə), the capital of Saskatchewan, in
the S part, in S Canada. 175,064.

Reims or **Rheims** (rēmz; *Fr.* ʀᴀɴs), a city in NE
France: cathedral; unconditional surrender of
Germany May 7, 1945. 181,985.

Rein/deer Lake/, a lake in central Canada, in
NE Saskatchewan and NW Manitoba. 2444 sq. mi.
(6330 sq. km).

Rem·scheid (rem/shīt), a city in North Rhine-
Westphalia, in W Germany, in the Ruhr region.
120,100.

Ren·frew (ren/frōō), a historic county in SW
Scotland. Also called **Ren/frew·shire/** (-shēr′,
-shər).

Rennes (ren), a city in NW France. 200,390.

Re·no (rē/nō), a city in W Nevada. 133,850.

Repub/lican Riv/er, a river flowing E from E
Colorado through Nebraska and Kansas into the
Kansas River. 422 mi. (680 km) long.

Resht (resht), ʀᴀsʜᴛ.

Re·sis·ten·ci·a (rā′zē sten/sē ə), a city in NE Argentina, on the Paraná River. 218,438.

Ré·u·nion (rē yōōn/yən, rā-), an island in the Indian Ocean, E of Madagascar: an overseas department of France. 574,800; 970 sq. mi. (2512 sq. km). *Cap.:* St. Denis.

Re·vere (ri vēr′), a city in E Massachusetts, on Massachusetts Bay, near Boston: seaside resort. 42,786.

Rey·kja·vik (rā′kyə vik, -vēk′), the capital of Iceland, on the SW coast. 93,245.

Rey·no·sa (rā nō′sə), a city in N Tamaulipas, in E Mexico, on the Rio Grande. 211,412.

Re·zai·yeh (ri zi′ə), former name of ORUMIYEH.

Rhae·ti·a (rē′shē ə, -shə), an ancient Roman province in central Europe, including what is now E Switzerland and a part of the Tyrol. —**Rhae′tian** (-shən, -shē ən), *adj., n.*

Rhae′tian Alps′, a chain of the Alps in E Switzerland and W Austria. Highest peak, Bernina, 13,295 ft. (4052 m).

Rheims (rēmz; *Fr.* RANS), REIMS.

Rhine (rin), a river flowing from SE Switzerland through Germany and the Netherlands into the North Sea. 820 mi. (1320 km) long. German, **Rhein** (RIN). French, **Rhin** (RAN). Dutch, **Rijn.** —**Rhen·ish** (ren′ish), *adj.*

Rhine·land (rin′land′, -lənd), **1.** that part of Germany W of the Rhine. **2.** RHINE PROVINCE. German, **Rhein·land** (rin′länt′).

Rhine′land-Palat′inate, a state in W Germany. 3,653,155; 7655 sq. mi. (19,825 sq. km). *Cap.:* Mainz. German **Rhein′land-Pfalz′** (pfälts).

Rhine′ Prov′ince, a former province of Prussia, mostly W of the Rhine: now divided between Rhineland-Palatinate and North Rhine-Westphalia. Also called **Rhineland.** German, **Rheinland.**

Rhode′ Is′land (rōd), a state in the NE United States, on the Atlantic coast: a part of New England. 1,003,464; 1214 sq. mi. (3145 sq. km). *Cap.:* Providence. *Abbr.:* RI, R.I. —**Rhode′ Is′lander,** *n.*

Rhodes (rōdz), **1.** a Greek island in the SE Aegean, off the SW coast of Turkey: largest of the Dodecanese Islands. 542 sq. mi. (1404 sq. km). **2.** a seaport on this island. 40,392. Greek, **Rho·dos** (RÔ′thôs). —**Rho′di·an,** *adj.*

Rho·de·sia (rō dē′zhə), **1.** a historical region in S Africa that comprised the British territories of Northern Rhodesia (now Zambia) and Southern

Rhodesia (now Zimbabwe). **2.** a former name (1964–80) of ZIMBABWE. **—Rho·de′sian,** *adj., n.*

Rhod·o·pe (rod′ə pē, ro dō′-), a mountain range in SW Bulgaria. Highest peak, 9595 ft. (2925 m).

Rhon·dda (ron′də; *Welsh* ʜʀon′thä), a city in Mid Glamorgan, in S Wales. 86,400.

Rhone or **Rhône** (rōn), a river flowing from the Alps in S Switzerland through the Lake of Geneva and SE France into the Mediterranean. 504 mi. (810 km) long.

Rhône-Alpes (rōn Alp′), a metropolitan region in SE France. 5,153,600; 16,872 sq. mi. (43,698 sq. km).

RI or **R.I.,** Rhode Island.

Ri·al·to (rē al′tō), **1.** a commercial center in Venice, Italy. **2.** a city in SW California, near Los Angeles. 72,388.

Rich·ard·son (rich′ərd sən), a city in NE Texas, near Dallas. 74,840.

Rich·e·lieu (rish′ə loo′; *Fr.* rēsh° lyœ′), a river in SE Canada, in Quebec, flowing N from Lake Champlain to the St. Lawrence. 210 mi. (340 km) long.

Rich·mond (rich′mənd), **1.** former name of STATEN ISLAND (def. 2). **2.** the capital of Virginia, in the E part on the James River: capital of the Confederacy 1861–65. 203,056. **3.** Also called **Rich′-mond-upon′-Thames′.** a borough of Greater London, England, on the Thames River. 163,000. **4.** a seaport in W California, on San Francisco Bay. 87,425. **5.** a city in E Indiana. 38,705.

Ri·ding (rī′ding), any of the three former administrative divisions of Yorkshire, England. Compare EAST RIDING, NORTH RIDING, WEST RIDING.

Rif (rif), a mountainous coastal region in N Morocco. Also called **Er Rif.**

Ri·ga (rē′gə), **1.** the capital of Latvia, on the Gulf of Riga. 915,000. **2. Gulf of,** an arm of the Baltic between Latvia and Estonia. 90 mi. (145 km) long.

Ri·je·ka (rē ek′ə, -yek′ä), a seaport in W Croatia, on the Adriatic. 193,044. Italian, **Fiume.**

Rijn (*Du.* rīn), RHINE.

Rijs·wijk (*Du.* rīs′vīk), a town in SW Netherlands, near The Hague. 48,657.

Rim·i·ni (rim′ə nē), a seaport in NE Italy, on the Adriatic. 130,787. Ancient, **Ariminum.**

Ri·o·bam·ba (rē′ō bäm′bə), a city in central Ecuador in the Andes, near the Chimborazo volcano. 149,757.

Ri·o Bran·co (rē′ō bräng′kō, -brang′-), the capital of Acre, in W Brazil. 119,815.

Rí·o Bra·vo (*Sp.* rē′ô brä′vô), Mexican name of RIO GRANDE (def. 1).

Ri·o de Ja·nei·ro (rē′ō dā zhə nâr′ō, jə-, dē), **1.** a state in SE Brazil. 11,489,797; 17,091 sq. mi. (44,268 sq. km). **2.** the capital of this state, on Guanabara Bay: former capital of Brazil. 5,184,292.

Rí·o de la Pla·ta (rē′ô dā lä plä′tä), PLATA, Río de la.

Rí·o de O·ro (rē′ō dā ôr′ō), the S part of Western Sahara.

Ri·o Grande (rē′ō grand′, gran′dē, grän′dā), **1.** Mexican, **Río Bravo.** a river flowing S from Colorado through central New Mexico and along the boundary between Texas and Mexico into the Gulf of Mexico. 1800 mi. (2900 km) long. **2.** a river flowing W from SE Brazil into the Paraná River. 650 mi. (1050 km) long. **3.** a seaport in SE Rio Grande do Sul state, in Brazil. 124,706.

Ri·o Gran·de (rē′ō grän′dā, -dē), a river in central Nicaragua, flowing NE to the Caribbean Sea. ab. 200 mi. (320 km) long.

Ri·o Gran′de do Nor′te (grän′dē dŏŏ nôr′tē), a state in NE Brazil. 2,194,500; 20,469 sq. mi. (53,015 sq. km). *Cap.:* Natal.

Ri·o Gran′de do Sul′ (dŏŏ sōōl′), a state in S Brazil. 8,749,400; 108,951 sq. mi. (282,184 sq. km). *Cap.:* Pôrto Alegre.

Rí·o Mu·ni (mōō′nē), former name of MBINI.

Ri·o Ne·gro (rē′ōō ne′grōō), Portuguese name of NEGRO River.

Rí·o Ne·gro (rē′ô ne′grô), Spanish name of NEGRO River.

Riv·er·side (riv′ər sid′), a city in SW California. 226,505.

Riv·i·er·a (riv′ē âr′ə), a resort area along the Mediterranean coast, extending from Saint-Tropez, in SE France, to La Spezia, in NW Italy.

Ri·yadh (rē yäd′), the capital of Saudi Arabia, in the E central part. 1,500,000.

Road′ Town′, a town on SE Tortola, in the NE

West Indies: capital of the British Virgin Islands. 3976.

Ro·a·noke (rō′ə nōk′), **1.** a city in SW Virginia. 96,397. **2.** a river flowing SE from W Virginia to Albemarle Sound in North Carolina. 380 mi. (610 km) long.

Ro′anoke Is′land, an island off the NE coast of North Carolina, S of Albemarle Sound: site of Raleigh's unsuccessful colonizing attempts 1585, 1587.

Rob·son (rob′sən), **Mount,** a mountain in E British Columbia, Canada: highest peak of the Rocky Mountains in Canada, 12,972 ft. (3954 m).

Ro·ca (rō′kə), **Cape,** a cape in W Portugal, near Lisbon: the western extremity of continental Europe.

Roch·dale (roch′dāl′), a borough of Greater Manchester, in N England: site of one of the earliest cooperative societies 1844. 211,500.

Roch·es·ter (roch′es tər, -ə stər), **1.** a city in W New York, on the Genesee River. 231,636. **2.** a town in SE Minnesota. 70,745. **3.** a city in N Kent, in SE England. 55,460.

Roch′ester Hills′, a city in SE Michigan. 61,766.

Rock·ford (rok′fərd), a city in N Illinois. 139,426.

Rock·hamp·ton (rok hamp′tən, -ham′-), a city in E Queensland, in E Australia. 60,406.

Rock′y Moun′tain Na′tional Park′, a national park in N Colorado. 405 sq. mi. (1050 sq. km).

Rock′y Moun′tains, a mountain system in W North America, extending NW from central New Mexico through W Canada to N Alaska. Highest peak in U.S., Mount Elbert, 14,431 ft. (4399 m); highest peak in Canada, Mount Robson, 12,972 ft. (3954 m). Also called **Rock′ies.**

Ro′man Em′pire, the lands and peoples subject to the authority of ancient Rome.

Ro·ma·ni·a (rō mā′nē ə, -mān′yə, rōō-) also **Rumania, Roumania,** a republic in SE Europe, bordering on the Black Sea. 22,823,479; 91,654 sq. mi. (237,385 sq. km). *Cap.:* Bucharest. Romanian, **Ro·mâ·nia** (Rô mu′nyä). —**Ro·ma′ni·an,** *adj., n.*

Rome (rōm), **1.** Italian, **Ro·ma** (Rô′mä). the capital of Italy, in the central part, on the Tiber: site of

Vatican City. 2,817,227. **2.** the ancient Italian kingdom, republic, and empire whose capital was the city of Rome. —**Ro/man,** *adj., n.*

Ron·ces·val·les (ron/səs vä/yes, ron/sə valz/), a village in N Spain, in the Pyrenees: defeat of part of Charlemagne's army and the death of Roland A.D. 788. French, **Ronce·vaux** (RôNS vō/).

Ron·dô·nia (ron dōn/yə, rōn-), a state in W Brazil. 981,800; 93,815 sq. mi. (242,980 sq. km). *Cap.:* Pôrto Velho. Formerly, **Guaporé.**

Ron/ne Ice/ Shelf/ (rô/nə), an ice barrier in Antarctica, in SW Weddell Sea.

Roo·de·poort-Ma·rais·burg (rōō/də pŏŏrt/mə-rā/bərg), a city in S Transvaal, in the NE Republic of South Africa. 139,810.

Roo·se·velt (rō/zə velt/, -vəlt, rōz/-; *spelling pron.* rōō/-), **Rio,** a river flowing N from W Brazil to the Madeira River. ab. 400 mi. (645 km) long.

Ro·rai·ma (rô ri/mə), a federal territory in N Brazil. 109,500; 88,844 sq. mi. (230,104 sq. km).

Ro·sa (rō/zə), **Mon·te** (mon/tē, -tā), a mountain between Switzerland and Italy, in the Pennine Alps: second highest peak of the Alps. 15,217 ft. (4638 m).

Ro·sa·ri·o (rō zär/ē ō/, -sär/-), a port in E Argentina, on the Paraná River. 954,606.

Ros·com·mon (ros kom/ən), a county in Connaught, in the N Republic of Ireland. 54,499; 950 sq. mi. (2460 sq. km). *Co. seat:* Roscommon.

Ro·seau (rō zō/), the capital of Dominica. 20,000.

Rose·mead (rōz/mēd/), a city in SW California, near Los Angeles. 51,638.

Ro·set·ta (rō zet/ə), a town in N Egypt, at the mouth of the Nile. 36,700.

Rose·ville (rōz/vil), a city in SE Michigan, near Detroit. 51,412.

Ross and Crom·ar·ty (rôs/ ən krom/ər tē, krum/-), a historic county in NW Scotland.

Ross/ Depend/ency, a territory in Antarctica, including Ross Island, the coasts along the Ross Sea, and adjacent islands: a dependency of New Zealand. ab. 175,000 sq. mi. (453,250 sq. km).

Ross/ Ice/ Shelf/, an ice barrier filling the S part of the Ross Sea.

Ross/ Is/land, an island in the W Ross Sea, off

the coast of Victoria Land: part of the Ross Dependency; location of Mt. Erebus.

Ros·si·ya (RU syē′yə), Russian name of RUSSIA.

Ross′ Sea′, an arm of the Antarctic Ocean, S of New Zealand, extending into Antarctica.

Ros·tock (ros′tok), a seaport in N Germany, on the Baltic. 253,990.

Ro·stov (rə stôf′, -stof′), a seaport in the SW Russian Federation in Europe, on the Don River, near the Sea of Azov. 1,020,000. Also called **Rostov′-on-Don′.**

Roth·er·ham (roᵺ′ər əm), a city in South Yorkshire, in N England. 251,700.

Rot·ter·dam (rot′ər dam′), a seaport in SW Netherlands. 574,299.

Rou·baix (rōō bā′), a city in N France, NE of Lille. 101,836.

Rou·en (rōō än′, -än′), a city in N France, on the Seine: execution of Joan of Arc 1431. 105,083.

Rou·ma·ni·a (rōō mā′nē ə, -mān′yə), ROMANIA. —**Rou·ma′ni·an,** *adj., n.*

Rou·me·li·a (rōō mē′lē ə, -mēl′yə), RUMELIA.

Rous·sil·lon (rōō′sē yôn′), a historical region in S France, bordering on the Pyrenees and the Mediterranean.

Ro·vno (rôv′nə, rov′-), a city in NW Ukraine, NE of Lvov. 233,000.

Ro·vu·ma (rōō vōō′mə), RUVUMA.

Rox·burgh (roks′bûr ō, -bur ō; *esp. Brit.* -brə), a historic county in SE Scotland. Also called **Rox′burgh·shire** (-shēr′, -shər).

Roy′al Oak′, a city in SE Michigan, near Detroit. 65,410.

Ru·an′da-Urun′di (rōō än′də), a former territory in central Africa, E of Zaire: administered by Belgium as a United Nations trust territory 1946–62; now divided into the independent states of Rwanda and Burundi.

Rub′ al Kha·li (rōōb′ al kä′lē), a desert in S Arabia, N of Hadhramaut and extending from Yemen to Oman. ab. 300,000 sq. mi. (777,000 sq. km). Also called **Empty Quarter, Great Sandy Desert.**

Ru·bi·con (rōō′bi kon′), a river in N Italy flowing E into the Adriatic. 15 mi. (24 km) long: in crossing this ancient boundary between Cisalpine Gaul and Italy, to march against Pompey in 49 B.C., Julius Caesar began a civil war.

Ru·dolf (rōō′dolf), **Lake,** former name of Lake TURKANA.

Rug·by (rug′bē), a city in E Warwickshire, in central England: boys' school, founded 1567. 86,400.

Ruhr (rōōr), **1.** a river in W Germany, flowing NW and W into the Rhine. 144 mi. (232 km) long. **2.** a mining and industrial region centered in the valley of the Ruhr River.

Ru·ma·ni·a (rōō mā′nē ə, -mān′yə), ROMANIA. —**Ru·ma′ni·an,** adj., n.

Ru·me·li·a or **Rou·me·li·a** (rōō mē′lē ə, -mēl′yə), a division of the former Turkish Empire, in the Balkan Peninsula: included Albania, Macedonia, and Thrace.

Run·ny·mede (run′i mēd′), a meadow on the S bank of the Thames, W of London, England: reputed site of the granting of the Magna Carta by King John, 1215.

Ru·se (rōō′sā), a city in N Bulgaria, on the Danube. 172,782.

Rush·more (rush′môr, -mōr), **Mount,** a peak in the Black Hills of South Dakota that is a memorial **(Mount Rushmore National Memorial)** having busts of Washington, Jefferson, Lincoln, and Theodore Roosevelt carved into its face. 5600 ft. (1707 m).

Rus·sia (rush′ə), **1.** RUSSIAN FEDERATION. **2.** Also called **Rus′sian Em′pire.** a former empire in E Europe and N and W Asia: overthrown by the Russian Revolution 1917. Cap.: St. Petersburg (1703–1917). **3.** UNION OF SOVIET SOCIALIST REPUBLICS. Russian, **Rossiya.** —**Rus′sian,** n., adj.

Rus′sian Federa′tion, a republic extending from E Europe to N and W Asia. 147,386,000; 6,593,000 sq. mi. (17,076,000 sq. km). Cap.: Moscow. Also called **Russia, Rus′sian Repub′lic.** Formerly (1918–91), **Rus′sian So′viet Fed′erated So′cialist Repub′lic.**

Rus′sian Tur′kestan. See under TURKESTAN.

Ru·the·ni·a (rōō thē′nē ə, -thēn′yə), a former province in E Czechoslovakia. —**Ru·the′ni·an,** adj., n.

Rut·land (rut′lənd), **1.** a city in W Vermont. 18,230. **2.** RUTLANDSHIRE.

Rut·land·shire (rut′lənd shēr′, -shər), a former

county, now part of Leicestershire, in central England. Also called **Rutland.**

Ru·vu·ma (rōō vōō′mə) also **Rovuma,** a river in SE Africa, flowing E along the Tanzania-Mozambique border to the Indian Ocean. ab. 450 mi. (725 km) long.

Ru·wen·zo·ri (rōō′wən zôr′ē, -zōr′ē), a mountain group in central Africa between Lake Albert and Lake Edward. Highest peak, Mt. Ngaliema. 16,763 ft. (5109 m).

Rwan·da (rōō än′də), a republic in central Africa, E of Zaire: formerly comprising the N part of the Belgian trust territory of Ruanda-Urundi; became independent 1962. 6,710,000; 10,169 sq. mi. (26,338 sq. km). *Cap.:* Kigali. —**Rwan′dan,** *n., adj.*

Rya·zan (rē′ə zän′, -zan′), a city in the W Russian Federation in Europe, SE of Moscow. 515,000.

Ry·binsk (rib′insk), a former name (1958–84) of ANDROPOV.

Ry′binsk Res′ervoir, a lake in the N central Russian Federation in Europe, formed by a dam on the upper Volga. 1768 sq. mi. (4579 sq. km).

Ryu′kyu Is′lands (rē ōō′kyōō), a chain of Japanese islands in the W Pacific between Japan and Taiwan. 1,235,000; 1205 sq. mi. (3120 sq. km). —**Ryu′kyu·an,** *n., adj.*

S

Saar (zär, sär), **1.** French, **Sarre.** a river in W Europe, flowing N from the Vosges mountains in NE France to the Moselle River in W Germany. 150 mi. (240 km) long. **2.** Also called **Saar′ Ba′sin.** a coal-producing region in W Germany, in the Saar River valley: under French economic control 1919–35, 1945–56. **3.** SAARLAND.

Saar·brück·en (zär brŏŏk′ən, sär-), a city in W Germany: the capital of Saarland. 187,400.

Saa·re·maa (sär′ə mä′), an island in the Baltic, at the mouth of the Gulf of Riga, belonging to Estonia. 1048 sq. mi. (2714 sq. km).

Saar·land (zär′land′, sär′-), a state in W Germany, in the Saar River valley. 1,054,142; 991 sq. mi. (2569 sq. km). *Cap.:* Saarbrücken. —**Saar′-land′er,** *n.*

Sa·ba (sä′bə), an island in the Netherlands Antilles, in the N Leeward Islands. 1011; 5 sq. mi. (13 sq. km).

Sa·ba·dell (sä′bə del′), a city in NE Spain, N of Barcelona. 185,960.

Sa·bah (sä′bä), a state in Malaysia, on the N tip of Borneo: formerly a British crown colony. 1,343,000; 28,460 sq. mi. (73,710 sq. km). *Cap.:* Kota Kinabalu. Formerly, **North Borneo.**

Sa·bine (sə bēn′), a river flowing SE and S from NE Texas, forming the boundary between Texas and Louisiana, and then through Sabine Lake to the Gulf of Mexico. ab. 500 mi. (800 km) long.

Sabine′ Lake′, a shallow lake on the boundary between Texas and Louisiana, formed by a widening of the Sabine River. ab. 17 mi. (27 km) long; 7 mi. (11 km) wide.

Sa·ble (sä′bəl), **Cape, 1.** a cape on a small island at the SW tip of Nova Scotia, Canada: lighthouse. **2.** a cape at the S tip of Florida.

Sach·sen (zäk′sən), German name of SAXONY.

Sach′sen-An′halt, German name of SAXONY-ANHALT.

Sac·ra·men·to (sak′rə men′tō), **1.** the capital of California, in the central part, on the Sacramento River. 369,365. **2.** a river flowing S from N California to San Francisco Bay. 382 mi. (615 km) long.

Sac·ra·men·to Moun·tains, a mountain range in S New Mexico and SW Texas: highest peak, 12,003 ft. (3660 m).

Sa·do·vá (sä′dō vä′), a village in NE Bohemia, in W Czechoslovakia: Prussian victory over Austrians 1866.

Sa·fi (saf′ē), a seaport in W central Morocco, on the Atlantic coast. 197,616. Also, **Saf′fi.**

Sa·fid Rud (sa fēd′ rōōd′), a river flowing from NW Iran into the Caspian Sea. 450 mi. (725 km) long.

Sa·ga·mi·ha·ra (sə gä′mē här′ə), a city on E central Honshu, in Japan, SW of Tokyo. 489,000.

Sag·i·naw (sag′ə nô′), a port in central Michigan, NW of Flint. 69,512.

Sag′inaw Bay′, an arm of Lake Huron, off the E coast of Michigan. 60 mi. (97 km) long.

Sa·guache (sə wäch′), SAWATCH.

Sag·ue·nay (sag′ə nā′), a river in SE Canada, in Quebec, flowing SE from Lake St. John to the St. Lawrence. 125 mi. (200 km) long.

Sa·gui·a el Ham·ra (sä′gē ə el ham′rə, häm′-), the northern part of Western Sahara.

Sa·gun·to (sə gōōn′tō), a city in E Spain, north of Valencia: besieged by Hannibal 219–218 B.C. 54,759. Ancient, **Sa·gun′tum** (-gun′təm).

Sa·har·a (sə har′ə, -hâr′ə, -här′ə), a desert in N Africa, extending from the Atlantic to the Nile valley. ab. 3,500,000 sq. mi. (9,065,000 sq. km). —**Sa·har′an, Sa·har′i·an,** adj.

Sa·ha·ran·pur (sə här′ən pŏŏr′), a city in NW Uttar Pradesh, in N India. 294,000.

Sa·hel (sə häl′, -hēl′), the arid area on the southern flank of the Sahara Desert that stretches across six countries from Senegal to Chad. —**Sa·hel′i·an,** adj., n.

Sa·i·da (sä′ē dä′), a seaport in SW Lebanon: the site of ancient Sidon. 24,740.

Sai·gon (sī gon′), former name of Ho CHI MINH CITY: capital of South Vietnam 1954–76.

Sai·maa (sī′mä), **Lake,** a lake in SE Finland. ab. 500 sq. mi. (1295 sq. km).

Saint′ John′, a seaport in S New Brunswick, in SE Canada, on the Bay of Fundy. 76,381.

Saint-Lou·is (Fr. san lwē′), a seaport in and the former capital of Senegal, at the mouth of the Senegal River. 96,594.

Saint-Tro·pez (saN trô pā/), a resort town in SE France, on the French Riviera. 4523.

Sai·pan (sī pan/), an island in and the capital of the Northern Mariana Islands in the W Pacific. 15,000; 71 sq. mi. (184 sq. km).

Sa·is (sā/is), an ancient city in N Egypt, on the Nile delta: an ancient capital of Egypt. —**Sa/ite,** *n., adj.*

Sa·kai (sä/kī/), a seaport on S Honshu, in S Japan, near Osaka. 810,120.

Sa·kha·lin (sak/ə lēn/, sak/ə lēn/), an island of the Russian Federation in the Sea of Okhotsk, N of Japan: formerly (1905–45) divided between the Soviet Union and Japan. 685,000; 29,100 sq. mi. (75,369 sq. km). Japanese, **Karafuto.**

Sak·ka·ra (sə kär/ə), SAQQARA.

Sa·la·do (sə lä/dō), **Río,** a river in N Argentina, flowing SE to the Paraná River. ab. 1200 mi. (1930 km) long.

Sal·a·man·ca (sal/ə mang/kə), a city in W Spain. 166,615.

Sa·lam·bri·a (sə lam/brē ə, sä/läm brē/ə), former name of PINIOS.

Sal·a·mis (sal/ə mis, sä/lä mēs/), an island off the SE coast of Greece, W of Athens, in the Saronic Gulf. 39 sq. mi. (101 sq. km).

Sa·lem (sā/ləm), **1.** a seaport in NE Massachusetts: founded 1626; execution of persons accused of witchcraft 1692. 38,091. **2.** the capital of Oregon, in the NW part, on the Willamette River. 107,786. **3.** a city in central Tamil Nadu, in S India. 515,000. **4.** an ancient city of Canaan, later identified with Jerusalem.

Sa·ler·no (sə lâr/nō, -lûr/-), a seaport in SW Italy. 153,807.

Sal·ford (sôl/fərd, sô/-, sal/-), a city in Greater Manchester, in N England. 266,500.

Sa·li·nas (sə lē/nəs), a city in W California. 108,777.

Salis·bur·y (sôlz/ber/ē, -bə rē, -brē), **1.** former name of HARARE. **2.** a city in Wiltshire, in S England: cathedral. 104,700.

Salis/bury Plain/, a plateau in S England, N of Salisbury: the site of Stonehenge.

Salm/on Riv/er Moun/tains, a range in central Idaho. Highest peak, 10,340 ft. (3150 m).

Sa·lon·i·ka (sə lon/i kə, sal/ə nē/kə), **1.** Also,

Sa·lon/i·ca, Sa·lo·ni·ki (*Gk.* sä′lô nē′kē). Official name, **Thessalonike**. Ancient, **Therma**. a seaport in S central Macedonia, in NE Greece, on the Gulf of Salonika. 339,496. **2. Gulf of,** an arm of the Aegean, in NE Greece. 70 mi. (113 km) long.

Sal·op (sal′əp), former name of SHROPSHIRE. —**Sa·lo·pi·an** (sə lō′pē ən), *adj., n.*

Sal·ta (säl′tə), a city in NW Argentina. 260,323.

Sal·ti·llo (säl tē′yō), the capital of Coahuila, in N Mexico. 233,600.

Salt′ Lake′ Cit′y, the capital of Utah, in the N part, near the Great Salt Lake. 159,936.

Sal′ton Sea′ (sôl′tn), a shallow saline lake in S California, in the Imperial Valley, formed by the diversion of water from the Colorado River into a salt-covered depression (**Sal′ton Sink′**). 236 ft. (72 m) below sea level.

Salt′ Riv′er, a river flowing W from E Arizona to the Gila River near Phoenix. 200 mi. (322 km) long.

Sal·va·dor (sal′və dôr′), **1.** EL SALVADOR. **2.** Formerly, **Bahia, São Salvador.** the capital of Bahia, in E Brazil. 1,525,831. —**Sal′va·do′ran, Sal′va·do′ri·an,** *adj., n.*

Sal·ween (sal′wēn), a river in SE Asia, flowing S from SW China through E Burma to the Bay of Bengal. 1750 mi. (2815 km) long.

Salz·burg (sôlz′bûrg; *Ger.* zälts′bŏŏRk), a city in W Austria: the birthplace of Mozart. 138,213.

Salz·git·ter (zälts′git/ər), a city in Lower Saxony in central Germany, SE of Hanover. 111,100.

Sa·ma·na Cay′ (sə mä′nə), a small, uninhabited island in the central Bahamas: now believed to be the first land in the New World seen by Christopher Columbus 1492. 9 mi. (14 km) long.

Sa·mar (sä′mär), an island in the E central Philippines. 5309 sq. mi. (13,750 sq. km).

Sa·ma·ra (sə mär′ə), a port in the SE Russian Federation in Europe, on the Volga. 1,257,000. Formerly (1935–91), **Kuibyshev.**

Sa·mar·i·a (sə mâr′ē ə), **1.** a district in ancient Palestine N of Judea: later part of the Roman province of Syria; taken by Jordan 1948; occupied by Israel 1967. **2.** the northern kingdom of the ancient Hebrews. **3.** the ancient capital of this kingdom.

Sam·ar·kand (sam′ər kand′), a city in SE Uzbekistan: taken by Alexander the Great 329 B.C.; Tamerlane's capital in the 14th century. 388,000. Ancient, **Maracanda.**

Sam·ni·um (sam′nē əm), an ancient country in central Italy.

Sa·mo·a (sə mō′ə), a group of islands in the S Pacific, N of Tonga: divided into American Samoa and Western Samoa. —**Sa·mo′an,** n., adj.

Sa·mos (sā′mos, sä′mōs), a Greek island in the E Aegean. 194 sq. mi. (502 sq. km). —**Sa·mi·an** (sā′mē ən), adj., n.

Sam·sun (säm sŏŏn′), a city in N Turkey, in Asia. 280,084.

San (sän), a river in central Europe, flowing from the Carpathian Mountains in W Ukraine through SE Poland into the Vistula. ab. 280 mi. (450 km) long.

Sa·n'a or **Sa·naa** (sä nä′), the political capital of the Republic of Yemen, in SW Arabia. 150,000.

San An·ge·lo (san an′jə lō′), a city in W Texas. 84,474.

San An·to·ni·o (san′ an tō′nē ō′), a city in S Texas: site of the Alamo. 935,933. —**San′ An·to′ni·an,** n., adj.

San Ber·nar·di·no (san′ bûr′nər dē′nō, -bûr′-nə-), a city in S California. 164,164.

San′ Bernardi′no Moun′tains, a mountain range in S California. Highest peak, 11,485 ft. (3500 m).

San Blas (sän bläs′), **Gulf of,** a gulf of the Caribbean on the N coast of Panama.

San Bue·na·ven·tu·ra (san bwä′nə ven tŏŏr′ə, -tyŏŏr′ə), official name of VENTURA.

San Cris·tó·bal (san′ kri stō′bəl), a city in SW Venezuela. 198,793.

San·da·kan (sän dä′kän, san dä′kən), a seaport in NE Sabah, in E Malaysia. 113,496.

Sand·hurst (sand′hûrst), a village in S England, near Reading, W of London: military college. 6445.

San Di·e·go (san′ dē ā′gō), a seaport in SW California. 1,110,549.

San Do·min·go (san′ də ming′gō), SANTO DO-MINGO (defs. 2, 3).

Sand·wich (sand′wich, san′-), a town in E Kent, in SE England. 4467.

Sand′wich Is′lands, former name of HAWAIIAN ISLANDS.

San·dy (san′dē), a city in central Utah. 75,058.

Sand′y Hook′, a peninsula in E New Jersey, at the entrance to lower New York Bay. 6 mi. (10 km) long.

San′ Fer·nan′do Val′ley (san′ fər nan′dō), a valley in SW California, NW of Los Angeles.

San·ford (san′fərd), **Mount,** a mountain in SE Alaska. 16,208 ft. (4940 m).

San Fran·cis·co (san′ frən sis′kō, fran-), a seaport in W central California, on San Francisco Bay. 723,959. —**San′ Fran·cis′can,** n., adj.

San′ Francis′co Bay′, an inlet of the Pacific in W central California: the harbor of San Francisco.

San′ Francis′co Peaks′, a mountain mass in N Arizona: highest point in the state, Humphrey's Peak, 12,611 ft. (3845 m).

San·gre de Cris·to (sang′grē də kris′tō), a mountain range in S Colorado and N New Mexico: a part of the Rocky Mountains. Highest peak, Blanca Peak, 14,390 ft. (4385 m).

San′i·bel Is′land (san′ə bəl, -bel′), an island in the Gulf of Mexico off the SW coast of Florida. 16 sq. mi. (41.5 sq. km).

San Ja·cin·to (san′ jə sin′tō, hə-), a river in E Texas, flowing SE to Galveston Bay: Texans defeated Mexicans near its mouth in 1836.

San Joa·quin (san′ wo kēn′), a river in California, flowing NW from the Sierra Nevada to the Sacramento River. 350 mi. (560 km) long.

San Jo·se (san′ hō zā′), a city in W California. 782,248.

San Jo·sé (san′ hō zā′), the capital of Costa Rica, in the central part. 241,464.

San Juan (san′ wän′, hwän′), **1.** the capital of Puerto Rico, on the NE coast. 431,227. **2.** a city in W Argentina. 290,479.

San′ Juan′ Hill′, a hill in SE Cuba, near Santiago de Cuba: captured by U.S. forces in battle during the Spanish-American War in 1898.

San′ Juan′ Is′lands, a group of islands in NW Washington between SE Vancouver Island and the mainland.

San′ Juan′ Moun′tains, a mountain range in SW Colorado and N New Mexico: a part of the

Rocky Mountains. Highest peak, Uncompahgre Peak, 14,309 ft. (4361 m).

Sankt Gal·len (Ger. zängkt′ gä′lən), St. GALLEN.

Sankt Mo·ritz (Ger. zängkt mō′Rits), St. MORITZ.

San Le·an·dro (san′ lē an′drō), a city in W California. 68,223.

San Lu·is Po·to·sí (sän′ loo ēs′ pô′tô sē′), **1.** a state in central Mexico. 1,527,000; 24,415 sq. mi. (63,235 sq. km). **2.** the capital of this state. 303,000.

San Ma·ri·no (san′ mə rē′nō), **1.** a small republic in E Italy. 22,746; 24 sq. mi. (61 sq. km). **2.** the capital of this republic. 4363. —**San′ Mar·i·nese′** (mar′ə nēz′, -nēs′), n., adj.

San Ma·te·o (san′ mə tā′ō), a city in W California. 85,486.

San Mi·guel (sän′ mē gel′), a city in E El Salvador. 131,977.

San′ Pa·blo Bay′ (san′ pä′blō), the N part of San Francisco Bay, in W California.

San Re·mo (san rē′mō, rā′-), a seaport in NW Italy, on the Riviera. 64,302.

San Sal·va·dor (san sal′və dôr′), **1.** an island in the E central Bahamas. 825; 60 sq. mi. (155 sq. km). **2.** the capital of El Salvador. 452,614.

San Se·bas·tián (sän′ sə bas′chən, sän seb′əs·tyän′), a seaport in N Spain. 180,043.

San Ste·fa·no (san stef′ə nō′), former name of YEŞILKÖY.

San·ta An·a (san′tə an′ə), a city in SW California, SE of Los Angeles. 293,742.

San′ta Bar′ba·ra (bär′bər ə, -brə), a city on the SW coast of California. 85,571.

San′ta Bar′bara Is′lands, a group of islands off the SW coast of California.

San′ta Catali′na, an island off the SW coast of California: resort. 132 sq. mi. (342 sq. km). Also called **Catalina Island, Catalina.**

San′ta Cat·a·ri′na (kat′ə rē′nə), a state in S Brazil. 4,235,800; 36,856 sq. mi. (95,455 sq. km). *Cap.:* Florianópolis.

San′ta Cla′ra (klar′ə), **1.** a city in central Cuba. 178,300. **2.** a city in central California, S of San Francisco. 93,613.

San′ta Cla·ri′ta (kla rē′tə), a city in SW California, N of Los Angeles. 110,642.

San′ta Cruz′ (krōōz), **1.** a city in central Bolivia.

441,717. **2.** an island in NW Santa Barbara Islands. **3.** St. Croix (def. 1).

San/ta Cruz/ de Tenerife/, a seaport on NE Tenerife island, in the W Canary Islands. 211,389.

San/ta Cruz/ Is/lands, a group of islands in the SW Pacific Ocean, part of the Solomon Islands. 380 sq. mi. (984 sq. km).

San/ta Fe/ (fā), the capital of New Mexico, in the N part: founded c1605. 55,859. —**San/ta Fe/an,** *n., adj.*

San/ta Fé/ (fā), a city in E Argentina. 287,240.

San/ta Fe/ Trail/, an important trade route between Independence, Missouri, and Santa Fe, New Mexico, used from about 1821 to 1880.

San/ta Is/a·bel (iz/ə bel/), former name of MaLABO.

San/ta Ma·ri/a (mə rē/ə), **1.** an active volcano in W Guatemala. 12,300 ft. (3750 m). **2.** a city in S Brazil. 151,202. **3.** a city in W California. 61,284.

San/ta Mar/ta (mär/tə), a seaport in NW Colombia. 218,205.

San/ta Mon/i·ca (mon/i kə), a city in SW California: suburb of Los Angeles. 86,905.

San·tan·der (san/tan där/, sän/tän-), a seaport in N Spain. 188,539.

San·ta·rém (san/tə rem/), a city in N Brazil, on the Amazon River. 111,706.

San/ta Ro/sa (rō/zə), a city in W California, N of San Francisco. 113,313.

San·tee (san tē/), **1.** a city in SW California. 52,902. **2.** a river flowing SE from central South Carolina to the Atlantic. 143 mi. (230 km) long.

San·ti·a·go (san/tē ä/gō), **1.** the capital of Chile, in the central part. 4,858,342. **2.** Also called **Santia/go de Com·pos·te/la** (də kom/pə stel/ə). a city in NW Spain. 104,045.

Santia/go de Cu/ba (də kyōō/bə), a seaport in SE Cuba. 358,800.

Santia/go del Es·te/ro (del e stâr/ō), a city in N Argentina. 148,357.

Santia/go de los Ca·ba·lle/ros (də lōs kä/bäl-yâr/ōs), a city in the N central Dominican Republic. 278,638.

San·to An·dré (sän/tōō än dre/), a city in E Brazil, near São Paulo. 549,556.

San·to Do·min·go (san/tō də ming/gō), **1.** the

capital of the Dominican Republic, on the S coast: first European settlement in the New World 1496. 1,313,172. **2.** a former name of DOMINICAN REPUBLIC. **3.** a former name of HISPANIOLA. Also, **San Domingo** (for defs. 2, 3).

San·to·rin (san'tə rēn') also **San·to·ri·ni** (-rē'nē), THERA.

San·tos (san'təs), a seaport in S Brazil. 410,933.

São Fran·cis·co (SOUN' frän sēs'kŏŏ), a river flowing NE and E through E Brazil into the Atlantic. 1800 mi. (2900 km) long.

São Mi·guel (SOUN' mē gel'), the largest island of the Azores. 150,000; 288 sq. mi. (746 sq. km).

Saône (sōn), a river flowing S from NE France to the Rhone. 270 mi. (435 km) long.

São Pau·lo (SOUN' pou'lō, -lŏŏ), **1.** a state in S Brazil. 30,942,600; 95,714 sq. mi. (247,898 sq. km). **2.** the capital of this state. 7,032,547.

São Sal·va·dor (SOUN' säl'vä dôr'), a former name of SALVADOR (def. 2).

São Tia·go (SOUN' tē ä'gŏ, -gŏŏ), the largest of the Cape Verde Islands, S of Cape Verde. ab. 383 sq. mi. (992 sq. km).

São To·mé (SOUN' tŏŏ mā'), **1.** an island in W Africa, off the W coast of Gabon: the larger component of the republic of São Tomé and Príncipe. 106,900; 326 sq. mi. (847 sq. km). **2.** a city on this island: capital of the republic. 35,000. —**São' To·me'an,** *n., adj.*

São' Tomé' and Prín'cipe or **Sao' Tomé' and Prin'cipe,** a republic in W Africa, comprising the islands of São Tomé and Príncipe, in the Gulf of Guinea, N of the equator: a former overseas province of Portugal; gained independence in 1975. 115,600; 387 sq. mi. (1001 sq. km). *Cap.:* São Tomé.

São Vi·cen·te (SOUN' vi seN'ti), an island city in SE Brazil. 116,075.

Sap·po·ro (sə pôr'ō, -pōr'ō), a city on W Hokkaido, in N Japan. 1,555,000.

Saq·qa·ra or **Sak·ka·ra** (sə kär'ə), a village in S Egypt, S of Cairo: site of the necropolis of Memphis. 12,700.

Sar·a·gos·sa (sar'ə gos'ə), a city in NE Spain, on the Ebro River. 596,080. Spanish, **Zaragoza.**

Sa·ra·je·vo (sar'ə yā'vō), the capital of Bosnia and Herzegovina, in the central part. 448,519.

Sar'a·nac Lakes' (sar'ə nak'), a group of three lakes in NE New York, in the Adirondack Mountains.

Sa·ransk (sə ränsk', -ransk'), the capital of the Mordovian Autonomous Republic in the Russian Federation in Europe. 312,000.

Sa·ra·so·ta (sar'ə sō'tə), a city in W Florida. 50,961.

Sar'a·to'ga Springs' (sar'ə tō'gə, sar'-), a city in E New York: resort. 25,001.

Sa·ra·tov (sə rä'tôf, -tof), a city in the SW Russian Federation in Europe, on the Volga. 905,000.

Sa·ra·wak (sə rä'wäk, -wä), a state in Malaysia, on NW Borneo: formerly a British crown colony and protectorate. 1,550,000; ab. 48,250 sq. mi. (124,449 sq. km). *Cap.:* Kuching.

Sar·din·i·a (sär din'ē ə, -din'yə), a large island in the Mediterranean, W of Italy: with small nearby islands it comprises a department of Italy. 1,594,175; 9301 sq. mi. (24,090 sq. km). *Cap.:* Cagliari. Italian, **Sar·de·gna** (sär de'nyä). —**Sar·din'i·an,** *adj., n.*

Sar·dis (sär'dis) also **Sar·des** (-dēz), an ancient city in W Asia Minor: the capital of Lydia. —**Sar'·di·an** (-dē ən), *n., adj.*

Sa·re'ra Bay' (sə rer'ə), a large bay on the NW coast of New Guinea, in Irian Jaya, in Indonesia. Formerly, **Geelvink Bay.**

Sar·gas'so Sea' (sär gas'ō), a relatively calm area of water in the N Atlantic, NE of the West Indies.

Sar·go·dha (sər gō'də), a city in NE Pakistan. 294,000.

Sark (särk), one of the Channel Islands, E of Guernsey. 2 sq. mi. (5 sq. km).

Sar·ma·ti·a (sär mā'shē ə, -shə), the ancient name of a region in E Europe, between the Vistula and the Volga rivers.

Sar·ni·a (sär'nē ə), a port in SE Ontario, in S Canada, on Lake Huron. 72,000.

Sa·ron'ic Gulf' (sə ron'ik), an inlet of the Aegean, on the SE coast of Greece, between Attica and the Peloponnesus.

Sa·ros (sär'ōs, -ôs), **Gulf of,** an inlet of the Aegean, N of the Gallipoli Peninsula.

Sarre (*Fr.* SAAR), SAAR.

Sa·se·bo (sä′sə bō′), a seaport on NW Kyushu, in SW Japan. 251,188.

Sask., Saskatchewan.

Sas·katch·e·wan (sa skach′ə won′, -wən), **1.** a province in W Canada. 1,009,613; 251,700 sq. mi. (651,900 sq. km). *Cap.:* Regina. **2.** a river in SW Canada, flowing E from the Rocky Mountains to Lake Winnipeg. 1205 mi. (1940 km) long.

Sas·ka·toon (sas′kə tōōn′), a city in S Saskatchewan, in SW Canada. 177,641.

Sas·sa·ri (sä′sə rē), a city in NW Sardinia. 120,497.

Sat·su·ma (sat sōō′mə, sat′sə mə), a former province on S Kyushu, in SW Japan: famous for its porcelain ware.

Sa·tu-Ma·re (sä′tōō mär′ā), a city in NW Romania. 128,115.

Sau′di Ara′bia (sou′dē, sô′-, sä ōō′-), a kingdom occupying most of Arabia. 12,566,000; ab. 830,000 sq. mi. (2,149,690 sq. km). *Cap.:* Riyadh. **—Sau′di,** *n., pl.* **-dis,** *adj.* **—Sau′di Ara′bian,** *n., adj.*

Sault Ste. (or **Sainte**) **Marie** (sōō′ sānt′ mə rē′), **1.** the rapids of the St. Marys River, between NE Michigan and Ontario, Canada. **2.** a city in S Ontario, in S Canada, near these rapids. 80,905.

Sault Ste. (or **Sainte**) **Marie Canals,** three ship canals, one in Canada and two in the U.S., that connect Lakes Superior and Huron. Also called **Soo Canals.**

Sau·ternes (sō tûrn′, sô-), a wine-producing district near Bordeaux.

Sa·va (sä′vä), a river flowing E from W Slovenia, through Croatia to the Danube at Belgrade, Yugoslavia. 450 mi. (725 km) long.

Sa·vai·i (sä vī′ē), an island in Western Samoa: largest of the Samoa group. 703 sq. mi. (1821 sq. km).

Sa·van·nah (sə van′ə), **1.** a seaport in E Georgia, near the mouth of the Savannah River. 137,560. **2.** a river flowing SE from E Georgia along most of the boundary between Georgia and South Carolina and into the Atlantic. 314 mi. (505 km) long.

Sa·vo·na (sə vō′nə), a city in N Italy on the Mediterranean. 79,393.

Sa·voy (sə voi′), a historic region in SE France,

adjacent to the Swiss-Italian border: formerly a duchy; later a part of the kingdom of Sardinia; ceded to France, 1860. French, **Sa·voie** (sa vwa′).
—**Sa·voy·ard** (sə voi′ərd, sav′oi ärd′), *n., adj.*

Savoy′ Alps′, a mountain range in SE France: a part of the Alps. Highest peak, Mont Blanc, 15,781 ft. (4810 m).

Sa·watch or **Sa·guache** (sə wäch′), a mountain range in central Colorado: a part of the Rocky Mountains. Highest peak, Mt. Elbert, 14,431 ft. (4399 m).

Sax·o·ny (sak′sə nē), **1.** a state in E central Germany. 4,900,000; 6561 sq. mi. (16,990 sq. km). *Cap.:* Dresden. **2.** a former state of the Weimar Republic in E central Germany. 5788 sq. mi. (14,990 sq. km). *Cap.:* Dresden. **3.** a medieval division of N Germany with varying boundaries: extended at its height from the Rhine to E of the Elbe. German, **Sachsen.** French, **Saxe** (saks). —**Sax′on,** *n., adj.*

Sax′ony-An′halt, a state in central Germany. 3,000,000; 9515 sq. mi. (24,644 sq. km). *Cap.:* Magdeburg. German, **Sachsen-Anhalt.**

Sa·yan′ Moun′tains (sä yän′), a mountain range in the S Russian Federation in central Asia. Highest peak, 11,447 ft. (3490 m).

SC or **S.C.,** South Carolina.

Sca·man·der (skə man′dər), ancient name of MENDERES (def. 2).

Scan·di·na·vi·a (skan′də nā′vē ə), **1.** Norway, Sweden, Denmark, and sometimes Finland, Iceland, and the Faeroe Islands. **2.** Also called **Scandina′vian Penin′sula.** the peninsula consisting of Norway and Sweden. —**Scan′di·na′vi·an,** *adj., n.*

Sca′pa Flow′ (skä′pə, skap′ə), a sea basin off the N coast of Scotland, in the Orkney Islands.

Scar·bor·ough (skär′bûr′ō, -bur′ō, -bər ə), a seaport in North Yorkshire, in NE England. 104,400.

Schaer·beek (*Flemish.* sᴋʜär′bāk), a city in central Belgium, near Brussels. 118,950.

Schaum·burg (shôm′bûrg), a city in NE Illinois. 68,586.

Schaum·burg-Lip·pe (shoum′bŏŏrk lip′ə), a former state in NW Germany.

Scheldt (skelt), a river in W Europe, flowing from

N France through W Belgium and SW Netherlands into the North Sea. 270 mi. (435 km) long. Flemish, **Schel·de** (sкнеl′də). French, **Escaut.**

Sche·nec·ta·dy (skə nek′tə dē), a city in E New York, on the Mohawk River. 65,566.

Schie·dam (skē däm′), a city in SW Netherlands. 71,280.

Schles·wig (shles′wig), a region in S Jutland, divided between Germany and Denmark: a former duchy. Danish, **Slesvig.**

Schles′wig-Hol′stein, a state in N Germany. 2,564,565; 6073 sq. mi. (15,728 sq. km). *Cap.:* Kiel.

Schuyl·kill (skool′kil, skoo′kəl), *n.* a river flowing SE from E Pennsylvania to the Delaware River at Philadelphia. 131 mi. (210 km) long.

Schwa·ben (shvä′bən), German name of SWABIA.

Schwarz·wald (shvärts′vält′), German name of the BLACK FOREST.

Schweiz (shvīts), German name of SWITZERLAND.

Schwe·rin (shvä rēn′), the capital of Mecklenburg-Western Pomerania in N Germany. 130,685.

Schwyz (shvēts), **1.** a canton in central Switzerland, bordering on the Lake of Lucerne. 104,800; 350 sq. mi. (900 sq. km). **2.** the capital of this canton, in the W part. 12,100.

Scil′ly Isles′ (sil′ē), a group of about 140 small islands, SW of Land's End, England. 6 sq. mi. (16 sq. km). Also called **Scil′ly Is′lands.**

Sci·o·to (sī ō′tə, -tō), a river in central Ohio, flowing S to the Ohio River. 237 mi. (382 km) long.

Scone (skoon, skōn), a village in central Scotland: site of coronation of Scottish kings.

Scot·land (skot′lənd), a division of the United Kingdom in the N part of Great Britain. 5,035,315; 30,412 sq. mi. (78,772 sq. km). *Cap.:* Edinburgh. —**Scot′tish,** *adj., n.*

Scotts·dale (skots′dāl′), a city in central Arizona, near Phoenix. 130,069.

Scran·ton (skran′tn), a city in NE Pennsylvania. 81,805.

Scu·ta·ri (skoo′tə rē), **Lake,** a lake between NW Albania and S Montenegro, Yugoslavia. ab. 135 sq. mi. (350 sq. km).

Scy·ros (skī′ros, skē′rōs), SKIROS.

Scyth·i·a (sith′ē ə), the ancient name of a region

in SE Europe and Asia, between the Black and Aral seas. —**Scyth/i·an,** *n., adj.*

SD or **S.D.,** South Dakota.

S. Dak., South Dakota.

Sea/ Is/lands, a group of islands in the Atlantic, along the coasts of South Carolina, Georgia, and N Florida.

Seal/ Beach/, a town in SW California: resort. 25,098.

Se·at·tle (sē at/l), a seaport in W Washington, on Puget Sound. 516,259.

Se·bas·to·pol (sə bas/tə pōl/), SEVASTOPOL.

Se·dan (si dan/, -dän/), a city in NE France, on the Meuse River: defeat and capture of Napoleon III 1870. 25,430.

Sedge·moor (sej/mŏŏr/), a plain in SW England, in central Somerset.

Se·go·vi·a (sə gō/vē ə), **1.** a city in central Spain. 41,880. **2.** Coco.

Seine (sān, sen), a river in France, flowing NW through Paris to the English Channel. 480 mi. (773 km) long.

Se·kon·di-Ta·ko·ra·di (sek/ən dē/tä/kə rä/dē), a seaport in SW Ghana. 93,400.

Se·lan·gor (sə lang/ər, -ôr, -läng/-), a state in Malaysia, on the SW Malay Peninsula. 1,515,536; 3160 sq. mi. (8184 sq. km).

Sel·en·ga (sel/eng gä/), a river in N central Asia, flowing E and N through the NW Mongolian People's Republic through the Buryat Autonomous Republic in the SE Russian Federation to Lake Baikal. ab. 700 mi. (1125 km) long.

Se·leu·cia (si lōō/shə), **1.** an ancient city in Iraq, on the Tigris River: capital of the Seleucid empire. **2.** an ancient city in Asia Minor, near the mouth of the Orontes River: the port of Antioch.

Sel·kirk (sel/kûrk), a historic county in SE Scotland. Also called **Sel/kirk·shire/** (-shēr/, -shər).

Sel/kirk Moun/tains, a mountain range in SW Canada, in SE British Columbia. Highest peak, 11,123 ft. (3390 m).

Sel·ma (sel/mə), a city in central Alabama, on the Alabama River: voting rights demonstrations led by Martin Luther King, Jr., 1965. 23,755.

Se·ma·rang (sə mär/äng), a seaport on N Java, in S Indonesia. 1,026,671.

Se·mi·pa·la·tinsk (sem′i pə lä′tinsk), a city in NE Kazakhstan, on the Irtysh River. 330,000.

Sen·dai (sen′dī′), a city on NE Honshu, in central Japan. 686,000.

Sen′e·ca Lake′ (sen′i kə), a lake in W New York: one of the Finger Lakes. 35 mi. (56 km) long.

Sen·e·gal (sen′i gôl′, -gäl′), **1.** a republic in W Africa: independent member of the French Community; formerly part of French West Africa. 6,980,000; 76,084 sq. mi. (197,057 sq. km). *Cap.:* Dakar. **2.** a river in W Africa, flowing NW from E Mali to the Atlantic. ab. 1000 mi. (1600 km) long. French, **Sé·né·gal** (sā nā gàl′). —**Sen′e·ga·lese′** (-gə lēz′, -lēs′), *adj., n., pl.* **-lese.**

Sen·e·gam·bi·a (sen′i gam′bē ə), **1.** a region in W Africa between the Senegal and Gambia rivers, now mostly in Senegal. **2.** a confederation of Senegal and the Gambia, formed in 1982. —**Sen′e·gam′bi·an,** *adj.*

Sen·lac (sen′lak), a hill in Sussex, in SE England: supposed site of the Battle of Hastings, 1066.

Sen·nar (sə när′), a region in the E Sudan between the White and Blue Nile rivers, S of Khartoum: a former kingdom.

Seoul (sōl), the capital of South Korea, in the W part. 9,645,824.

Se·pik (sā′pik), a river in N Papua New Guinea, on the NE part of the island of New Guinea. 600 mi. (966 km) long.

Sequoi′a Na′tional Park′, a national park in central California, in the Sierra Nevada: giant sequoia trees, mountains. 604 sq. mi. (1565 sq. km).

Se·ram (si ram′, sā′räm), CERAM.

Ser·bi·a (sûr′bē ə), a constituent republic of Yugoslavia, in the N part. 9,660,000; 34,116 sq. mi. (88,360 sq. km). *Cap.:* Belgrade. —**Serb,** *n.* —**Ser′bi·an,** *adj., n.*

Serbs′, Cro′ats, and Slo′venes, Kingdom of the, former name (1918–29) of YUGOSLAVIA.

Ser·en·dip (ser′ən dip′) also **Ser·en·dib** (-dēb′), Arabic name of SRI LANKA.

Ser·en·get·i (ser′ən get′ē), a plain in N Tanzania, SE of Lake Victoria, including a major wildlife reserve (**Serenget′i Na′tional Park′**).

Ser·gi·pe (sər zhē′pə), a state in NE Brazil.

1,345,100; 8490 sq. mi. (21,990 sq. km). *Cap.:* Aracajú.

Ser·gi·yev Po·sad (sûr′gē əf pə säd′), a city in the NW Russian Federation in Europe, NE of Moscow. 111,000. Formerly (1930–91), **Zagorsk.**

Ser·ra do Mar (ser′ə də mär′), a mountain range on the SE coast of Brazil. Highest point, 7420 ft. (2262 m).

Ses·tos (ses′tos), an ancient Thracian town on the Hellespont opposite Abydos: Xerxes crossed the Hellespont here when he began his invasion of Greece.

Se·tú·bal (si tōō′bäl), **1. Bay of,** an inlet of the Atlantic, in W Portugal. **2.** a seaport on this bay, near Lisbon. 64,531.

Se·vas·to·pol (sə vas′tə pōl′) also **Sebastopol,** a seaport in the S Crimea, in S Ukraine. 350,000.

Sev′en Hills′ of Rome′, the seven hills (the Aventine, Caelian, Capitoline, Esquiline, Palatine, Quirinal, and Viminal) on and about which the ancient city of Rome was built.

Sev′en Pines′, FAIR OAKS.

Sev·ern (sev′ərn), a river in Great Britain, flowing from central Wales through W England into the Bristol Channel. 210 mi. (338 km) long.

Se·ve·ro·dvinsk (sev′ər ə dvinsk′), a city in the N Russian Federation in Europe, on Dvina Bay, E of Archangel. 239,000.

Se·ville (sə vil′), a port in SW Spain, on the Guadalquivir River. 668,356. Spanish, **Se·vi·lla** (se-vē′lyä). —**Se·vil′lian** (-yən), *adj., n.*

Sè·vres (se′vʀᵉ), a suburb of Paris in N France. 21,296.

Sew′ard Penin′sula (sōō′ərd), a peninsula in W Alaska, on Bering Strait.

Sey·chelles (sā shel′, -shelz′), a republic consisting of 115 islands in the Indian Ocean, NE of Madagascar: a member of the Commonwealth of Nations. 67,000; 175 sq. mi. (455 sq. km). *Cap.:* Victoria.

Sey·han (sā hän′), **1.** ADANA. **2.** a river in S central Turkey, flowing S from the Anatolia plateau to the Mediterranean Sea. 748 mi. (1204 km) long.

Sfax (sfäks), a seaport in E Tunisia, in N Africa. 231,911.

's Gra·ven·ha·ge (sꭓʀä′vən hä′ꭓə), a Dutch name of The HAGUE.

Shaan·xi (shän′shē′) also **Shensi,** a province in N central China. 30,430,000; 75,598 sq. mi. (195,799 sq. km). *Cap.:* Xian.

Sha·ba (shä′bə), a province in SE Zaire: important mining area. 3,874,019; 191,878 sq. mi. (496,964 sq. km). *Cap.:* Lubumbashi.

Sha·che (shä′chœ′) also **Soche,** a city in W Xinjiang Uygur, in W China, in a large oasis of the Tarim Basin. Also called **Yarkand.**

Shah·ja·han·pur (shä′jə hän′pŏōr′), a city in central Uttar Pradesh, in N India. 205,325.

Shakh·ty (shäкн′ti), a city in the SW Russian Federation in Europe, in the Donets Basin. 225,000.

Shan·dong (shän′dông′) also **Shantung, 1.** a maritime province in E China. 77,760,000; 59,189 sq. mi. (153,299 sq. km). *Cap.:* Jinan. **2.** a peninsula in the E part of this province, extending into the Yellow Sea.

Shang·hai (shang hī′), a seaport and municipality in Jiangsu province, in E China, near the mouth of the Chang Jiang. 6,980,000 (municipality 12,300,000).

Shan·non (shan′ən), a river flowing SW from N Ireland to the Atlantic: the principal river of Ireland. 240 mi. (386 km) long.

Shan′ State′ (shän, shan), a state in E Burma, along the Salween River. 3,718,805; ab. 56,000 sq. mi. (145,040 sq. km).

Shan·tou (shän′tō′) also **Swatow,** a seaport in E Guangdong province, in SE China. 722,805.

Shan·tung (shan′tung′), SHANDONG.

Shan·xi or **Shan·si** (shän′shē′), a province in N China. 26,550,000; 60,656 sq. mi. (157,099 sq. km). *Cap.:* Taiyuan.

Shao·xing or **Shao·hsing** (shou′shing′), a city in NE Zhejiang province, in E China. 1,107,175.

Shao·yang (shou′yäng′), a city in central Hunan province, in E China. 399,255. Formerly, **Paoking.**

Sha·ri or **Cha·ri** (shär′ē), a river in N central Africa, flowing NW from the Central African Republic into Lake Chad. 1400 mi. (2254 km) long.

Shar·on (shar′ən), a fertile coastal plain in ancient Palestine: now a coastal region N of Tel Aviv in Israel.

Sharps·burg (shärps′bûrg), a town in NW Maryland: nearby is the site of the Civil War battle of Antietam 1862. 659.

Shas·ta (shas′tə), **Mount,** the peak of an extinct volcano in N California, in the Cascade Range. 14,161 ft. (4315 m).

Shatt-al-A·rab (shat′al ar′əb, shät′-), a river in SE Iraq, formed by the junction of the Tigris and Euphrates rivers, flowing SE to the Persian Gulf. 123 mi. (198 km) long.

Shcher·ba·kov (sher′bə kôf′, -kof′), a former name (1946–57) of ANDROPOV.

She·be·li or **Shi·be·li** (shi bā′lē), a river in E Africa, flowing SE from central Ethiopia to the Juba River in Somalia. 1130 mi. (1820 km) long.

She·chem (shē′kəm, shek′əm; *Heb.* shə ка̄em′), **1.** a town in ancient Palestine, in Samaria near modern Nablus. **2.** Hebrew name of NABLUS.

Shef·field (shef′ēld), a city in South Yorkshire, in N England. 559,800.

Shen′an·do′ah Na′tional Park′ (shen′ən dō′-ə, shen′-), a national park in N Virginia, including part of the Blue Ridge mountain range. 302 sq. mi. (782 sq. km).

Shen·si (*Chin.* shun′shē′), SHAANXI.

Shen·yang (shun′yäng′), the capital of Liaoning province, in NE China: cultural capital of Manchuria. 4,200,000. Formerly, **Fengtien, Mukden.**

Sher·brooke (shûr′brŏŏk), a city in S Quebec, in SE Canada. 74,438.

's Her·to·gen·bosch (seR′tō кнəн bôs′), the capital of North Brabant, in the S Netherlands. 89,988.

Sher′wood For′est (shûr′wŏŏd), a forest in central England, chiefly in Nottinghamshire: the traditional haunt of Robin Hood.

Shet·land (shet′lənd), a region in NE Scotland, comprising the Shetland Islands. Formerly, **Zetland.**

Shet′land Is′lands (shet′lənd), a group of islands NE of the Orkney Islands: northernmost part of Great Britain. 22,429; 550 sq. mi. (1425 sq. km). **—Shet′land·er,** *n.*

Shi·be·li (shi bā′lē), SHEBELI.

Shi·jia·zhuang or **Shih·chia·chuang** (shœ′jyä′jwäng′), the capital of Hebei province, in NE China, SW of Beijing. 1,160,000.

Shi·ko·ku (shē/kô kōō/), an island in SW Japan, S of Honshu. 7249 sq. mi. (18,775 sq. km).

Shil·long (shi lông/), the capital of Meghalaya state, in NE India: resort. 109,244.

Shi·loh (shī/lō), **1.** a military national park in SW Tennessee: Civil War battle 1862. **2.** an ancient town in central Palestine, west of the Jordan River.

Shim·la (shim/lə) also **Simla,** the capital of Himachal Pradesh, in N India. 70,604.

Shi·mo·no·se·ki (shim/ə nə sā/kē), a seaport on SW Honshu, in SW Japan. 261,000.

Shi·nar (shī/när), a land mentioned in the Bible, often identified with Sumer.

Shi·raz (shi räz/), a city in SW Iran. 848,289.

Shi·ré (shē/rā), a river in SE Africa, flowing S from Lake Malawi to the Zambezi River. 370 mi. (596 km) long.

Shi·zu·o·ka (shē/zōō ō/kə), a city on S Honshu, in central Japan. 468,000.

Shko·dër (shkō/dər), a city in NW Albania, on Lake Scutari: a former capital of Albania. 71,000.

Sho·la·pur (shō/lə pŏŏr/), a city in S Maharashtra, in SW India. 514,000.

Sho·sho·ne (shō shō/nē), a river in NW Wyoming, flowing NE into the Bighorn River. 120 mi. (193 km) long.

Shosho/ne Falls/, falls of the Snake River, in S Idaho. 210 ft. (64 m) high.

Shreve·port (shrēv/pôrt/, -pōrt/), a city in NW Louisiana, on the Red River. 198,525.

Shrews·bur·y (shrōōz/ber/ē, -bə rē, shrōz/-), a city in Shropshire, in W England. 90,500.

Shrop·shire (shrop/shēr, -shər), a county in W England. 396,500; 1348 sq. mi. (3490 sq. km). Formerly, **Salop.**

Shu·shan (shōō/shan, -shän), Biblical name of SUSA.

Si·al·kot (sē äl/kōt/), a city in NE Pakistan. 296,000.

Si·am (si am/, sī/am), **1.** former name of THAILAND (def. 1). **2. Gulf of,** THAILAND (def. 2).

Si·an (*Chin.* shē/än/), XIAN.

Siang·tan (*Chin.* shyäng/tän/), XIANGTAN.

Šiau·liai (shou lyī/), a city in N Lithuania, N of Kaunas. 145,000.

Si·be·ri·a (sī bēr/ē ə), a part of the Russian Federation in N Asia, extending from the Ural Mountains to the Pacific. —**Si·be′ri·an,** *adj., n.*

Si·biu (sē byŏŏ′), a city in central Romania. 176,928.

Si·chuan (sich′wän′, sich′ōō än′) also **Szechwan, Szechuan,** a province in S central China. 103,200,000; 219,691 sq. mi. (569,000 sq. km). *Cap.:* Chengdu.

Sic·i·lies, Two (sis′ə lēz), **Two SICILIES.**

Sic·i·ly (sis′ə lē), the largest island in the Mediterranean, constituting a region of Italy, and separated from the SW tip of the mainland by the Strait of Messina. 5,141,343; 9924 sq. mi. (25,705 sq. km). Italian, **Si·ci·lia** (sē chē′lyä). Ancient, **Si·ci·lia** (si sil′yə, -sil′ē ə), **Trinacria.** —**Si·cil·ian** (si sil′yən), *adj., n.*

Sic·y·on (sish′ē on′, sis′-), an ancient city in S Greece, near Corinth.

Si·di-bel-Ab·bès (sē/dē bel ə bes′), a city in NW Algeria. 186,978.

Si·don (sīd′n), a city of ancient Phoenicia: site of modern Saida. —**Si·do′ni·an** (-dō′nē ən), *adj., n.*

Sid·ra (sid′rä), **Gulf of,** an inlet of the Mediterranean, on the N coast of Libya.

Si·e·na (sē en′ə), a city in Tuscany, in central Italy. 64,745. —**Si′en·ese′** (-ə nēz′, -nēs′), *adj., n., pl.* **-ese.**

Si·er·ra Le·o·ne (sē er′ə lē ō′nē, lē ōn′), a republic in W Africa: member of the Commonwealth of Nations; former British colony and protectorate. 3,880,000; 27,925 sq. mi. (72,326 sq. km). *Cap.:* Freetown.

Sier′ra Ma′dre (mä′drä), a mountain system in Mexico, comprising three ranges bordering the central plateau. Highest peak, Orizaba, 18,546 ft. (5653 m).

Sier′ra Nevad′a, 1. a mountain range in E California. Highest peak, Mt. Whitney, 14,495 ft. (4418 m). **2.** a mountain range in S Spain. Highest peak, Mulhacén, 11,411 ft. (3478 m).

Si·kang (*Chin.* shē′käng′), **XIKANG.**

Si Kiang (*Chin.* shē′ kyäng′), **XI JIANG.**

Sik·kim (sik′im), a state in NE India, in the Himalayas between Nepal and Bhutan. 315,682; 2818 sq. mi. (7298 sq. km). *Cap.:* Gangtok. —**Sik′kim·ese′,** *n., pl.* **-ese,** *adj.*

Si·le·sia (si lē′zhə, -shə, sī-), a region in central

Europe along both banks of the upper Oder River, mainly in SW Poland and N Czechoslovakia. —**Si·le′sian,** *adj., n.*

Sil·ver Spring′, a town in central Maryland, near Washington, D.C. 72,893.

Sim·birsk (sim bērsk′), former name of ULYA-NOVSK.

Sim·fe·ro·pol (sim′fə rō′pəl), a city in S Ukraine, on S Crimea. 338,000.

Si·mi′ Val′ley (si mē′, sē′mē), a city in SW California. 100,217.

Sim·la (sim′lə), SHIMLA.

Sim·plon (sim′plon), **1.** a mountain pass in S Switzerland, in the Lepontine Alps. 6592 ft. (2010 m) high. **2.** a tunnel between Switzerland and Italy, near this pass. 12 mi. (20 km) long.

Si·nai (si′ni, si′nē i′), **1.** Also called **Si′nai Penin′sula.** a peninsula in NE Egypt, at the N end of the Red Sea between the Gulf of Suez and the Gulf of Aqaba. **2. Mount,** the mountain in S Sinai, of uncertain identity, on which Moses received the Law. —**Si′na·it′ic** (-nē it′ik), **Si·na·ic** (si nā′ik), *adj.*

Si·na·lo·a (sēn′l ō′ə, sin′-), a state in W Mexico, bordering on the Gulf of California. 2,367,567; 22,582 sq. mi. (58,485 sq. km). *Cap.:* Culiacán.

Sind (sind), a province in SE Pakistan, in the lower Indus valley. 21,682,000; 54,407 sq. mi. (140,914 sq. km). *Cap.:* Karachi.

Sin·ga·pore (sing′gə pôr′, -pōr′, sing′ə-), **1.** an island off the S tip of the Malay Peninsula. **2.** a republic comprising this island and a few adjacent islets: member of the Commonwealth of Nations; formerly a British crown colony (1946–59) and a state of Malaysia (1963–65); independent since 1965. 2,610,000; 240 sq. mi. (623 sq. km). **3.** the capital of this republic, a port on the S coast. 206,500. —**Sin′ga·po′re·an,** *n., adj.*

Si·ning (*Chin.* shē′ning′), XINING.

Sin·kiang Ui·ghur (*Chin.* shin′jyäng′ wē′gər), XINJIANG UYGUR.

Sin·siang (*Chin.* shin′shyäng′), XINXIANG.

Sint Maar·ten (sint mär′tn), Dutch name of ST. MARTIN.

Sin·ui·ju (shin′wē′jōō′), a city in W North Korea, on the Yalu River. 500,000.

Sion (*Fr.* syôN), the capital of Valais, in SW Switzerland. 23,100.

Sioux′ Cit′y (soo̅), a port in W Iowa, on the Missouri River. 80,505.

Sioux′ Falls′, a city in SE South Dakota. 100,814.

Sip·par (si pär′), an ancient Babylonian city on the Euphrates, in SE Iraq.

Si·ra·cu·sa (*It.* sē′rä koo̅′zä), Syracuse (def. 2).

Si·ret (si ret′), a river in NE Romania, flowing SE from the Carpathian Mountains in Ukraine to the Danube in Romania. 270 mi. (435 km) long.

Sit·ka (sit′kə), a town in SE Alaska, on an island in the Alexander Archipelago. 7803. —**Sit′kan**, *n., adj.*

Sit·twe (sit′wā), a seaport in W Burma. 107,907. Formerly, **Akyab.**

Si·vas (sē väs′), a city in central Turkey. 197,266.

Sjael·land (shel′län), Danish name of Zealand.

Skag·er·rak (skag′ə rak′, skä′gə räk′), an arm of the North Sea, between Denmark and Norway.

Skag·way (skag′wā′), a town in SE Alaska, near the famous White and Chilkoot passes to the Klondike gold fields: railway terminus. 768.

Skaw (skô), **The,** a cape at the N tip of Denmark. Also called **Ska·gen** (skä′gən).

Skik·da (skēk′də), a seaport in NE Algeria. 141,159.

Ski·ros or **Sky·ros** or **Scy·ros** (ski′ros, skē′rōs), a Greek island in the W Aegean: the largest island of the Northern Sporades. 81 sq. mi. (210 sq. km).

Sko·kie (skō′kē), a city in NE Illinois, near Chicago. 59,432.

Skop·je (skôp′ye), the capital of Macedonia, in the N part. 504,932. Serbo-Croatian, **Skop·lje** (skôp′lye).

Skye (ski), an island in the Hebrides, in NW Scotland. 7372; 670 sq. mi. (1735 sq. km).

Slave′ Coast′, the coast of W equatorial Africa, between the Benin and Volta rivers: a center of slavery traffic 16th–19th centuries.

Slave′ Riv′er, a river in NE Alberta and the Northwest Territories, in Canada: flowing from Lake Athabaska NW to Great Slave Lake. 258 mi. (415 km) long.

Slav·kov (släf′kôf), Czech name of Austerlitz.

Sla·vo·ni·a (slə vō′nē ə), a historic region in N Croatia. —**Sla·vo′ni·an,** *adj., n.*

Sla·vyansk (sləv yänsk′), a city in E central Ukraine, NW of Donetsk. 140,000.

Sles·vig (*Dan.* sles′viкн), Schleswig.

Sli·go (slī′gō), a county in Connaught province, in the NW Republic of Ireland. 55,425; 694 sq. mi. (1795 sq. km).

Slo·va·ki·a (slō vä′kē ə, -vak′ē ə), a region comprising a constituent republic (**Slo′vak Repub′-lic**) of Czechoslovakia, in the E part. 5,177,441; 18,934 sq. mi. (49,039 sq. km). *Cap.*: Bratislava. Czech and Slovak, **Slo·ven·sko** (slô′ven skô/). —**Slo′vak,** *n., adj.* —**Slo·va′ki·an,** *adj., n.*

Slo·ve·ni·a (slō vē′nē ə, -vēn′yə), a republic in SE Europe: formerly part of Yugoslavia. 1,930,000; 7819 sq. mi. (20,250 sq. km). *Cap.*: Ljubljana. —**Slo·vene** (slō vēn′, slō′vēn), **Slo·ve′ni·an,** *n., adj.*

Smok′y Hill′, a river flowing E from E Colorado to the Republican River in central Kansas. 540 mi. (870 km) long.

Smok′y Moun′tains, Great Smoky Mountains.

Smo·lensk (smō lensk′), a city in the W Russian Federation in Europe, on the upper Dnieper. 338,000.

Smyr·na (smûr′nə), **1.** former name of Izmir. **2. Gulf of,** former name of the Gulf of Izmir.

Snake′ Riv′er, a river flowing from NW Wyoming through S Idaho into the Columbia River in SE Washington. 1038 mi. (1670 km) long.

Sno·qual′mie Falls′ (snō kwol′mē), falls on a river (**Snoqualmie**) in W Washington. 270 ft. (82 m) high.

Snow·don (snōd′n), a mountain in NW Wales: highest peak in Wales. 3560 ft. (1085 m).

So·che (*Chin.* sô′chu′), Shache.

So·chi (sō′chē), a seaport in the SW Russian Federation in Europe, on the Black Sea: resort. 317,000.

Soci′ety Is′lands, a group of islands in the S Pacific: a part of French Polynesia; largest island, Tahiti. 142,270; 650 sq. km. (1683 sq. km). *Cap.*: Papeete.

So·co·tra or **So·ko·tra** (sō kō′trə, sok′ə trə), an island of the Republic of Yemen in the Indian

Ocean, S of Arabia. 1382 sq. mi. (3579 sq. km).
—**So·co/tran,** *n., adj.*

Sod·om (sod/əm), an ancient city destroyed, with
Gomorrah, because of its wickedness. —**Sod/om·**
ite/, *n.*

So·fi·a or **So·fi·ya** (sō/fē ə, sō fē/ə), the capital
of Bulgaria, in the W part. 1,128,859.

Sog·di·a·na (sog/dē ā/nə, -an/ə), an ancient region in Central Asia between the Amu Darya and
the Syr Darya.

So·ho (sō/hō; *for 1 also* sō hō/), **1.** a district in
central London, England. **2.** SoHo.

So·Ho or **So·ho** (sō/hō), a district on the lower W
side of Manhattan: art galleries and studios.

So·lent (sō/lənt), **The,** a channel between the
Isle of Wight and the mainland of S England. 2–5
mi. (3.2–8 km) wide.

So·leure (sô lœr/), French name of SOLOTHURN.

So·li·mões (*Port.* sô/li moINs/), the upper Amazon in Brazil, from the Negro River to the border
of Peru.

So·ling·en (zō/ling ən), a city in W Germany, in
the Ruhr region. 159,100.

So·lo (sō/lō), former name of SURAKARTA.

Sol/o·mon Is/lands (sol/ə mən), **1.** an archipelago in the W Pacific Ocean, E of New Guinea: politically divided between Papua New Guinea and
the Solomon Islands. **2.** an independent country
comprising the larger, SE part of this archipelago:
a former British protectorate; gained independence in 1978. 285,796; 11,458 sq. mi. (29,676
sq. km). *Cap.:* Honiara.

So·lo·thurn (zō/lə tŏŏrn/, -tûrn/), **1.** a canton in
NW Switzerland. 224,800; 305 sq. mi. (790 sq.
km). **2.** the capital of this canton, on the Aare
River. 19,000. French, **Soleure.**

Sol/way Firth/ (sol/wā), an arm of the Irish Sea
between SW Scotland and NW England. 38 mi.
(61 km) long.

So·ma·li·a (sō mä/lē ə, -mäl/yə), a republic on
the E coast of Africa, formed by the merger of
British Somaliland and Italian Somaliland in
1960. 6,260,000; 246,198 sq. mi. (637,653 sq.
km). *Cap.:* Mogadishu. Official name, **So·ma/li**
Democrat/ic Repub/lic (sō mä/lē, sə-). —**So·**
ma/li·an, *adj., n.*

So·ma·li·land (sō mä/lē land/, sə-), a coastal region in E Africa, including Djibouti, Somalia, and the Ogaden part of Ethiopia.

Som·er·set·shire (sum/ər set shēr/, -shər, -sit-), a county in SW England. 452,300; 1335 sq. mi. (3455 sq. km). Also called **Som/er·set/**.

Som·er·ville (sum/ər vil/), a city in E Massachusetts, near Boston. 76,210.

Somme (som, sôm), a river in N France, flowing NW to the English Channel. 150 mi. (241 km) long.

Song Hong (sông/ hông/), Vietnamese name of RED RIVER.

Song·hua (sông/hwä/), a river in NE China, flowing through Manchuria into the Amur. 800 mi. (1287 km) long. Also called **Song/hua Jiang/** (jyäng), **Sungari.**

So·no·ra (sə nôr/ə, -nōr/ə), a state in NW Mexico. 1,799,646; 70,484 sq. mi. (182,555 sq. km). *Cap.:* Hermosillo. —**So·no/ran,** *n., adj.*

Soo/ Canals/ (sōō), SAULT STE. MARIE CANALS.

Soo·chow (*Chin.* sōō/jō/), SUZHOU.

So·ra·ta (sô rä/tə), **Mount,** a mountain in W Bolivia, in the Andes, near Lake Titicaca: two peaks, Ancohuma, 21,490 ft. (6550 m), and Illampu, 20,958 ft. (6388 m).

So·ro·ca·ba (sôr/ōō kä/bä), a city in SE Brazil, W of São Paulo. 254,672.

Sor·ren·to (sə ren/tō), a seaport and resort in SW Italy, on the Bay of Naples. 15,133. —**Sor·ren·tine** (sôr/ən tēn/, sə ren/tēn), *adj.*

Sos·no·wiec (sos nōv/yets), a city in S Poland. 255,000.

Sound (sound), **The,** English name of ØRESUND.

South (south), **the,** the general area south of Pennsylvania and the Ohio River and east of the Mississippi, consisting mainly of those states that formed the Confederacy. —**South·ern** (suth/ərn), *adj.* —**South/ern·er,** *n.*

South/ Af/rica, Republic of, a country in S Africa: member of the Commonwealth of Nations until 1961. 29,600,000; 472,000 sq. mi. (1,222,480 sq. km). *Caps.:* Pretoria and Cape Town. Formerly, **Union of South Africa.** —**South/ Af/rican,** *adj., n.*

South/ Amer/ica, a continent in the S part of the Western Hemisphere. ab. 6,900,000 sq. mi.

(17,871,000 sq. km). —**South′ Amer′ican,** *n.,* *adj.*

South·amp·ton (south amp′tən, -hamp′-), a seaport in Hampshire county in S England. 215,400.

Southamp′ton Is′land, an island in N Canada, in the Northwest Territories at the entrance to Hudson Bay. 15,913 sq. mi. (44,124 sq. km).

South′ Ara′bia, Protectorate of, a former protectorate of Great Britain in S Arabia, now part of Yemen.

South′ Austral′ia, a state in S Australia. 1,388,100; 380,070 sq. mi. (984,380 sq. km). *Cap.:* Adelaide. —**South′ Austral′ian,** *n., adj.*

South′ Bend′, a city in N Indiana. 105,511.

South′ Caroli′na, a state in the SE United States, on the Atlantic coast. 3,486,703; 31,055 sq. mi. (80,430 sq. km). *Cap.:* Columbia. *Abbr.:* SC, S.C. —**South′ Carolin′ian,** *n., adj.*

South′ Chi′na Sea′, a part of the W Pacific, bounded by SE China, Vietnam, the Malay Peninsula, Borneo, and the Philippines.

South′ Dako′ta, a state in the N central United States. 696,004; 77,047 sq. mi. (199,550 sq. km). *Cap.:* Pierre. *Abbr.:* SD, S. Dak. —**South′ Dako′tan,** *n., adj.*

South′ Downs′, a range of low hills, from Hampshire to East Sussex, in S England.

South·east (south′ēst′), **the,** the southeast region of the United States. —**South′east′ern,** *adj.* —**South′east′ern·er,** *n.*

South′east A′sia, a region including Indochina, the Malay Peninsula, and the Malay Archipelago. —**South′east A′sian,** *n., adj.*

South′end-on-Sea′ (south′end′), a seaport in SE Essex, in SE England, on the Thames estuary. 159,400.

South′ern Alps′, a mountain range in New Zealand, on South Island. Highest peak, Mt. Cook, 12,349 ft. (3764 m).

South′ern Hem′isphere, the half of the earth between the South Pole and the equator.

South′ern Rhode′sia, a former name (until 1964) of ZIMBABWE. —**South′ern Rhode′sian,** *n., adj.*

South′ern Spor′ades, See under SPORADES.

South·field (south′fēld′), a city in SE Michigan, W of Detroit. 75,728.

South′ Gate′, a city in SW California, near Los Angeles. 86,284.

South′ Geor′gia, an island in the S Atlantic, about 800 mi. (1287 km) SE of the Falkland Islands: a British dependent territory. 1450 sq. mi. (3755 sq. km). —**South′ Geor′gian,** n., adj.

South′ Glamor′gan, a county in SE Wales. 399,500; 161 sq. mi. (416 sq. km).

South′ Hol′land, a province in the SW Netherlands. 3,208,414; 1287 sq. mi. (2906 sq. km). *Cap.:* The Hague. Dutch, **Zuid-Holland.**

South′ Is′land, the largest island of New Zealand. 863,603; 57,843 sq. mi. (149,813 sq. km).

South′ Kore′a, a country in E Asia: formed 1948 after the division of the former country of Korea at 38° N. 42,082,000; 38,232 sq. mi. (99,022 sq. km). *Cap.:* Seoul. Compare KOREA. Official name, **Republic of Korea.** —**South′ Kore′an,** n., adj.

South′ Ork′ney Is′lands, a group of islands in the British Antarctic Territory, N of the Antarctic Peninsula: formerly a dependency of the Falkland Islands; claimed by Argentina.

South′ Osse′tian Auton′omous Re′gion, an autonomous region in the Georgian Republic, in the N part. 99,000; 1506 sq. mi. (3900 sq. km). *Cap.:* Tskhinvali.

South′ Platte′, a river flowing NE from central Colorado to the Platte River in W Nebraska. 424 mi. (683 km) long.

South′ Pole′, the southern end of the earth's axis of rotation, the southernmost point on earth.

South·port (south′pôrt′, -pōrt′), a seaport in Merseyside, in W England: resort. 89,745.

South′ San′ Francis′co, a city in central California. 54,312.

South′ Sea′ Is′lands, the islands in the S Pacific Ocean. Compare OCEANIA. —**South′ Sea′ Is′lander,** n.

South′ Seas′, the seas south of the equator.

South′ Shet′land Is′lands, a group of islands in the British Antarctic Territory, N of the Antarctic Peninsula: formerly a dependency of the Falkland Islands; claimed by Argentina and Chile.

South′ Shields′, a seaport in Tyne and Wear, in

NE England, at the mouth of the Tyne River. 100,513.

South′ Vietnam′, a former country in SE Asia that comprised Vietnam south of the 17th parallel: a separate state 1954–75; now part of reunified Vietnam. *Cap.:* Saigon. Compare NORTH VIETNAM, VIETNAM.

South·wark (suᵺ′ərk), a borough of Greater London, England, S of the Thames. 216,800.

South·west (south′west′), **the,** the southwest region of the United States. —**South′west′ern,** *adj.* —**South′west′ern·er,** *n.*

South′ West′ A′frica, a former name (1920–68) of NAMIBIA.

South′ Yem′en, YEMEN (def. 3).

South′ York′shire, a metropolitan county in N England. 1,317,500; 603 sq. mi. (1561 sq. km).

So′vi·et Rus′sia (sō′vē et′, -it, sō′vē et′), **1.** UNION OF SOVIET SOCIALIST REPUBLICS. **2.** RUSSIAN SOVIET FEDERATED SOCIALIST REPUBLIC.

So′viet Un′ion, UNION OF SOVIET SOCIALIST REPUBLICS.

So·we·to (sə wet′ō, -wā′tō), a group of townships housing black South Africans SW of Johannesburg in NE South Africa. ab. 2,000,000; 26 sq. mi. (67 sq. km).

Spa (spä), a resort town in E Belgium, SE of Liège: famous mineral springs. 9391.

Spain (spān), a kingdom in SW Europe, on the Iberian Peninsula. 39,000,000; 194,988 sq. mi. (505,019 sq. km). *Cap.:* Madrid. Spanish, **Es·paña.** —**Span·iard** (span′yərd), *n.* —**Span′ish,** *adj., n.*

Span·dau (spän′dou, shpän′-), a district of Berlin, in E Germany: site of prison for Nazi war criminals.

Span′ish Amer′ica, the Spanish-speaking countries S of the U.S.: Mexico, Central America (except Belize), South America (except Brazil, French Guiana, Guyana, and Suriname), and most of the West Indies.

Span′ish Guin′ea, former name of EQUATORIAL GUINEA.

Span′ish Main′, 1. the mainland of America adjacent to the Caribbean Sea, esp. the area between the mouth of the Orinoco River and the Isthmus of Panama. **2.** the Caribbean Sea: the route of the Spanish treasure galleons and a former haunt of pirates.

Span'ish Moroc'co, See under Morocco.

Span'ish Sahar'a, former name of Western Sahara.

Sparks (spärks), a city in W Nevada, E of Reno. 53,367.

Spar·ta (spär'tə), an ancient city in S Greece: the capital of Laconia and the chief city of the Peloponnesus, at one time the dominant city of Greece. Also called **Lacedaemon.** —**Spar'tan,** adj., n.

Spe·zia (spāt'sē ə), La Spezia.

Spice' Is'lands, former name of the Moluccas.

Spits·ber·gen (spits'bûr'gən), a group of islands in the Arctic Ocean, N of and belonging to Norway: a major part of Svalbard. 23,641 sq. mi. (61,229 sq. km).

Split (split), a seaport in S Croatia, on the Adriatic: Roman ruins. 180,571.

Spo·kane (spō kan'), a city in E Washington. 177,196.

Spor·a·des (spôr'ə dēz'), two groups of Greek islands in the Aegean: the one **(Northern Sporades)** off the E coast of Greece; the other **(Southern Sporades),** including the Dodecanese, off the SW coast of Asia Minor.

Spree (sprā, shprā), a river in E Germany, flowing N through Berlin to the Havel River. 220 mi. (354 km) long.

Spring·field (spring'fēld'), **1.** a city in S Massachusetts, on the Connecticut River. 156,983. **2.** a city in SW Missouri. 140,494. **3.** the capital of Illinois, in the central part. 105,227. **4.** a city in W Ohio. 70,487.

Springs (springz), a city in S Transvaal, in the E Republic of South Africa, E of Johannesburg. 142,812.

Spuy'ten Duy'vil Creek' (spit'n di'vəl), a channel in New York City at the N end of Manhattan Island, connecting the Hudson and Harlem rivers.

Sri Lan·ka (srē' läng'kə, lang'kə, shrē'), an island republic in the Indian Ocean, S of India: a member of the Commonwealth of Nations. 16,600,000; 25,332 sq. mi. (65,610 sq. km).

Cap.: Colombo. Formerly, **Ceylon.** —**Sri′ Lan′-kan,** *adj., n.*

Sri·na·gar (srē nug′ər, shrē-), the summer capital of Jammu and Kashmir, on the Jhelum River. 520,000.

Staf·fa (staf′ə), an island in W Scotland, in the Hebrides: site of Fingal's Cave.

Staf·ford (staf′ərd), **1.** a city in Staffordshire, in central England. 117,900. **2.** STAFFORDSHIRE.

Staf·ford·shire (staf′ərd shēr′, -shər), a county in central England. 1,027,500; 1154 sq. mi. (2715 sq. km). Also called **Stafford, Staffs** (stafs).

Sta·gi·ra (stə ji′rə) also **Sta·gi·ros** (-rəs, -ros), an ancient town in NE Greece, in Macedonia on the E Chalcidice peninsula: birthplace of Aristotle. —**Stag·i·rite** (staj′ə rit′), *n.*

Staked′ Plain′, LLANO ESTACADO.

Sta·kha·nov (stə kä′nəf), a city in E Ukraine, W of Lugansk. 108,000. Formerly, **Kadiyevka.**

St. Al·bans (ôl′bənz), a city in W Hertfordshire, in SE England. 128,600.

Sta·lin (stä′lin, -lēn, stal′in), **1.** a former name of DONETSK. **2.** former name of VARNA.

Sta·li·na·bad (stä′lə nə bäd′), a former name of DUSHANBE.

Sta·lin·grad (stä′lin grad′), a former name of VOLGOGRAD.

Sta·li·no (stä′lə nō′), a former name of DONETSK.

Sta·linsk (stä′linsk), former name of NOVOKUZNETSK.

Stam·bul or **Stam·boul** (stäm bool′), the oldest section of Istanbul.

Stam·ford (stam′fərd), a city in SW Connecticut. 108,056.

Stan·ley (stan′lē), **1.** the capital and principal harbor of the Falkland Islands, in the E part. 1200. **2. Mount,** former name of Mount NGALIEMA.

Stan′ley Falls′, former name of BOYOMA FALLS.

Stan′ley Pool′, MALEBO POOL.

Stan·ley·ville (stan′lē vil′), former name of KISANGANI.

Sta·no·voi (stan′ə voi′), a mountain range in the E Russian Federation in Asia: a watershed between the Pacific and Arctic oceans; highest peak, 8143 ft. (2480 m).

Sta·ra Za·go·ra (stär′ə zä gôr′ə), a city in central Bulgaria. 156,441.

Stat′en Is′land (stat′n), **1.** an island facing New York Bay. **2.** Formerly, **Richmond.** a borough of New York City including this island. 378,977; 64½ sq. mi. (167 sq. km).

States′ of the Church′, PAPAL STATES.

Sta·tia (stā′shə), ST. EUSTATIUS.

St. Au·gus·tine (ô′gə stēn′), a seacoast city in NE Florida: founded by the Spanish 1565; oldest city in the U.S. 11,692.

Sta·vang·er (stä väng′ər), a seaport in SW Norway. 96,439.

Stav·ro·pol (stav rō′pəl), **1.** a territory of the Russian Federation in Europe, N of the Caucasus. 2,306,000; 29,600 sq. mi. (76,960 sq. km). **2.** the capital of this territory. 306,000. **3.** former name of TOLYATTI.

St. Ber·nard (sänt′ bər närd′), **1. Great,** a mountain pass between SW Switzerland and NW Italy, in the Pennine Alps. 8108 ft. (2470 m) high. **2. Little,** a mountain pass between SE France and NW Italy, in the Alps, S of Mont Blanc. 7177 ft. (2185 m) high.

St. Cath·ar·ines (kath′ər inz, kath′rinz), a city in SE Ontario, in SE Canada. 123,455.

St. Charles (chärlz), a city in E Missouri, on the Missouri River. 54,555.

St. Chris·to·pher (kris′tə fər), ST. KITTS.

St. Christopher-Nevis, ST. KITTS-NEVIS.

St. Clair (klâr), **1.** a river in the N central U.S. and S Canada, flowing S from Lake Huron to Lake St. Clair, forming part of the boundary between Michigan and Ontario. 41 mi. (66 km) long. **2. Lake,** a lake between SE Michigan and Ontario, Canada. 460 sq. mi. (1190 sq. km).

St. Clair Shores, a city in SE Michigan, near Detroit. 68,107.

St. Cloud (sänt′ kloud′; for 2 also saN klŏŏ′), **1.** a city in central Minnesota, on the Mississippi. 42,566. **2.** a suburb of Paris in N France, on the Seine: former royal palace. 28,350.

St. Croix (kroi), **1.** Also called **Santa Cruz.** a U.S. island in the N Lesser Antilles: the largest of the Virgin Islands. 55,300; 82 sq. mi. (212 sq. km). **2.** a river flowing from NW Wisconsin along the boundary between Wisconsin and Minnesota into the Mississippi. 164 mi. (264 km) long. **3.** a river in the northeast U.S. and SE Canada, forming a part of the boundary between Maine and New Brunswick, flowing into Passamaquoddy Bay. 75 mi. (121 km) long.

St. Den·is (sănt′ den′is; *Fr.* saɴ də nē′), **1.** a suburb of Paris in N France: famous abbey, the burial place of many French kings. 96,759. **2.** the capital of Réunion Island, in the Indian Ocean. 109,072.

Steam′boat Springs′, a town in NW Colorado: ski resort. 6695.

Ste.-Foy (sănt′fwä′; *Fr.* saɴt fwA′), a city in S Quebec, in E Canada, near Quebec. 68,883.

Stei·er·mark (shti′ər märk′), German name of STYRIA.

St. E·li·as (i lī′əs), **Mount,** a mountain on the boundary between Alaska and Canada, in the St. Elias Mountains. 18,008 ft. (5490 m).

St. Elias Mountains, a mountain range between SE Alaska and the SW Yukon territory. Highest peak, Mt. Logan, 19,850 ft. (6050 m).

Ste·pa·na·kert (step′ə nə kärt′), the capital of the Nagorno-Karabakh Autonomous Region, within Azerbaijan. 33,000.

Step·ney (step′nē), a former borough of Greater London, England, now part of Tower Hamlets.

Steppes (steps), **The,** the vast grasslands in the S and E European and W and SW Asian parts of Russia.

Ster′ling Heights′, a city in SE Michigan, near Detroit. 117,810.

Ster·li·ta·mak (stûr′lit ə mak′), a city in the Russian Federation in Europe, W of the Southern Urals. 251,000.

St.-É·tienne (saɴ tā tyen′), a city in SE France. 206,688.

Stet·tin (shte tēn′), German name of SZCZECIN.

St. Eu·sta·ti·us (yōō stā′shē əs, -shəs), an island in the Netherlands Antilles, in the E West Indies. 1421; 7 sq. mi. (18 sq. km). Also called **Statia.**

Stew′art Is′land (stōō′ərt, styōō′-), one of the islands of New Zealand, S of South Island. 670 sq. mi. (1735 sq. km).

St. Gal′len (gä′lən), **1.** a canton in NE Switzerland. 407,000; 777 sq. mi. (2010 sq. km). **2.** the capital of this canton. 73,200. French, **St. Gall** (san gȧl′). German, **Sankt Gallen.**

St. George's (jôr′jiz), the capital of Grenada, in the SW part. 6657.

St. George's Channel, a channel between Wales and Ireland, connecting the Irish Sea and the Atlantic.

St.-Ger·main-en-Laye (san zheʀ ma nän lā′), a city in N France, near Paris. 40,471. Also called **St.-Ger·main** (san zheʀ man′).

St. Got·thard (sänt′ got′ərd), **1.** a mountain range in S Switzerland: a part of the Alps. Highest peak, 10,490 ft. (3195 m). **2.** a mountain pass over this range. 6935 ft. (2115 m) high. French, **St. Go·thard** (san gô taʀ′).

St. He·le·na (hə lē′nə), **1.** a British island in the S Atlantic: Napoleon's place of exile 1815–21. 47 sq. mi. (122 sq. km). **2.** a British colony comprising this island, Ascension Island, and the Tristan da Cunha group. 5,564; 126 sq. mi. (326 sq. km). *Cap.:* Jamestown.

St. Hel·ens (hel′ənz), **1.** a city in Merseyside, in NW England, near Liverpool. 187,300. **2. Mount,** an active volcano in SW Washington, part of the Cascade Range: major eruptions 1980. 8364 ft. (2549 m).

St. Hel·ier (hel′yər), a seaport on and capital of the island of Jersey in the English Channel: resort. 28,135.

Stir·ling (stûr′ling), **1.** Also called **Stir·ling·shire′** (-shēr′, -shər). a historic county in central Scotland. **2.** a city in and the administrative center of the Central region, on the Forth River. 38,638.

St. James-As·sin·i·boi·a (ə sin′ə boi′ə), a city in SE Manitoba, in S central Canada: suburb of Winnipeg. 71,431.

St. John, 1. an island of the Virgin Islands of the United States, in the E West Indies. 2800; ab. 20 sq. mi. (52 sq. km). **2. Lake,** a lake in SE Canada, in Quebec province, draining into the Saguenay River. 365 sq. mi. (945 sq. km). **3.** a river in the NE United States·and SE Canada, flowing NE and E from Maine to New Brunswick province and then S to the Bay of Fundy. 450 mi. (725 km) long. **4.** a seaport in S New Brunswick, in SE Canada, on the Bay of Fundy, at the mouth of the St. John River. 80,521. **5.** St. John's.

St. Johns (sānt′ jonz′), a river flowing N and E through NE Florida into the Atlantic. 276 mi. (444 km) long.

St. John's (or **John**), **1.** the capital of Newfoundland, on the SE part of the island. 96,216. **2.** a seaport on and the capital of Antigua and Barbuda, on NW Antigua, in the E West Indies. 30,000.

St. Jo·seph (jō′zəf, -səf), a city in NW Missouri, on the Missouri River. 71,852.

St. Kitts (kits), one of the Leeward Islands, in the E West Indies: part of St. Kitts-Nevis; formerly a British colony. 68 sq. mi. (176 sq. km). Also called **St. Christopher.**

St. Kitts-Nevis, a twin-island state in the Leeward Islands, in the E West Indies, consisting of St. Kitts and Nevis: formerly a British colony; gained independence 1983. 44,400; 104 sq. mi. (269 sq. km). *Cap.:* Basseterre. Also called **St. Christopher-Nevis.**

St. Kitts-Nevis-Anguilla, a former British colony (1967–71) in the Leeward Islands, in the E West Indies: comprising St. Kitts, Nevis, Anguilla, and adjacent small islands. Compare Sᴛ. Kɪᴛᴛs-Nᴇᴠɪs.

St. Lau·rent (san lô rän′), a city in S Quebec, in E Canada, W of Montreal. 65,900.

St. Lawrence (sānt), **1.** a river in SE Canada, flowing NE from Lake Ontario, forming part of the boundary between New York and Ontario, and emptying into the Gulf of St. Lawrence. 760 mi. (1225 km) long. **2. Gulf of,** an arm of the Atlantic between SE Canada and Newfoundland.

St. Lawrence Seaway, a series of channels, locks, and canals between Montreal and the mouth of Lake Ontario, a distance of 182 miles (293 km), enabling most deep-draft vessels to travel from the Atlantic Ocean, up the St. Lawrence River, to all the Great Lakes ports: developed jointly by the U.S. and Canada.

St.-Lé·o·nard (sānt/len′ərd; *Fr.* san lā ô naʀ′), a city in S Quebec, in E Canada: suburb of Montreal. 79,429.

St. Lou·is (sānt′ lōō′is), a port in E Missouri, on the Mississippi. 396,685.

St. Lu·cia (lōō′shə, -sē ə), one of the Windward Islands, in the E West Indies: a former British colony; gained independence 1979. 146,000; 238

sq. mi. (616 sq. km). *Cap.:* Castries. —**St. Lu·cian,** *n., adj.*

St. Ma·lo (sän ma lō′), **1.** a seaport in NW France, on the Gulf of St. Malo: resort. 46,270. **2. Gulf of,** an arm of the English Channel in NW France. 60 mi. (97 km) wide.

St. Mar·tin (sänt′ mär′tn, sənt), an island in the N Leeward Islands, in the E West Indies, divided in two parts: the N section is a dependency of Guadeloupe. 8072; 20 sq. mi. (52 sq. km); the S section is an administrative part of the Netherlands Antilles. 14,639; 17 sq. mi. (44 sq. km). Dutch, **Sint Maarten.**

St. Mar·ys (mâr′ēz), a river in the N central U.S. and S Canada, forming the boundary between NE Michigan and Ontario, flowing SE from Lake Superior into Lake Huron. 63 mi. (101 km) long. Compare SAULT STE. MARIE.

St. Mo·ritz (sän′ mō rits′, mô-, mə-), a resort town in SE Switzerland. 5900; 6037 ft. (1840 m) above sea level. German, **Sankt Moritz.**

St. Na·zaire (san na zar′), a seaport in W France, on the Loire estuary. 69,769.

Stock·holm (stok′hōm, -hōlm), the chief seaport in and the capital of Sweden, in the SE part. 666,810; with suburbs 1,606,157.

Stock·port (stok′pôrt′, -pōrt′), a borough of Greater Manchester, in NW England. 293,400.

Stock·ton (stok′tən), a city in central California, on the San Joaquin River. 210,943.

Stock′ton-on-Tees′, a seaport in Cleveland, in NE England, on the Tees River. 175,900.

Stoke′-on-Trent′ or **Stoke′-upon-Trent′** (stōk), a city in N Staffordshire, in central England, on the Trent River. 255,800.

Stone·henge (stōn′henj), a prehistoric megalithic monument on Salisbury Plain, in S England, dating to late Neolithic and early Bronze Age times (3rd to 2nd millennium B.C.): believed to have had religious or astronomical functions.

Stone′ Moun′tain, a dome-shaped granite outcrop in NW Georgia, near Atlanta: sculptures of Confederate heroes. 1686 ft. (514 m) high.

Ston′y Tungu′ska. See under TUNGUSKA.

St.-Ouen (san twän′), a suburb of Paris in N France. 52,000.

St. Paul, a port in and the capital of Minnesota, in the SE part, on the Mississippi. 272,235.

St. Petersburg. 1. Formerly, **Leningrad** (1924–91); **Petrograd** (1914–24). a seaport in the NW Russian Federation in Europe, on the Gulf of Finland: capital of the Russian Empire (1712–1917). 5,020,000. **2.** a city in W Florida, on Tampa Bay. 238,629.

St. Pierre and Miq·ue·lon (sānt/ pyâr/; *Fr.* saN pyeR/; mik/ə lon/; *Fr.* mēk/ə lôn/), two small groups of islands off the S coast of Newfoundland: an overseas territory of France. 6041; 93 sq. mi. (240 sq. km).

St. Quen·tin (sānt/ kwen/tn; *Fr.* saN kän taN/), a city in N France, on the Somme. 69,153.

Straits/ Set/tlements, a former British crown colony in SE Asia: included the settlements of Singapore, Penang, Malacca, and Labuan.

Stral·sund (sträl/zŏŏnt, shträl/-), a seaport in NE Germany. 75,408.

Stras·bourg (stras/bûrg, sträz/bŏŏrg), a city in NE France, near the Rhine. 257,303. German, **Strass·burg** (shträs/bŏŏrk).

Strat·ford (strat/fərd). **1.** a town in SW Connecticut. 50,541. **2.** a city in SE Ontario, in S Canada: summer Shakespeare festival. 25,657.

Strat/ford-upon-A/von or **Strat/ford-on-A/von,** a town in SW Warwickshire, in central England, on the Avon River: birthplace and burial place of Shakespeare. 107,200.

Strath·clyde (strath klīd/), a region in SW Scotland. 2,504,909; 5300 sq. mi. (13,727 sq. km).

Strom·bo·li (strom/bə lē), **1.** an island off the NE coast of Sicily, in the Lipari group. **2.** an active volcano on this island. 3040 ft. (927 m).

Stru·ma (strŏŏ/mä), a river in S Europe, flowing SE through SW Bulgaria and NE Greece into the Aegean. 225 mi. (362 km) long.

St. Thom·as (tom/əs), **1.** an island in the Virgin Islands of the U.S., in the E West Indies. 52,660; 32 sq. mi. (83 sq. km). **2.** former name of CHARLOTTE AMALIE.

Stu·art (stŏŏ/ərt, styŏŏ/-), former name of ALICE SPRINGS.

Stutt·gart (stut/gärt, stŏŏt/-, shtŏŏt/-), the capital of Baden-Württemberg, in SW Germany. 552,300.

St. Vin·cent (vin/sənt), **1.** an island in the S

Windward Islands, in the SE West Indies: part of the state of St. Vincent and the Grenadines. 133 sq. mi. (345 sq. km). **2. Cape,** the SW tip of Portugal.

St. Vin·cent and the Gren·adines, a country in the S Windward Islands, in the SE West Indies, comprising St. Vincent island and the N Grenadines: a former British colony; gained independence 1979. 112,614; 150 sq. mi. (388 sq. km). *Cap.*: Kingstown.

Styr (stēr), a river in NW Ukraine, flowing N to the Pripet River. 300 mi. (480 km) long.

Styr·i·a (stēr′ē ə), a province in SE Austria: formerly a duchy. 1,187,512; 6327 sq. mi. (16,385 sq. km). *Cap.*: Graz. German, **Steiermark.**

Su·bo·ti·ca (sōō′bə tit′sə), a city in N Serbia, in N Yugoslavia. 154,611.

Sü·chow (*Chin.* sy′jō′), Xuzhou.

Su·cre (sōō′krā), the official capital of Bolivia, in the S part. 86,609.

Su·dan (sōō dan′), **1.** a region in N Africa, S of the Sahara and Libyan deserts, extending from the Atlantic to the Red Sea. **2. Republic of the.** Formerly, **Anglo-Egyptian Sudan.** a republic in NE Africa, S of Egypt and bordering on the Red Sea: a former condominium of Egypt and Great Britain; gained independence 1956. 25,560,000; 967,500 sq. mi. (2,505,825 sq. km). *Cap.*: Khartoum. —**Su′da·nese′,** *n.*, *pl.* **-nese,** *adj.*

Sud·bur·y (sud′ber′ē, -bə rē), a city in S Ontario, in S Canada. 88,717.

Su·de·ten (sōō dāt′n), **1.** Also, **Su·de·tes** (sōō-dē′tēz); *Czech,* **Su·de·ty** (sōō′de ti). a mountain range in E central Europe, extending along the N boundary of Czechoslovakia between the Elbe and Oder rivers. Highest peak, 5259 ft. (1603 m). **2.** Sudetenland.

Su·de·ten·land (sōō dāt′n land′, -länt′), a mountainous region in N Czechoslovakia, including the Sudeten and the Erzgebirge: annexed by Germany 1938; returned to Czechoslovakia 1945. Also called **Sudeten.**

Su·ez (sōō ez′, sōō′ez), **1.** a seaport in NE Egypt, near the S end of the Suez Canal. 275,000. **2. Gulf of,** a NW arm of the Red Sea, W of the Sinai Peninsula. **3. Isthmus of,** an isthmus in NE Egypt, joining Africa and Asia. 72 mi. (116 km) wide.

Su′ez Canal′, a canal in NE Egypt, crossing the Isthmus of Suez and connecting the Mediterranean and Red seas. 107 mi. (172 km) long.

Suf·folk (suf′ək), **1.** a county in E England. 635,100; 1470 sq. mi. (3805 sq. km). **2.** a city in SE Virginia. 52,141.

Sug·ar·loaf Moun′tain (shŏŏg′ər lōf′), a mountain in SE Brazil in Rio de Janeiro, at the entrance to Guanabara Bay. 1296 ft. (395 m). Portuguese, **Pão de Açúcar.**

Suisse (swēs), French name of SWITZERLAND.

Su·i·ta (sŏŏ ē′tä), a city on S Honshu, in Japan: a suburb of Osaka. 343,000.

Su·khu·mi (sŏŏ kōō′mē, sŏŏ′kə-), the capital of Abkhazia, in the NW Georgian Republic, on the Black Sea. 122,000.

Suk·kur (suk′ər), a city in SE Pakistan, on the Indus River. 191,000.

Su·la·we·si (sŏŏ′lä wā′sē), an island in central Indonesia, E of Borneo. 10,409,533 with adjacent islands; 72,986 sq. mi. (189,034 sq. km). Formerly, **Celebes.**

Su′lu Archipel′ago (sŏŏ′lōō), an island group in the SW Philippines, extending SW from Mindanao to Borneo. 555,239; 1086 sq. mi. (2813 sq. km).

Su′lu Sea′, a sea in the W Pacific, between the SW Philippines and Borneo.

Su·ma·tra (sŏŏ mä′trə), a large island in the W part of Indonesia. 28,016,160; 164,147 sq. mi. (425,141 sq. km). **—Su·ma′tran,** *adj., n.*

Sum·ba (sŏŏm′bä), one of the Lesser Sunda Islands, in Indonesia, S of Flores. 4306 sq. mi. (11,153 sq. km).

Sum·ba·wa (sŏŏm bä′wä), one of the Lesser Sunda Islands, in Indonesia. 5965 sq. mi. (15,449 sq. km).

Su·mer (sŏŏ′mər), an ancient region in S Mesopotamia containing a number of independent cities and city-states, fl. c3200–2000 B.C. **—Su·me′ri·an** (-mēr′ē ən, -mer′-), *n., adj.*

Sum·ga·it (sŏŏm′gä ēt′), a city in SE Azerbaijan, on the Caspian Sea. 234,000.

Sum·ter (sum′tər, sump′-), Fort. FORT SUMTER.

Sun′belt′ or **Sun′ Belt′,** (*sometimes l.c.*) the southern and southwestern region of the U.S.

Sun·da Is′lands (sun′də), a chain of islands in Indonesia, in the Malay Archipelago, including

Borneo, Sumatra, Java, and Sulawesi **(Greater Sunda Islands)** and a group of smaller islands extending E from Java to Timor **(Lesser Sunda Islands).**

Sun·der·land (sun′dər lənd), a seaport in Tyne and Wear, in NE England. 298,000.

Sunds·vall (sunts′väl), a seaport in E Sweden, on the Gulf of Bothnia. 92,721.

Sun·ga·ri (sŏŏng′gə rē), SONGHUA.

Sun·ny·vale (sun′ē vāl′), a city in central California, S of San Francisco. 117,229.

Sun·rise (sun′rīz′), a city in SE Florida. 64,407.

Sun′ Val′ley, a village in S central Idaho: winter resort. 938.

Suo·mi (swô′mi), Finnish name of FINLAND.

Su·pe·ri·or (sə pēr′ē ər, sŏŏ-), **Lake,** a lake in the N central U.S. and S Canada: the northernmost of the Great Lakes; the largest body of fresh water in the world. 31,820 sq. mi. (82,415 sq. km).

Sur (sŏŏr), a town in S Lebanon, on the Mediterranean Sea: site of the ancient port of Tyre.

Su·ra·ba·ya or **Su·ra·ba·ja** (sŏŏr′ə bä′yə), a seaport on NE Java, Indonesia. 2,027,913.

Su·ra·kar·ta (sŏŏr′ə kär′tə), a city on central Java, in central Indonesia. 469,888. Formerly, **Solo.**

Su·rat (sŏŏ rat′, sŏŏr′ət), a seaport in S Gujarat, in W India: first British trading post in India 1612. 913,000.

Su·ri·ba·chi (sŏŏr′ə bä′chē), an extinct volcano on Iwo Jima island: World War II battle 1945.

Su·ri·na·me (sŏŏr′ə nä′mə) also **Su′ri·nam′** (-näm′, -nam′), a republic on the NE coast of South America: formerly a territory of the Netherlands; gained independence 1975. 415,000; 63,251 sq. mi. (163,820 sq. km). *Cap.:* Paramaribo. Formerly, **Dutch Guiana, Netherlands Guiana.** —**Su′ri·nam′er,** *n.* —**Su′ri·na·mese′** (-nə-mēz′, -mēs′), *n., pl.* **-mese,** *adj.*

Sur·rey (sûr′ē, sur′ē), a county in SE England, bordering S London. 1,000,700; 648 sq. mi. (1680 sq. km).

Surt·sey (sûrt′sē, sŏŏrt′sā), an island S of and belonging to Iceland: formed by an undersea volcano 1963. ab. one mi. (1.5 km) in diameter; ab. 500 ft. (150 m) high.

Su·sa (soo′sə, -sä), a ruined city in W Iran: the capital of ancient Elam. Biblical name, **Shushan.** —**Su′si·an** (-zē ən), *n., adj.*

Sus·que·han·na (sus′kwə han′ə), a river flowing S from central New York through E Pennsylvania and NE Maryland into Chesapeake Bay. 444 mi. (715 km) long.

Sus·sex (sus′iks), **1.** a former county in SE England: divided into East Sussex and West Sussex. **2.** a kingdom of the Anglo-Saxon heptarchy in SE England.

Suth·er·land (suth′ər lənd), Also called **Suth′er·land·shire′** (-shēr′, -shər). a historic county in N Scotland.

Suth′erland Falls′, a waterfall in New Zealand, on SW South Island. 1904 ft. (580 m) high.

Sut·lej (sut′lej), a river in S Asia, flowing W and SW from Tibet through NW India into the Indus River in Pakistan. 900 mi. (1450 km) long.

Sut′ter's Mill′ (sut′ərz), the location of John Sutter's mill in California, NE of Sacramento, near which gold was discovered, precipitating the gold rush of 1849.

Sut·ton (sut′n), a borough of Greater London, England. 165,800.

Su·va (soo′vä), the capital of Fiji, on Viti Levu island. 71,608.

Su·wan·nee (sə won′ē, -wô′nē, swon′ē, swô′nē) also **Swanee,** a river in SE Georgia and N Florida, flowing SW to the Gulf of Mexico. 240 mi. (386 km) long.

Su·wŏn (soo′wun′), a city in NW South Korea, S of Seoul. 430,827.

Su·zhou (sy′jō′) also **Soochow,** a city in S Jiangsu province, in E China. 673,308.

Sval·bard (sväl′bär), a group of islands in the Arctic Ocean, N of and belonging to Norway: includes the Spitsbergen group. 23,958 sq. mi. (62,050 sq. km).

Sverd·lovsk (sverd lôfsk′, -lofsk′, sferd-), former name (1924–91) of EKATERINBURG.

Sver·drup (sver′drəp, sfer′-), Also called **Sver′-drup Is′lands.** a group of islands in the N Northwest Territories of Canada, in the Arctic.

Sve·ri·ge (sve′rē ye), Swedish name of SWEDEN.

Sviz·ze·ra (zvēt′tse rä), Italian name of SWITZER-LAND.

Swa·bi·a (swā′bē ə), a region and medieval duchy in SW Germany, now part of the states of Baden-Württemberg and Bavaria in S Germany. German, **Schwaben.** —**Swa′bi·an,** adj., n.

Swa·nee (swon′ē, swô′nē), SUWANNEE.

Swan·sea (swon′sē, -zē), a seaport in West Glamorgan, in S Wales. 190,500.

Swat (swät), a former princely state in NW India: now a part of Pakistan.

Swa·tow (swä′tou′), SHANTOU.

Swa·zi·land (swä′zē land′), a kingdom in SE Africa between Mozambique and the Republic of South Africa: formerly a British protectorate. 676,049; 6704 sq. mi. (17,363 sq. km). *Cap.:* Mbabane.

Swe·den (swēd′n), a kingdom in N Europe, in the E part of the Scandinavian Peninsula. 8,414,083; 173,394 sq. mi. (449,090 sq. km). *Cap.:* Stockholm. Swedish, **Sverige.** —**Swede,** n. —**Swed′ish,** adj.

Switz·er·land (swit′sər lənd), a republic in central Europe. 6,620,000; 15,944 sq. mi. (41,295 sq. km). *Cap.:* Bern. French, **Suisse.** German, **Schweiz.** Italian, **Svizzera.** Latin, **Helvetia.** —**Swiss** (swis), adj., n.

Syb·a·ris (sib′ə ris), an ancient Greek city in S Italy: noted for its wealth and luxury; destroyed 510 B.C.

Syd·ney (sid′nē), the capital of New South Wales, in SE Australia. 3,430,600.

Sy·e·ne (si ē′nē), ancient name of ASWAN.

Syk·tyv·kar (sik′tif kär′), the capital of the Komi Autonomous Republic in the NW Russian Federation in Europe. 233,000.

Syr·a·cuse (sir′ə kyoōs′, -kyoōz′), **1.** a city in central New York. 163,860. **2.** Italian, **Siracusa.** a seaport in SE Sicily: ancient city founded by the Carthaginians 734 B.C.; battles 413 B.C., 212 B.C. 121,134. —**Syr′a·cu′san,** adj., n.

Syr Dar·ya (sēr′ där′yə), a river in central Asia, flowing NW from the Tien Shan Mountains in Kyrgyzstan, through Uzbekistan and Kazakhstan to the Aral Sea. 1300 mi. (2100 km) long. Ancient, **Jaxartes.**

Syr·i·a (sēr′ē ə), **1.** Official name, **Syr′ian Ar′ab Repub′lic.** a republic in SW Asia at the E end of the Mediterranean. 11,400,000; 71,227 sq. mi.

(184,478 sq. km). *Cap.:* Damascus. **2.** an ancient country in W Asia, including modern Syria, Lebanon, Israel, and adjacent areas: a part of the Roman Empire 64 B.C.–A.D. 636. —**Syr′i‧an,** *n., adj.*

Syr′ian Des′ert, a desert in N Saudi Arabia, SE Syria, W Iraq, and NE Jordan. ab. 125,000 sq. mi. (323,750 sq. km).

Syz‧ran (siz′ran), a city in the E Russian Federation in Europe, on the Volga. 174,000.

Szcze‧cin (shchet′chēn, -sēn), a seaport in NW Poland. 391,000. German, **Stettin.**

Sze‧chwan or **Sze‧chuan** (sech′wän′, -ōō än′), SICHUAN.

Sze‧ged (seg′ed), a city in S Hungary, on the Tisza River. 188,000.

Szé‧kes‧fe‧hér‧vár (sā′kesh fā hâr vär′, -fâr-vär′), a city in W central Hungary. 113,000.

T

Ta·bas·co (tə bas′kō), a state in SE Mexico, on the Gulf of Campeche. 1,299,507; 9783 sq. mi. (25,338 sq. km). *Cap.:* Villahermosa.

Ta′ble Moun′tain, a mountain in the SW Republic of South Africa, near Cape Town. 3550 ft. (1080 m).

Ta·bor (tā′bər), **Mount,** a mountain in N Israel. 1929 ft. (588 m).

Ta·briz (tä brēz′, tə-), a city in NW Iran. 971,482.

Ta·clo·ban (tä klō′bän), a seaport on NE Leyte, in the central Philippines. 138,000.

Tac·na (tak′nə, täk′-), a city in S Peru. 137,500.

Tac′na-Ari′ca, a maritime region in W South America: now divided between Peru and Chile.

Ta·co·ma (tə kō′mə), a seaport in W Washington, on Puget Sound. 176,664. —**Ta·co′man,** *n.*

Tad·mor (tad′môr, täd′-), Biblical name of PALMYRA.

Ta·dzhik·i·stan (tə jik′ə stan′, -stän′, -jē′kə-), TAJIKISTAN.

Tae·gu (tī′gōō′), a city in SE South Korea. 2,030,649.

Tae·jon (tī′jon′), a city in W South Korea. 866,303.

Ta·gan·rog (taq′ən rog′), a seaport in the S Russian Federation in Europe, on the Sea of Azov. 295,000.

Ta·gus (tā′gəs), a river in SW Europe, flowing W through central Spain and Portugal to the Atlantic at Lisbon. 566 mi. (910 km) long. Spanish, **Tajo.** Portuguese, **Tejo.**

Ta·hi·ti (tə hē′tē, tä-), the principal island of the Society Islands, in the S Pacific. 115,820; 402 sq. mi. (1041 sq. km). *Cap.:* Papeete. —**Ta·hi′tian,** *adj., n.*

Ta·hoe (tä′hō), **Lake,** a lake in E California and W Nevada, in the Sierra Nevada Mountains: resort. ab. 200 sq. mi. (520 sq. km).

T'ai·chou or **Tai·chow** (*Chin.* tī′jō′), TAIZHOU.

Tai·chung (tī′jŏŏng′), a city in W Taiwan. 607,238.

Ta·if (tä′if), a city in W Saudi Arabia. 300,000.

Tai·myr′ (or **Tai·mir′**) **Penin′sula** (tī mēr′), a

peninsula in the N Russian Federation in Asia, between the Kara and Laptev seas.

Tai·nan (ti/nän/), a city in SW Taiwan. 594,739.

Tai·pei (ti/pā/, -bā/), a city in the N part of Taiwan: the capital of the Republic of China. 2,640,000.

Tai·wan (ti/wän/), an island off the SE coast of China: seat of the Republic of China since 1949. 13,900 sq. mi. (36,000 sq. km). Also called **Formosa.** —**Tai/wan·ese/** (-wä nēz/, -nēs/), adj., n., pl. -**ese.**

Tai/wan Strait/, an arm of the Pacific Ocean between China and Taiwan, connecting the East and South China seas. Formerly, **Formosa Strait.**

Tai·yuan (ti/yyän/), the capital of Shanxi province, in N China. 1,880,000. Formerly, **Yangkü.**

Ta·iz or **Ta·'izz** (ta iz/), a city in S Yemen. 178,043.

Tai·zhou or **T'ai·chou** or **Tai·chow** (ti/jō/), a city in central Jiangsu province, in E China. 275,000.

Ta·jik·i·stan or **Ta·dzhik·i·stan** (tə jik/ə stan/, -stän/, -jē/kə-), a republic in central Asia, N of Afghanistan. 5,112,000; 55,240 sq. mi. (143,100 sq. km). Cap.: Dushanbe. Formerly (1929–91), **Ta·jik/ So/viet So/cialist Repub/lic.** —**Ta·jik/, Ta·dzhik/** n., adj.

Taj Ma·hal (täzh/ mə häl/, täj/), a white marble mausoleum built at Agra, India, by Shah Jahan (fl. 1628–58).

Ta·jo (tä/hô), Spanish name of TAGUS.

Ta·ka·ma·tsu (tä/kə mät/sōō), a seaport on NE Shikoku, in SW Japan. 326,000.

Ta·ka·tsu·ki (tä/kət sōō/kē, tə kät/sōō kē), a city on S Honshu, in Japan: a suburb of Osaka. 350,000.

Ta·kla·ma·kan or **Ta·kli·ma·kan** (tä/klə mə-kän/), a desert in S central Xinjiang Uygur, in W China. ab. 125,000 sq. mi. (323,750 sq. km).

Tal·ca (täl/kə), a city in central Chile. 164,482.

Tal·ca·hua·no (täl/kə wä/nō, -hwä/-), a seaport in central Chile. 231,356.

Ta·lien (dä/lyen/), DALIAN.

Tal·la·has·see (tal/ə has/ē), the capital of Florida, in the N part. 124,773.

Tal·linn or **Tal·lin** (tä′lin, tal′in), the capital of Estonia, on the Gulf of Finland. 499,800.

Ta·mau·li·pas (tä′mou lē′päs), a state in NE Mexico, bordering on the Gulf of Mexico. 1,901,000; 30,731 sq. mi. (79,595 sq. km). *Cap.:* Ciudad Victoria.

Tam·bo·ra (täm′bər ə, täm bôr′ə, -bōr′ə), an active volcano in Indonesia, on N Sumbawa. 9042 ft. (2756 m).

Tam·bov (täm bôf′, -bôv′), a city in the Russian Federation, SE of Moscow. 305,000.

Tam·il Na·du (tam′il nä′dōō, tum′-, tä′məl), a state in S India. 48,408,077; 50,215 sq. mi. (130,058 sq. km). *Cap.:* Madras. Formerly, **Madras.**

Tam·pa (tam′pə), a city in W Florida, on Tampa Bay. 280,015. —**Tam′pan,** *n., adj.*

Tam′pa Bay′, an inlet of the Gulf of Mexico, in W Florida.

Tam·pe·re (täm′pə rā′), a city in SW Finland. 170,533.

Tam·pi·co (tam pē′kō), a seaport in SE Tamaulipas, in E Mexico. 267,957. —**Tam·pi′can,** *n.*

Ta·na (tä′nä, -nə), **1.** a river in E Africa, in Kenya, flowing SE to the Indian Ocean. 500 mi. (800 km) long. **2. Lake.** Also, **Tsana.** a lake in NW Ethiopia: the source of the Blue Nile. 1100 sq. mi. (2850 sq. km).

Tan·a·gra (tan′ə grə, tə nag′rə), a town in ancient Greece, in Boeotia: Spartan victory over the Athenians 457 B.C.

Tan·a·na (tan′ə nä′, -nô′), a river flowing NW from E Alaska to the Yukon River. ab. 650 mi. (1045 km) long.

Ta·na·na·rive (tə nan′ə rēv′), former name of Antananarivo.

Tan·ga (tang′gə), a seaport in NE Tanzania. 103,409.

Tan·gan·yi·ka (tan′gən yē′kə, -gə nē′-, tang′-), **1.** a former country in E Africa: formed the larger part of German East Africa; British trusteeship **(Tan′ganyi′ka Ter′ritory)** 1946–61; became independent 1961; now part of Tanzania. 361,800 sq. mi. (937,062 sq. km). **2. Lake,** a lake in central Africa, between Zaire and Tanzania: longest freshwater lake in the world. 12,700 sq. mi. (32,893 sq. km). —**Tan′gan·yi′kan,** *adj., n.*

Tan·gier (tan jēr′) also **Tan·giers** (-jērz′), a seaport in N Morocco, on the W Strait of Gibraltar: capital of the former Tangier Zone. 266,346. French, **Tan·ger** (tän zhä′).

Tangier′ Zone′, a former internationalized zone on the Strait of Gibraltar: became a part of Morocco 1956.

Tan·gle·wood (tang′gəl wŏŏd′), See under LENOX.

Tang·shan (täng′shän′), a city in NE Hebei province, in NE China. 1,390,000.

Ta·nis (tā′nis), an ancient city in Lower Egypt, in the Nile delta.

Tan·jore (tan jôr′, -jōr′), former name of THANJAVUR.

Tan·ta (tän′tə), a city in N Egypt, in the Nile delta. 373,500.

Tan·za·ni·a (tan′zə nē′ə), a republic in E Africa formed in 1964 by the merger of Tanganyika and Zanzibar. 23,200,000; 363,950 sq. mi. (942,623 sq. km). *Cap.*: Dodoma. —**Tan′za·ni′an,** *n., adj.*

Taos (tous), a town in N New Mexico: resort. 4065.

Ta·pa·jós (tä′pə zhôs′, tap′ə-), a river flowing NE through central Brazil to the Amazon. 500 mi. (800 km) long.

Tar·a (tar′ə), a village in the NE Republic of Ireland, NW of Dublin: traditional residence of ancient Irish kings **(Hill of Tara).**

Ta·ran·to (tär′ən tō′, tar′-, tə ran′tō), **1.** Ancient, **Ta·ren·tum** (tə ren′təm). a seaport in SE Italy, on the Gulf of Taranto: founded by the Greeks in the 8th century B.C. 244,249. **2. Gulf of,** an arm of the Ionian Sea, in S Italy.

Ta·ra·wa (tə rä′wə, tar′ə wä′), an island in the central Pacific: capital of Kiribati. 24,598; 14 sq. mi. (36 sq. km).

Ta·rim (tä′rēm′), a river in NW China, in Xinjiang Uygur. ab. 1300 mi. (2090 km) long.

Ta′rim Ba·sin′, a region in W China between the Tien Shan and Kunlun mountain ranges. ab. 350,000 sq. mi. (906,000 sq. km).

Tar·lac (tär′läk), a city on N central Luzon, in the N Philippines. 160,595.

Tar·nów (tär′nŏŏf), a city in SE Poland, E of Kraków. 107,139.

Tar·pe′ian Rock′ (tär pē′ən), a rock on the

Capitoline Hill in Rome from which criminals and traitors were hurled.

Tar·ra·sa (tə rä′sə, -sä), a city in NE Spain, N of Barcelona. 159,530.

Tar·sus (tär′səs), a city in S Turkey, near the Cilician Gates: important seaport of ancient Cilicia; birthplace of Saint Paul. 121,074.

Tar·ta·ry (tär′tə rē), TATARY.

Tar·tu (tär′tōō), a city in SE Estonia. 115,000.

Tash·kent (täsh kent′, tash-), the capital of Uzbekistan, in the NE part. 2,073,000.

Tas·ma·ni·a (taz mā′nē ə, -mān′yə), an island S of Australia: a state of the commonwealth of Australia. 436,353; 26,382 sq. mi. (68,330 sq. km). *Cap.:* Hobart. Formerly, **Van Diemen's Land.** —**Tas·ma′ni·an,** *adj., n.*

Tas′man Sea′ (taz′mən), a part of the Pacific Ocean between SE Australia and New Zealand.

Ta′tar Auton′omous Repub′lic (tä′tər), an autonomous republic in the E Russian Federation in Europe. 3,640,000; ab. 26,255 sq. mi. (68,000 sq. km). *Cap.:* Kazan.

Ta·ta·ry (tä′tə rē) also **Tartary,** a historic region of indefinite extent in E Europe and Asia: designates the area overrun by the Tartars in the Middle Ages, from the Dnieper River to the Pacific.

Ta′tra Moun′tains (tä′trə), a mountain range in NE Czechoslovakia and S Poland: a part of the Carpathian Mountains. Highest peak, Gerlachovka, 8737 ft. (2663 m).

Ta·tung (*Chin.* dä′tōōng′), DATONG.

Taun′ton Deane′ (tôn′tn dēn′, tän′-), a city in Somersetshire, in SW England. 92,900. Formerly, **Taun′ton.**

Tau·po (tou′pō), **Lake,** a lake in N New Zealand, in central North Island: largest lake in New Zealand. ab. 234 sq. mi. (605 sq. km).

Tau·rus (tôr′əs), a mountain range in S Turkey. Highest peak, 12,251 ft. (3734 m).

Tay (tā), **1.** a river flowing through central Scotland into the Firth of Tay. 118 mi. (190 km) long. **2. Firth of,** an estuary of the North Sea, off the coast of central Scotland. 25 mi. (40 km) long.

Tay·lor (tā′lər), a city in SE Michigan. 70,811.

Tay·side (tā′sid′), a region in E Scotland. 401,987; 1100 sq. mi. (2849 sq. km).

Tbi·li·si (tə bə lē′sē, -bil′ə-), the capital of the

Georgian Republic, in the SE part, on the Kura. 1,194,000. Formerly, **Tiflis.**

Tchad (*Fr.* chàd), CHAD.

Tees (tēz), a river in N England, flowing E along the boundary between Durham and Yorkshire to the North Sea. 70 mi. (113 km) long.

Te·gu·ci·gal·pa (tə gōō'si gäl'pə, -gäl'pä), the capital of Honduras, in the S part. 604,600.

Te·he·ran or **Teh·ran** (te ran', -rän', tā'ə-), the capital of Iran, in the N part. 6,042,584.

Te·huan·te·pec (tə wän'tə pek'), **1.** Isthmus of, an isthmus in S Mexico, between the Gulf of Tehuantepec and the Gulf of Campeche. 125 mi. (200 km) wide at its narrowest point. **2.** Gulf of, an inlet of the Pacific, off the S coast of Mexico.

Tei·de or **Tey·de** (tā'dā), **Pi·co de** (pē'kō dā), a volcanic peak in the Canary Islands, on Tenerife. 12,190 ft. (3716 m). Also called **Pico de Tener·ife** (or **Teneriffe**).

Te·jo (te'zhōō), Portuguese name of TAGUS.

Te·la·nai·pu·ra (tel'ə ni pōōr'ə), former name of JAMBI (def. 2).

Tel A·viv (tel' ə vēv'), a city in W central Israel, on the Mediterranean Sea. 334,900. Official name, **Tel' Aviv'-Jaf'fa** (-yä'fə), **Tel' Aviv'-Ya'fo** (-yä'fō). —**Tel' A·viv'an,** *n.*

Tel el A·mar·na (tel' el ə mär'nə), a village in central Egypt, on the Nile: site of ancient Egyptian city.

Tel·loh (te lō'), a village in SE Iraq, between the lower Tigris and Euphrates: site of the ancient Sumerian city of Lagash.

Te·luk·be·tung (tə lŏŏk'bə tŏŏng'), a port on SE Sumatra, in Indonesia. 284,275.

Tem·es·vár (te'mesh vär'), Hungarian name of TIMIŞOARA.

Te·mir·tau (tā'mēr tou'), a city in E central Kazakhstan, NW of Karaganda. 228,000.

Tem·pe (tem'pē), **1.** Vale of, a valley in E Greece, in Thessaly, between Mounts Olympus and Ossa. **2.** a city in central Arizona, near Phoenix. 141,865.

Te·mu·co (tā mōō'kō), a city in S Chile. 217,789.

Ten·e·dos (ten'i dos', -dōs'), ancient name of BOZCAADA.

Ten·er·ife or **Ten·er·iffe** (ten'ə rēf', -rif', -rē'fä), **1.** the largest of the Canary Islands, off the NW coast of Africa. 794 sq. mi. (2055 sq. km). **2. Pi·co de** (pē′kō dä), TEIDE, Pico de.

Ten·gri Khan (teng′grē kän′, κHän′), KHAN TEN-GRI.

Tenn., Tennessee.

Ten·nes·see (ten′ə sē′), **1.** a state in the SE United States. 4,877,185; 42,246 sq. mi. (109,415 sq. km). *Cap.:* Nashville. *Abbr.:* TN, Tenn. **2.** a river flowing from E Tennessee through N Alabama, W Tennessee, and SW Kentucky into the Ohio near Paducah. 652 mi. (1050 km) long. —**Ten′nes·se′an,** *adj., n.*

Te·noch·ti·tlán (tā nôch′tē tlän′), the capital of the Aztec empire: now the site of Mexico City.

Te·o·ti·hua·cán (tā′ō tē′wä kän′), the ruins of an ancient city in central Mexico, near Mexico City, fl. A.D. c200–c750.

Te·pic (tā pēk′), the capital of Nayarit, in W central Mexico. 177,007.

Ter·cei·ra (tər sâr′ə, -sēr′ə), an island in the Azores, in the N Atlantic. 153 sq. mi. (395 sq. km).

Te·re·si·na (tir′ə zē′nə), the capital of Piauí, in NE Brazil, on the Parnaíba River. 388,922.

Ter·na·te (tər nä′tē, -tā), an island in E Indonesia, W of Halmahera. 53 sq. mi. (137 sq. km).

Ter·ni (târ′nē), a city in central Italy. 110,704.

Ter·no·pol (tər nō′pəl), a city in W Ukraine. 175,000.

Ter·re Haute (ter′ə hōt′, hut′, hōt′), a city in W Indiana, on the Wabash River. 57,483.

Tes·sin (*Fr.* te saN′; *Ger.* te sēn′), French and German name of TICINO.

Te′ton Range′ (tē′ton), a mountain range in NW Wyoming: a part of the Rocky Mountains. Highest peak, Grand Teton, 13,766 ft. (4196 m).

Te·tuán (te twän′) also **Té·touan** (*Fr.* tā twän′), a seaport in N Morocco, on the Mediterranean. 704,205.

Teu′to·burg For′est (tōō′tə bûrg′, tyōō′-), a chain of wooded hills in NW Germany, in Westphalia, taken to be the site of a Roman defeat by Germanic tribes A.D. 9. German, **Teu·to·bur·ger Wald** (toi′tō bûr′gər vält′, -bōōr′-).

Te·ve·re (te′ve re), Italian name of the TIBER.

Tewkes·bur·y (tōōks′ber′ē, -bə rē, -brē, tyōōks′-), a town in N Gloucestershire, in W England: final defeat of the Lancastrians in the Wars of the Roses 1471. 79,500.

Tex., Texas.

Tex·as (tek′səs), a state in the S United States. 16,986,510; 267,339 sq. mi. (692,410 sq. km). *Cap.:* Austin. *Abbr.:* Tex., TX **—Tex′an,** *adj., n.*

Tey·de (tā′dā), **Pi·co de** (pē′kō dā), TEIDE, Pico de.

Thai·land (tī′land′, -lənd), **1.** Formerly, **Siam.** a kingdom in SE Asia. 53,900,000; 198,242 sq. mi. (513,445 sq. km). *Cap.:* Bangkok. **2. Gulf of.** Also called **Gulf of Siam.** an arm of the South China Sea, S of Thailand. **—Thai,** *n., adj.*

Thames (temz; *for 3 also* thāmz, tāmz), **1.** a river in S England, flowing E through London to the North Sea. 209 mi. (336 km) long. **2.** a river in SE Canada, in Ontario province, flowing SW to Lake St. Clair. 160 mi. (260 km) long. **3.** an estuary in SE Connecticut, flowing S to Long Island Sound. 15 mi. (24 km) long.

Than·ja·vur (tun′jə voor′), a city in E Tamil Nadu, in SE India. 183,464. Formerly, **Tanjore.**

Thap·sus (thap′səs), an ancient town on the coast of Tunisia: decisive victory of Caesar 46 B.C.

Thar′ Des′ert (tûr, tär), a desert in NW India and S Pakistan. ab. 100,000 sq. mi. (259,000 sq. km). Also called **Indian Desert.**

Tha·sos (thā′sōs, thā′sos), a Greek island in the N Aegean. ab. 170 sq. mi. (440 sq. km).

The·ba·id (thē′bā id, -bē-), the ancient region surrounding Thebes, in Egypt.

Thebes (thēbz), **1.** an ancient city in S Egypt, on the Nile, on the site of the modern towns of Karnak and Luxor. **2.** a city of ancient Greece, in Boeotia. **—The′ban,** *adj., n.*

The·ra or **Thi·ra** (thēr′ə), a Greek island in the S Aegean, in the Cyclades group. 30 sq. mi. (78 sq. km). Also called **Santorin, Santorini.**

Ther·ma (thûr′mə), ancient name of SALONIKA.

Ther·mop·y·lae (thər mop′ə lē′), a pass in E Greece, in Locris, near an arm of the Aegean: Persian defeat of the Spartans 480 B.C.

Thes·sa·lo·ni·ke (*Gk.* the′sä lô nē′kē) also **Thes·sa·lon·i·ca** (thes′ə lon′i kə, -ə lō ni′kə), official name of SALONIKA. **—Thes′sa·lo′ni·an** (-lō′nē ən), *adj., n.*

Thes·sa·ly (thes′ə lē), a region in E Greece, between the Pindus mountains and the Aegean. 695,654; 5208 sq. mi. (14,490 sq. km). —**Thes·sa·li·an** (the sə'lē ən, -sāl'yən), *adj., n.*

Thet'ford Mines' (thet'fərd), a city in S Quebec, in E Canada: asbestos mining. 19,965.

Thim·phu (tim pōō') also **Thim·bu** (-bōō'), the capital of Bhutan, in the W part. 15,000.

Thi·ra (thēr'ə), THERA.

Tho·hoy·an·dou (tō hoi'an dōō'), the capital of Venda, in NE South Africa. 40,000.

Thorn·ton (thôrn'tn), a city in NE central Colorado. 55,031.

Thors·havn (tôrs houn'), the capital of the Faeroe Islands, in the N Atlantic. 11,618.

Thou'sand Is'lands, a group of about 1500 islands in S Ontario, Canada, and N New York State, in the St. Lawrence River at the outlet of Lake Ontario: summer resorts.

Thou'sand Oaks', a town in S California. 104,352.

Thrace (thrās), **1.** an ancient region of varying extent in the E Balkan Peninsula: later a Roman province; now in Bulgaria, Turkey, and Greece. **2.** a modern region corresponding to the S part of the Roman province: now divided between Greece (**Western Thrace**) and Turkey (**Eastern Thrace**). —**Thra·cian** (thrā'shən), *adj., n.*

Three' Mile' Is'land, an island in the Susquehanna River, SE of Harrisburg, Pennsylvania: nuclear plant accident in 1979.

Three' Riv'ers, a city in S Quebec, in SE Canada, on the St. Lawrence. 55,240. French, **Trois-Rivières.**

Thu·le (tōō'lē), a settlement in NW Greenland: site of U.S. air base. 749.

Thun (tōōn), **Lake of,** a lake in central Switzerland, formed by a widening in the course of the Aare River. 10 mi. (16 km) long. German, **Thuner See** (tōō'nər zā').

Thun'der Bay', a port in W Ontario, in S Canada, on Lake Superior: created in 1970 by the merger of twin cities (**Fort William** and **Port Arthur**) and two adjoining townships. 112,272.

Thur·gau (tōōr'gou), a canton in NE Switzerland. 194,600; 388 sq. mi. (1005 sq. km). *Cap.:* Frauenfeld.

Thu·rin·gi·a (thōō rin'jē ə, -jə), a state in central

Germany. 2,500,000; 5985 sq. mi. (15,500 sq. km). *Cap.:* Erfurt. German, **Thü·ring·en** (tyˈRĭng-ən). —**Thu·rin′gi·an,** *adj., n.*

Thurin′gian For′est, a forested mountain region in central Germany: a resort area. German, **Thü·ring·er Wald** (tyˈRĭng ər vält′).

Thurs′day Is′land, an island in Torres Strait between NE Australia and New Guinea: part of Queensland. 1½ sq. mi. (4 sq. km).

Thy·a·ti·ra (thīˈə tīˈrə), ancient name of AKHISAR.

Tian′an·men Square′ (tyänˈän men′) also **Tien′an·men Square** (tyenˈ-), a large plaza in Beijing, noted esp. as the site of major student demonstrations in 1989 suppressed by the government.

Tian·jin (tyänˈjinˈ) also **Tien·tsin** (tinˈtsinˈ), a port in E Hebei province, in NE China. 6,280,000.

Ti·ber (tīˈbər), a river in central Italy, flowing through Rome into the Mediterranean. 244 mi. (395 km) long. Italian, **Tevere.**

Ti·be·ri·as (ti bērˈē əs), Lake, GALILEE, Sea of.

Ti·bet (ti betˈ), an autonomous region in SW China, on a plateau N of the Himalayas: average elevation ab. 16,000 ft. (4877 m). 2,030,000; 471,660 sq. mi. (1,221,600 sq. km). *Cap.:* Lhasa. Chinese, **Xizang.** —**Ti·bet′an,** *adj., n.*

Ti·bur (tīˈbər), ancient name of TIVOLI (def. 1).

Ti·ci·no (ti chēˈnō), a canton in S Switzerland. 279,100; 1086 sq. mi. (2813 sq. km). *Cap.:* Bellinzona. French and German, **Tessin.**

Ti·con·der·o·ga (tīˈkon də rōˈgə), a village in NE New York, on Lake Champlain: site of fort captured by the English 1759 and by Americans under Ethan Allen 1775. 2938.

Tien Shan (tyenˈ shänˈ) also **Tian′ Shan′** (tyän), a mountain range in central Asia, in Kyrgyzstan and China. Highest peak, Pobeda Peak, 24,406 ft. (7439 m).

Tier·ra del Fue·go (tē erˈə del fwāˈgō), a group of islands at the S tip of South America, separated from the mainland by the Strait of Magellan: jointly owned by Argentina and Chile; boundary disputed. 27,476 sq. mi. (71,165 sq. km).

Tif·lis (tifˈlis), former name of TBILISI.

Ti·gris (tīˈgris), a river in SW Asia, flowing SE from SE Turkey through Iraq, joining the Euphrates to form the Shatt-al-Arab. 1150 mi. (1850 km) long.

Ti·jua·na (tē'ə wä'nə, tē hwä'nä), a city in N Baja California Norte, in NW Mexico, on the Mexico–U.S. border. 461,257.

Ti·kal (tē käl'), an ancient Mayan city in N Guatemala occupied c200 B.C. to A.D. 900.

Til·burg (til'bûrg), a city in the S Netherlands. 153,117.

Tim·buk·tu (tim'buk tōō', tim buk'tōō), former name of TOMBOUCTOU.

Times' Square', a wide intersection extending from 43rd Street to 47th Street in central Manhattan, New York City, where Broadway and Seventh Avenue intersect: theater and entertainment area.

Ti·mi·şoa·ra (tē'mē shwär'ə), a city in W Romania. 318,955. Hungarian, **Temesvár.**

Tim·mins (tim'inz), a city in E Ontario, in S Canada: gold-mining center. 46,114.

Ti·mor (tē'môr, tē môr'), an island in the S part of Indonesia: largest and easternmost of the Lesser Sunda Islands. 13,095 sq. mi. (33,913 sq. km). —**Ti'mo·rese'** (-mô rēz', -rēs'), adj., n., pl. **-rese**.

Ti'mor Sea', an arm of the Indian Ocean, between Timor and NW Australia.

Tin·tag'el Head' (tin taj'əl), a cape in SW England, on the W coast of Cornwall.

Tip·pe·ca·noe (tip'ē kə nōō'), a river in N Indiana, flowing SW to the Wabash. 200 mi. (320 km) long.

Tip·per·ar·y (tip'ə râr'ē), a county in Munster province, in the S Republic of Ireland. 135,204; 1643 sq. mi. (4255 sq. km).

Ti·ran (ti rän'), **Strait of,** a navigable waterway between the N Red Sea and the Gulf of Aqaba.

Ti·ra·në or **Ti·ra·na** (ti rä'nə), the capital of Albania, in the central part. 210,000.

Ti·ra·spol (ti ras'pəl), a city in E Moldova, NW of Odessa. 158,000.

Tîr·gu Mu·reş (tēr'gōō mōōr'esh), a city in central Romania. 157,411.

Ti·rich Mir (tē'rich mēr'), a mountain in N Pakistan, on the border of Afghanistan: highest peak of the Hindu Kush. 25,230 ft. (7690 m).

Ti·rol (ti rōl', ti-, ti'rōl), TYROL.

Tir·u·chi·ra·pal·li or **Tir·uch·chi·rap·pal·li**

(tir′ōō chi rop′ə lē), a city in central Tamil Nadu, in S India.

Tir·yns (tir′inz), an ancient city in Greece, in the Peloponnesus.

Ti·sza (tis′ò), a river in S central Europe, flowing from the Carpathian Mountains through E Hungary and NE Yugoslavia into the Danube N of Belgrade. 800 mi. (1290 km) long.

Ti·ti·ca·ca (tit′i kä′kə, -kä), **Lake,** a lake on the boundary between S Peru and W Bolivia, in the Andes. 3200 sq. mi. (8290 sq. km); 12,508 ft. (3812 m) above sea level.

Ti·to·grad (tē′tō grad′, -gräd′), the former (1945–92) name of **Podgorica.**

Tiv·o·li (tiv′ə lē), **1.** Ancient, **Tibur.** a town in central Italy, E of Rome. 50,969. **2.** a park and entertainment center in Copenhagen, Denmark.

Tlax·ca·la (tläs kä′lä), **1.** a state in SE central Mexico. 665,606; 1554 sq. mi. (4025 sq. km). **2.** the capital of this state. 12,000.

Tlem·cen (tlem sen′), a city in NW Algeria. 146,089.

TN, Tennessee.

To·a·ma·si·na (tō′ə mə sē′nə), a seaport on E Madagascar. 139,000.

To·ba·go (tə bā′gō), an island in the SE West Indies, off the NE coast of Venezuela: formerly a British colony, now part of Trinidad and Tobago. 45,000; 117 sq. mi. (303 sq. km). —**To·ba·go·ni·an** (tō′bə gō′nē ən, -gōn′yən), *n.*

To·bol (tə bôl′), a river rising in Kazakhstan, flowing NE through the Russian Federation in Asia to the Irtysh River. 800 mi. (1290 km) long.

To·can·tins (tō′kən tēnz′, -kän tēns′), a river in E Brazil, flowing N to the Pará River. 1700 mi. (2735 km) long.

To·go (tō′gō), **Republic of,** a country in W Africa, on the Gulf of Guinea: formerly a French mandate; gained independence in 1960. 3,246,000; 21,830 sq. mi. (56,540 sq. km). *Cap.:* Lomé. —**To′go·lese′** (-gə lēz′, -lēs′), *adj., n., pl.* **-lese.**

To·go·land (tō′gō land′), a region in W Africa, on the Gulf of Guinea: a German protectorate until 1919, then divided between Great Britain and Great Britain; the French part is now the Republic of Togo; the British part is now part of Ghana. —**To′go·land′er,** *n.*

To·ku·shi·ma (tō′kŏŏ shē′mə), a seaport on NE Shikoku, in SW Japan. 256,000.

To·kyo (tō′kē ō′), the capital of Japan, on Tokyo Bay in SE Honshu. 11,618,281. Formerly, **Edo, Yedo.** —**To′ky·o·ite′,** *n.*

To′kyo Bay′, an inlet of the Pacific, in SE Honshu in Japan. 30 mi. (48 km) long; 20 mi. (32 km) wide.

To·le·do (tə lē′dō; *for 2 also* -lā′-), **1.** a port in NW Ohio, on Lake Erie. 332,943. **2.** a city in central Spain, on the Tagus River. 57,769.

To·li·ma (tə lē′mə), a volcano in W Colombia, in the Andes. 18,438 ft. (5620 m).

To·lu·ca (tə lōō′kə), **1.** the capital of Mexico state, in S central Mexico. 357,071. **2.** an extinct volcano in central Mexico, in Mexico state. 15,026 ft. (4580 m).

To·lyat·ti (tôl yä′tē), a city in the SW Russian Federation in Europe, on the Volga River. 630,000. Formerly, **Stavropol.**

Tom·big·bee (tom big′bē), a river flowing S through NE Mississippi and SW Alabama to the Mobile River. 525 mi. (845 km) long.

Tom·bouc·tou (*Fr.* tôn bŏŏk tōō′), a town in central Mali, in W Africa, near the Niger River. 20,483. Formerly, **Timbuktu.**

Tomsk (tomsk), a city in the central Russian Federation in Asia, E of the Ob River. 502,000.

Ton·ga (tong′gə), a kingdom consisting of three groups of islands in the SW Pacific, E of Fiji: a former British protectorate. 100,105; 289 sq. mi. (748 sq. km). *Cap.:* Nukualofa. Also called **Ton′ga Is′lands, Friendly Islands.** —**Ton′gan,** *n., adj.*

Tong·hua (tông′hwä′) also **T'unghua, Tunghwa,** a city in SE Jilin province, in NE China. 354,842.

Ton·kin (ton′kin′, tong′-) also **Tong·king** (tong′king′), **1.** a former state in N French Indochina, now part of Vietnam. **2. Gulf of,** an arm of the South China Sea, W of Hainan. 300 mi. (485 km) long.

Ton·le Sap (ton′lā säp′), a lake in W Cambodia, draining into the Mekong River.

Too·woom·ba (tə wŏŏm′bə), a city in SE Queensland, in E Australia. 79,137.

To·pe·ka (tə pē′kə), the capital of Kansas, in the NE part, on the Kansas River. 119,883.

Tor·bay (tôr′bā′, -bā′), a borough in S Devonshire, in SW England: seaside resort. 117,700.

Tor·de·si·llas (tôr′də sēl′yäs, -sē′-), a town in NW Spain, SW of Valladolid: treaty (1494) defining the colonial spheres of Spain and Portugal. 6604.

To·ri·no (tô rē′nô), Italian name of TURIN.

To·ron·to (tə ron′tō), the capital of Ontario, in SE Canada, on Lake Ontario. 612,289. —**To·ron·to·ni·an** (tôr′ən tō′nē ən, tor′-, tə ron-), adj., n.

Tor·rance (tôr′əns, tor′-), a city in SW California, SW of Los Angeles. 133,107.

Tor·re del Gre·co (tôr′ā del grek′ō, grä′kō, tôr′ē), a city in SW Italy, near Naples. 104,646.

Tor·rens (tôr′ənz, tor′-), **Lake**, a salt lake in Australia, in E South Australia. 2400 sq. mi. (6220 sq. km); 25 ft. (8 m) below sea level.

Tor·re·ón (tôr′rā ōn′, -rē-), a city in N Mexico. 363,886.

Tor′res Strait′ (tôr′iz, tor′-), a strait between NE Australia and S New Guinea. 80 mi. (130 km) wide.

Tórs·havn (tôrs houn′), THORSHAVN.

Tor·to·la (tôr tō′lə), the principal island of the British Virgin Islands, in the NE West Indies. 9730; 21 sq. mi. (54 sq. km).

Tor·tu·ga (tôr tōō′gə), an island off the N coast of and belonging to Haiti: formerly a pirate stronghold. 70 sq. mi. (180 sq. km).

To·ruń (tô′rōōn′yə), a city in N Poland, on the Vistula. 186,000.

Tos·ca·na (tôs kä′nä), Italian name of TUSCANY.

Tot·ten·ham (tot′n əm, tot′nəm), a former borough, now part of Haringey, in SE England, N of London.

Tou·lon (tōō lôn′), a seaport in SE France. 181,985.

Tou·louse (tōō lōōz′), a city in S France, on the Garonne River. 354,289.

Tou·raine (tōō ren′, -rān′), a region and former province in W France. Cap.: Tours.

Tou·rane (tōō rän′), former name of DANANG.

Tour·coing (tōōr kwaN′), a city in N France. 97,121.

Tour·nai or **Tour·nay** (tōōr nā′), a city in W Belgium, on the Scheldt River. 66,749.

Tours (tŏŏr), a city in W France, on the Loire River. 136,483.

Tow·er Ham·lets (tou′ər), a borough of Greater London, England. 159,000.

Towns·ville (tounz′vil), a seaport on the E coast of Queensland, in E Australia. 108,342.

Tow·son (tou′sən), a town in central Maryland, near Baltimore. 51,083.

To·ya·ma (tô′yə mä′, tô yä′mə), a city on W Honshu, in central Japan. 314,000.

To·yo·ha·shi (tô′yə hä′shē), a seaport on S Honshu, in central Japan. 323,000.

To·yo·na·ka (tô′yə nä′kə), a city on S Honshu, in Japan, N of Osaka. 407,000.

To·yo·ta (tô′yə tä′, tô yō′tə), a city on S Honshu, in Japan. 311,000.

Trab·zon (träb zôn′), Turkish name of TREBIZOND.

Tra·fal·gar (trə fal′gər; *Sp.* trä′fäl gär′), **Cape,** a cape on the SW coast of Spain, W of Gibraltar: British naval victory over the French and Spanish fleets 1805.

Tra·lee (trə lē′), the county seat of Kerry, in the SW Republic of Ireland. 16,988.

Trans A·lai (trans′ ə lī′, tranz′), a mountain range in central Asia, between Kyrgyzstan and Tajikistan. Highest peak, Lenin Peak, 23,382 ft. (7127 m).

Trans·al·pine Gaul′ (trans al′pin, -pin, tranz-), See under GAUL.

Trans·cau·ca·sia (trans′kô kā′zhə, -shə), a region in SE Europe, S of the Caucasus Mountains, between the Black and Caspian seas: includes the republics of Armenia, Azerbaijan, and Georgia. —**Trans·cau·ca·sian** (-kā′zhən, -shən, -kazh′ən, -kash′-), *adj., n.*

Trans·jor·dan (trans jôr′dn, tranz-), an area E of the Jordan River, in SW Asia: a British mandate (1921–23); an emirate (1923–49); now the major part of the kingdom of Jordan.

Trans·kei (trans kā′, -kī′), a self-governing black homeland in SE South Africa, on the Indian Ocean: granted independence in 1976. 2,876,122; 16,910 sq. mi. (43,798 sq. km). *Cap.:* Umtata. —**Trans·kei′an,** *adj., n.*

Trans·vaal (trans väl′, tranz-), a province in the NE Republic of South Africa. 11,885,000; 110,450

sq. mi. (286,066 sq. km). *Cap.*, Pretoria
—**Trans·vaal′er**, *n.* —**Trans·vaal′i·an**, *adj.*

Tran·syl·va·nia (tran′sil vān′yə, -vā′nē ə), a region in central Romania: formerly part of Hungary. 24,027 sq. mi. (62,230 sq. km). —**Tran′syl·va′ni·an**, *adj., n.*

Tran′syl·va′nian Alps′, a mountain range in S Romania, forming a SW extension of the Carpathian Mountains. Highest peak, Moldoveanul, 8343 ft. (2543 m).

Tra·pa·ni (trä′pə nē), a seaport in NW Sicily. 71,430.

Tra·si·me·no (traz′ə mā′nō), a lake in central Italy, in Umbria near Perugia: Romans defeated by Hannibal 217 B.C. ab. 50 sq. mi. (130 sq. km). Latin, **Tras·i·me·nus** (tras′ə mē′nəs).

Trav·an·core (trav′ən kôr′, -kōr′), a former state in SW India: now a part of Kerala state.

Treb·bi·a (treb′ē ə), a river in N Italy, flowing N into the Po at Piacenza: Romans defeated by Hannibal near here 218 B.C. 70 mi. (113 km) long.

Treb·i·zond (treb′ə zond′), **1.** a medieval empire in NE Asia Minor 1204–1461. **2.** Turkish, **Trab·zon.** a seaport in NE Turkey, on the Black Sea: an ancient Greek colony; capital of the medieval empire of Trebizond. 155,960.

Treng·ga·nu (treng gä′nōō), a state in Malaysia, on the E central Malay Peninsula. 540,627; 5050 sq. mi. (13,080 sq. km).

Trent (trent), **1.** a river in central England, flowing NE from Staffordshire to the Humber. 170 mi. (275 km) long. **2.** Italian, **Tren·to.** (ʀɛn′tô). a city in N Italy, on the Adige River. 100,677.

Tren·ti·no-Al·to A·di·ge (tren tē′nō äl′tō ä′di-jā′), a region in NE Italy. 881,986; 5256 sq. mi. (13,615 sq. km).

Tren·ton (tren′tn), the capital of New Jersey, in the W part, on the Delaware River. 88,675. —**Tren·to′ni·an** (-tō′nē ən), *n.*

Tre·vi·so (trä vē′zō), a city in NE Italy. 90,632.

Trier (trēr), a city in W Germany, on the Moselle River. 93,472.

Tri·este (trē est′, -es′tā, -tē), **1.** a seaport in NE Italy, on the Gulf of Trieste. 237,191. **2. Free Territory of,** an area bordering the N Adriatic: designated a free territory by the U.N. 1947; N zone, including the city of Trieste, turned over to Italy in 1954; S zone incorporated into Yugoslavia; now part of Slovenia. **3. Gulf of,** an inlet at the N end of the Adriatic, in NE Italy.

Tri·na·cri·a (tri nä′krē ə, -nak′rē ə, tri-), an ancient name of SICILY. —**Tri·na′cri·an,** adj.

Trin·i·dad (trin′i dad′), an island in the SE West Indies, off the NE coast of Venezuela: formerly a British colony, now part of Trinidad and Tobago. 1,198,000; 1864 sq. mi. (4828 sq. km). —**Trin′i·da/di·an** (-dā′dē ən, -dad′ē-), adj., n.

Trin′idad and Toba′go, a republic in the West Indies, comprising the islands of Trinidad and Tobago: member of the Commonwealth of Nations. 1,243,000; 1980 sq. mi. (5128 sq. km). Cap.: Port-of-Spain.

Trip·o·li (trip′ə lē), **1.** Also, **Trip·o·li·ta·ni·a** (trip′ə li tā′nē ə, -tān′yə, tri pol′i-). one of the Barbary States of N Africa: later a province of Turkey; now a part of Libya. **2.** the capital of Libya, in the NW part. 858,000. **3.** a seaport in NW Lebanon, on the Mediterranean. 175,000. —**Tri·pol·i·tan** (tri pol′i tn), n., adj.

Trip·u·ra (trip′ər ə), a state in E India. 2,060,189; 4033 sq. mi. (10,445 sq. km). Cap.: Agartala.

Tris·tan da Cu·nha (tris′tən də kōō′nə, kōōn′yə), a group of volcanic islands in the S Atlantic, belonging to St. Helena. 40 sq. mi. (104 sq. km).

Tri·van·drum (tri van′drəm), the capital of Kerala state, in S India: Vishnu pilgrimage center. 520,000.

Tro·ad (trō′ad), **The,** a region in NW Asia Minor surrounding ancient Troy. Also called **Tro′as** (-as).

Tro·bri·and Is′lands (trō′brē änd′, -and′), a group of islands in the SW Pacific, off SE New Guinea: part of Papua New Guinea. 170 sq. mi. (440 sq. km).

Trois-Ri·vières (trwä rē vyer′), French name of THREE RIVERS.

Trond·heim (tron′hām), a seaport in central Norway, on Trondheim Fjord. 134,889.

Trond′heim Fjord′, an inlet of the North Sea, extending into N Norway. 80 mi. (129 km) long.

Tros·sachs (tros′əks), a valley in central Scotland, in Perth county, near Loch Katrine.

Trou·ville (trōō vēl′), a seaport in NW France, on the English Channel: resort. 6577. Also called **Trouville′-sur-Mer′** (sûr mâr′).

Troy (troi), **1.** Latin, **Ilium.** Greek, **Ilion.** an ancient ruined city in NW Asia Minor: the seventh of nine settlements on the site is commonly identified as the Troy of the *Iliad.* **2.** a city in SE Michigan, near Detroit. 72,884. **3.** a city in E New York, on the Hudson River. 54,269. —**Tro·jan** (trō′jən), *adj., n.*

Troyes (tRwä), a city in NE France, on the Seine. 64,769.

Tru·cial O·man (trōō′shəl ō män′), a former name of UNITED ARAB EMIRATES. Also called **Tru′cial Coast′, Tru′cial States′.**

Truck·ee (truk′ē), a river in E California and W Nevada, rising in Lake Tahoe and flowing E and NE for about 125 mi. (201 km).

Tru·ji·llo (trōō hē′ō), a seaport in NW Peru. 491,100.

Truk′ Is′lands (truk, trōōk), a group of islands in the W Pacific, in the Caroline Islands: part of the Federated States of Micronesia. 46,159; 49 sq. mi. (127 sq. km).

Tsa·na (tsä′nä, -nə), **Lake, TANA, Lake.**

Tsa·ri·tsyn (zə rēt′sin, tsə-), a former name of VOLGOGRAD.

Tse·li·no·grad (tsə lin′ə grad′, -gräd′, -lē′nə-), a city in N central Kazakhstan. 276,000.

Tsi·nan (*Chin.* jē′nän′), JINAN.

Tsing·hai (*Chin.* ching′hī′), QINGHAI.

Tsing·tao (*Chin.* ching′dou′), QINGDAO.

Tsing·yuan (*Chin.* ching′yyän′), former name of BAODING.

Tsi·tsi·har (tsē′tsē′här′; *Chin.* chē′chē′här′), QIQIHAR.

Tskhin·va·li (skin′və lē, tskin′-), the capital of the South Ossetian Autonomous Region, in the N Georgian Republic. 34,000.

Tsu·shi·ma (tsōō′shē mä′), two adjacent Japanese islands between Korea and Kyushu. 58,672; 271 sq. mi. (702 sq. km).

Tsu′shima Strait′, a channel between the Tsushima islands and Kyushu island, connecting the

Sea of Japan and the East China Sea: sometimes considered part of the Korea Strait.

Tu·a·mo/tu Archipel/ago (tōō/ə mō/tōō), a group of islands in the S Pacific: part of French Polynesia. 11,793; 332 sq. mi. (860 sq. km).

Tuc·son (tōō/son, tōō son/), a city in S Arizona. 405,390.

Tu·cu·mán (tōō/kōō män/), a city in NW Argentina. 496,914.

Tu·la (tōō/lə), **1.** a city in the W Russian Federation, S of Moscow. 540,000. **2.** a city in SW Hidalgo, in central Mexico, NW of Mexico City: site of Toltec ruins. 36,460.

Tul·sa (tul/sə), a city in NE Oklahoma, on the Arkansas River. 367,302. —**Tul/san,** *n., adj.*

Tu·men (tv/mœn), a river in E Asia, flowing NE along the China–North Korea border and then SE along the border between China and Russia to the Sea of Japan. ab. 325 mi. (525 km) long.

Tun/bridge Wells/ (tun/brij), a city in SW Kent, in SE England: mineral springs. 99,100.

Tung·hwa or **T'ung·hua** (*Chin.* tōōng/hwä/), Tonghua.

Tung·ting (dōōng/ting/), Dongting.

Tun·gu·ska (tōōng gōō/skə), any of three tributaries (**Lower Tunguska, Stony Tunguska,** and **Upper Tunguska**) of the Yenisei River in the central Russian Federation in Asia.

Tu·nis (tōō/nis, tyōō/-), **1.** the capital of Tunisia, in the NE part. 596,654. **2.** one of the former Barbary States in N Africa: constitutes modern Tunisia.

Tu·ni·sia (tōō nē/zhə, -shə, -nizh/ə, -nish/ə, tyōō-), a republic in N Africa, on the Mediterranean: a French protectorate until 1956. 7,320,000; 48,330 sq. mi. (125,175 sq. km). *Cap.:* Tunis. —**Tu·ni/sian,** *adj., n.*

Tu·pun·ga·to (tōō/pōōng gä/tō), a mountain between Argentina and Chile, in the Andes. ab. 22,310 ft. (6800 m).

Tu·rin (tōōr/in, tyōōr/-, tōō rin/, tyōō-), a city in NW Italy, on the Po River. 1,025,390. Italian, **Torino.**

Tur·ka·na (tōōr kä/nə), **Lake,** a lake in E Africa, in N Kenya. 3500 sq. mi. (9100 sq. km). Formerly, **Lake Rudolf.**

Tur·ke·stan (tûr/kə stan/, -stän/), a vast region

in central Asia, from the Caspian Sea to the Gobi desert: includes the Xinjiang Uygur region in W China **(Chinese Turkestan),** a strip of N Afghanistan, and the area **(Russian Turkestan)** comprising the republics of Kazakhstan, Kyrgyzstan, Tajikistan, Turkmenistan, and Uzbekistan.

Tur·key (tûr′kē), a republic in W Asia and SE Europe. 50,664,458; 300,948 sq. mi. (779,452 sq. km). *Cap.:* Ankara. Compare OTTOMAN EMPIRE. —**Turk,** *n.* —**Turk′ish,** *adj.*

Turk·me·ni·stan (tûrk′me ne stan′, -stän′), a republic in central Asia, bordering the Caspian Sea, Iran, and Afghanistan. 3,534,000; 188,417 sq. mi. (488,000 sq. km). *Cap.:* Ashkhabad. Formerly, **Turk′men So′viet So′cialist Repub′lic.** —**Turk′man,** *n.,* *pl.* -**men.** —**Turk·me′ni·an** (-mē′nē ən),, *adj.*

Turks′ and Cai′cos Is′lands (tûrks′, ki′kōs, kā′-), two groups of islands in the SE Bahamas: British crown colonies. 7436; ab. 166 sq. mi. (430 sq. km). *Cap.:* Grand Turk.

Tur·ku (tōōr′kōō), a seaport in SW Finland. 163,400. Swedish, **Åbo.**

Tus·ca·loo·sa (tus′ke lōō′se), a city in W Alabama. 77,759.

Tus·ca·ny (tus′ke nē), a region in W central Italy: formerly a grand duchy. 3,578,814; 8879 sq. mi. (22,995 sq. km). Italian, **Toscana.** —**Tus′can,** *adj., n.*

Tus·cu·lum (tus′kye lem), an ancient city of Latium, SE of Rome: Roman villas, esp. that of Cicero. —**Tus′cu·lan,** *adj.*

Tus·tin (tus′tin), a city in SW California. 50,689.

Tu·tu·i·la (tōō/tōō ē′le), the largest of the islands of American Samoa: harbor at Pago Pago. 30,626; 53 sq. mi. (137 sq. km). —**Tu′tu·i′lan,** *adj., n.*

Tu′va Auton′omous Repub′lic (tōō′ve), an autonomous republic in the Russian Federation in Asia: formerly an independent republic in Mongolia. 309,000; 65,810 sq. mi. (170,500 sq. km). *Cap.:* Kyzyl.

Tu·va·lu (tōō′ve lōō′, tōō vä′lōō), a parliamentary state consisting of a group of islands in the central Pacific, S of the equator: a former British colony; gained independence 1978. 8229; 10 sq. mi. (26 sq. km). *Cap.:* Funafuti. Formerly, **Ellice Islands.** —**Tu′va·lu′an,** *adj., n.*

Tux·tla Gu·tiér·rez (tōōs′tlä gōō tyer′res), the capital of Chiapas, in SE Mexico. 166,476. Also called **Tux′tla.**

Tver (tvâr), a city in the W Russian Federation in Europe, on the Volga. 447,000. Formerly (1934–90), **Kalinin.**

Tweed (twēd), a river flowing E from S Scotland along part of the NE boundary of England into the North Sea. 97 mi. (156 km) long.

Tweed·dale (twēd′dāl′), Peebles.

Twick·en·ham (twik′ə nəm), a former borough, now part of Richmond-upon-Thames, in SE England.

Twin′ Cit′ies, the cities of St. Paul and Minneapolis.

Two′ Sic′ilies, a former kingdom in Sicily and S Italy that existed intermittently from 1130 to 1861.

TX, Texas.

Ty·ler (tī′lər), a city in E Texas. 75,450.

Tyne (tīn), a river in NE England, in Northumberland, flowing E into the North Sea. ab. 30 mi. (48 km) long.

Tyne′ and Wear′ (wēr), a metropolitan county in NE England. 1,135,800.

Tyne·mouth (tīn′məth, tīn′-), a seaport in Tyne and Wear, in NE England, at the mouth of the Tyne River. 72,000.

Tyre (tīªr), an ancient seaport and trading center of Phoenicia: site of modern Sur.

Ty·ree (tī rē′), **Mount,** a mountain in Antarctica, near Ronne Ice Shelf. ab. 16,290 ft. (4965 m).

Ty·rol or **Ti·rol** (tī rōl′, ti-, tī′rōl; Ger. tē rōl′), an alpine region in W Austria and N Italy: a former Austrian crown land. —**Ty·ro′le·an,** adj., n.

Ty·rone (tī rōn′), a county in W Northern Ireland. 143,900; 1211 sq. mi. (3136 sq. km).

Tyr·rhe′ni·an Sea′ (ti rē′nē ən), a part of the Mediterranean, bounded by W Italy, Corsica, Sardinia, and Sicily.

Tyu·men (tyōō men′), a city in the SW Russian Federation in Asia. 456,000.

Tze·kung or **Tzu·kung** (Chin. dzu′gŏŏng′), Zi-gong.

Tze·po or **Tzu·po** (Chin. dzu′bô′), Zibo.

U

U.A.E. or **UAE**, United Arab Emirates.

U.A.R. or **UAR**, United Arab Republic.

U·ban·gi (yŏŏ bang/gē, ŏŏ bäng/-), a river in W central Africa, forming part of the boundary between Zaire and the Central African Republic, flowing W and S into the Congo (Zaire) River. 700 mi. (1125 km) long.

Uban/gi-Sha/ri (shär/ē), former name of the CENTRAL AFRICAN REPUBLIC.

U·ca·ya·li (ŏŏ/kä yä/lē), a river in central Peru, flowing N and joining the Marañón to form the Amazon. 1200 mi. (1930 km) long.

U·di·ne (ŏŏ/dē ne), a city in NE Italy. 103,504.

Ud·murt/ Auton/omous Repub/lic (ŏŏd-mŏŏrt/), an autonomous republic in the Russian Federation in Europe. 1,609,000; 16,250 sq. mi. (42,088 sq. km). Cap.: Izhevsk.

Ue·le (wā/lə), a river in central Africa flowing W from NE Zaire to the Ubangi River. 700 mi. (1125 km) long.

U·fa (ŏŏ fä/), the capital of the Bashkir Autonomous Republic, in the W Russian Federation in Europe. 1,083,000.

U·gan·da (yŏŏ gan/də, ŏŏ gän/-), a republic in E Africa, between NE Zaire and Kenya: member of the Commonwealth of Nations; formerly a British protectorate. 15,500,000; 91,343 sq. mi. (236,860 sq. km). Cap.: Kampala. —**U·gan/dan**, adj., n.

U·ga·rit (ŏŏ/gə rēt/, yŏŏ/-), an ancient city on the site of modern Ras Shamra, in NW Syria, fl. 2nd millennium B.C. —**U/ga·rit/ic** (-rit/ik), adj.

U·in/ta Moun/tains (yŏŏ in/tə), a mountain range in NE Utah, part of the Rocky Mountains. Highest peak, Kings Peak, 13,498 ft. (4115 m).

Uj·jain or **U·jain** (ŏŏ/jin), a city in W Madhya Pradesh, in W central India: one of the seven holy cities of India. 282,000.

U·jung Pan·dang (ŏŏ jŏŏng/ pän däng/), a seaport on SW Sulawesi, in central Indonesia. 709,038. Formerly, **Macassar, Makassar.**

U.K. or **UK**, United Kingdom.

U·kraine (yŏŏ krān/, -krin/, yŏŏ/krān), a republic in SE Europe. 51,704,000; 233,090 sq. mi.

(603,700 sq. km). *Cap.:* Kiev. Formerly, **Ukrain/‑ian So/viet So/cialist Repub/lic.** —**U·krain/i·an,** *adj., n.*

U·lan Ba·tor (ŏŏ/län bä/tôr), the capital of the Mongolian People's Republic, in the N central part. 500,000. Formerly, **Urga.**

U·lan U·de (ŏŏ län/ ŏŏ dā/), the capital of the Buryat Autonomous Republic, in the SE Russian Federation in Asia, on the Selenga River. 353,000. Formerly, **Verkhneudinsk.**

Ulm (ŏŏlm), a city in E Baden‑Württemberg, in S Germany, on the Danube. 103,600.

Ul·ster (ul/stər), **1.** a former province in Ireland, now comprising Northern Ireland and a part of the Republic of Ireland. **2.** a province in N Republic of Ireland. 235,641; 3123 sq. mi. (8090 sq. km). **3.** NORTHERN IRELAND. —**Ul/ster·ite/,** *n.*

Ul·ya·novsk (ŏŏl yä/nôfsk, ‑nofsk, ‑nəfsk), a city in the W Russian Federation, on the Volga: birthplace of Lenin. 625,000. Formerly, **Simbirsk.**

Um·bri·a (um/brē ə), a region in central Italy. 818,226; 3270 sq. mi. (8470 sq. km). —**Um/bri·an,** *adj., n.*

Um·ta·ta (ŏŏm tä/tə), the capital of Transkei, in SE Africa. 24,805.

U·na·las·ka (ŏŏ/nə las/kə, un/ə las/‑), an island off the coast of SW Alaska, one of the Aleutian Islands. ab. 75 mi. (120 km) long.

Un·ga·va (ung gä/və, ‑gä/‑), a region in N Quebec, in E Canada, comprising the larger part of the peninsula of Labrador.

Unga/va Bay/, an inlet of the Hudson Strait in NE Quebec province, in E Canada, between Ungava Peninsula and N Labrador.

Unga/va Penin/sula, a peninsula in N Quebec, in E Canada, between Hudson Bay and Ungava Bay.

Un·ion (yōōn/yən), a township in NE New Jersey. 50,024.

Un/ion Cit/y, 1. a city in NE New Jersey. 58,012. **2.** a city in W California. 53,762.

Un/ion of South/ Af/rica, former name of Republic of SOUTH AFRICA.

Un/ion of So/viet So/cialist Repub/lics, a former federal union of 15 constituent republics, in E Europe and N Asia, comprising the larger part of the earlier Russian Empire: dissolved in

December 1991. *Cap.:* Moscow. Also called **So-viet Union.** *Abbr.:* U.S.S.R., USSR

Unit′ed Ar′ab Em′irates, an independent fed-eration in E Arabia, formed in 1971, now com-prising seven emirates on the S coast of the Per-sian Gulf, formerly under British protection. 1,600,000; ab. 32,300 sq. mi. (83,657 sq. km). *Cap.:* Abu Dhabi. *Abbr.:* U.A.E. Formerly, **Trucial Coast, Trucial Oman, Trucial States.**

Unit′ed Ar′ab Repub′lic, 1. a name given the union of Egypt and Syria from 1958 to 1961. **2.** the official name of Egypt from 1961 to 1971. *Abbr.:* U.A.R. Compare EGYPT.

Unit′ed King′dom, a kingdom in NW Europe, consisting of Great Britain and Northern Ireland: formerly comprising Great Britain and Ireland 1801–1922. 55,900,000; 94,242 sq. mi. (244,100 sq. km). *Cap.:* London. *Abbr.:* U.K. Official name, **Unit′ed King′dom of Great′ Brit′ain and North′ern Ire′land.**

Unit′ed States′, a republic in the N Western Hemisphere comprising 48 conterminous states, the District of Columbia, and Alaska in North America, and Hawaii in the N Pacific. 248,709,873; conterminous United States, 3,022,387 sq. mi. (7,827,982 sq. km); with Alaska and Hawaii, 3,615,122 sq. mi. (9,363,166 sq. km). *Cap.:* Washington, D.C. *Abbr.:* U.S., US Also called **United States of America, Amer-ica.**

Unit′ed States′ of Amer′ica, UNITED STATES. *Abbr.:* U.S.A., USA

Un·ter·wal·den (*Ger.* ŏŏn′tər väl′dən), a former canton in central Switzerland: now divided into Nidwalden and Obwalden.

Up·land (up′lənd), a city in SW California, E of Los Angeles. 63,374.

U·po·lu (ōō pō′lōō), an island in Western Samoa, in the S Pacific. 114,980; 430 sq. mi. (1113 sq. km).

Up′per Can′ada, a former British province in Canada 1791–1840: now the S part of Ontario province.

Up′per Dar′by (där′bē), a town in SE Pennsyl-vania, near Philadelphia. 84,054.

Up′per Palat′inate, See under PALATINATE.

Up′per Penin′sula, the peninsula between

lakes Superior and Michigan constituting the N part of Michigan. Also called **Up/per Mich/igan.**

Up/per Tungu/ska, See under TUNGUSKA.

Up/per Vol/ta, former name of BURKINA FASO. —**Up/per Vol/tan,** *adj., n.*

Upp·sa·la or **Up·sa·la** (up/sä lə, -sə-, ŏŏp/-), a city in SE Sweden. 159,962.

Ur (ûr, ŏŏr), an ancient Sumerian city on the Euphrates, in what is now S Iraq.

U·ral (yŏŏr/əl), a river in the Russian Federation, flowing S from the S Ural Mountains to the Caspian Sea. 1400 mi. (2255 km) long.

U/ral Moun/tains, a mountain range in the W Russian Federation, extending N and S from the Arctic Ocean to near the Caspian Sea, forming a natural boundary between Europe and Asia. Highest peak, 6214 ft. (1894 m). Also called **U/rals.**

U·ralsk (yŏŏ ralsk/), a city in W Kazakhstan, on the Ural River. 201,000.

U·ra·wa (ŏŏ rä/wə), a city on E Honshu, in Japan. 383,000.

Ur·fa (ŏŏr fä/), a city in SE Turkey, E of the Euphrates River: on the site of ancient Edessa. 148,434.

Ur·ga (ŏŏr/gä), former name of ULAN BATOR.

U·ri (ŏŏr/ē), a canton in central Switzerland. 33,400; 415 sq. mi. (1075 sq. km). *Cap.:* Altdorf.

Ur·mi·a (ŏŏr/mē ə), **Lake,** a salt lake in NW Iran. ab. 2000 sq. mi. (5180 sq. km).

U·rua·pan (ŏŏr wä/pän), a city in central Michoacán, in SW Mexico: near Paricutín volcano. 147,000.

U·ru·bam·ba (ŏŏr/ə bäm/bə), a river rising in SE Peru, flowing NW through the Andes Mountains to join the Apurímac River and form the Ucayali River. 450 mi. (725 km) long.

U·ru·guay (yŏŏr/ə gwā/, -gwi/, ŏŏr/-), **1.** a republic in SE South America. 3,080,000; 72,172 sq. mi. (186,925 sq. km). *Cap.:* Montevideo. **2.** a river in SE South America, flowing from S Brazil into the Río de la Plata. 981 mi. (1580 km) long. —**U/ru·guay/an,** *adj., n.*

U·ruk (ŏŏ/rŏŏk), an ancient Sumerian city near the Euphrates, in what is now S Iraq. Biblical name, **Erech.**

Ü·rüm·qi or **U·rum·chi** (Y/RYM/chē/; *Eng.* ŏŏrōōm/chē), the capital of Xinjiang Uygur region, in NW China. 1,000,000.

U·run·di (ōō rōōn/dē), former name of BURUNDI. Compare RUANDA-URUNDI.

U.S. or **US**, United States.

U.S.A. or **USA**, United States of America.

Us·hua·ia (ōō swī/ə), a city in S Argentina, on the S coast of Tierra del Fuego: the southernmost city in the world. 10,998.

Üs·kü·dar (ōōs/kə där/), a section of Istanbul, Turkey, on the Asian shore of the Bosporus.

Us·pa·lla/ta Pass/ (ōōs/pä yä/tə), a mountain pass in S South America, in the Andes, connecting Mendoza, Argentina, and Santiago, Chile. ab. 12,650 ft. (3855 m) high.

U.S.S.R. or **USSR**, Union of Soviet Socialist Republics.

Us·su·ri (ōō sōōr/ē), a river in E Asia, forming part of the boundary between E Manchuria and the SE Russian Federation in Asia, flowing N to the Amur River. 500 mi. (805 km) long.

Ust-Ka·me·no·gorsk (ōōst/kə mē/nə gôrsk/), a city in E Kazakhstan, on the Irtysh River. 321,000.

U·sum·bu·ra (ōō/sōōm bōōr/ə), former name of BUJUMBURA.

UT or **Ut.**, Utah.

U·tah (yōō/tô, -tä), a state in the W United States. 1,722,850; 84,916 sq. mi. (219,930 sq. km). *Cap.:* Salt Lake City. *Abbr.:* UT, Ut. —**U·tah·an, U·tahn** (yōō/tôn, -tän), *adj., n.*

U·ti·ca (yōō/ti kə), **1.** an ancient city on the N coast of Africa, NW of Carthage. **2.** a city in central New York, on the Mohawk River. 68,637.

U·trecht (yōō/trekt, -treкнt), **1.** a province in the central Netherlands. 965,229; 511 sq. mi. (1325 sq. km). **2.** the capital of this province. 230,373.

U·tsu·no·mi·ya (ōō tsōō/nō mē/yä), a city on central Honshu, in central Japan. 412,000.

Ut·tar Pra·desh (ŏŏt/ər prə däsh/, -desh/), a state in N India. 110,858,019; 113,409 sq. mi. (293,730 sq. km). *Cap.:* Lucknow.

Ux·mal (ōōs mäl/), an ancient ruined city in SE Mexico, in Yucatán: a center of later Mayan civilization.

Uz·bek·i·stan (ŏŏz bek/ə stan/, -stän/, uz-), a republic in S central Asia. 19,906,000; 172,741 sq. mi. (447,400 sq. km). *Cap.:* Tashkent. For-

merly, **Uz′bek So′viet So′cialist Repub′lic.**
—**Uz′bek,** *n.*

V

VA or **Va.,** Virginia.

Vaal (väl), a river in the Republic of South Africa, flowing SW from the Transvaal to the Orange River. 700 mi. (1125 km) long.

Vac·a·ville (vak′ə vil′), a city in central California. 71,479.

Va·do·da·ra (vä′dō där′ə), **1.** a former state in W India. **2.** a city in E Gujarat state, in W India: former capital of the state of Vadodara. 744,000. Formerly, **Baroda.**

Va·duz (fä dōōts′), the capital of Liechtenstein, on the upper Rhine. 4891.

Va·lais (va lā′), a canton in SW Switzerland. 235,500; 2021 sq. mi. (5235 sq. km). *Cap.:* Sion.

Val·dai′ Hills′ (väl dī′), a region of hills and plateaus in the W Russian Federation in Europe, at the source of the Volga River: highest point, 1140 ft. (347 m).

Val·dez (val dēz′), an ice-free port in S Alaska, at the N end of the Gulf of Alaska: S terminus of the Trans-Alaska Pipeline. 3079.

Val·di·vi·a (val dē′vē ə), a seaport in S Chile. 117,205.

Va·lence (va läns′), a city in SE France. 70,307.

Va·len·ci·a (və len′shē ə, -shə, -sē ə), **1.** a region in E Spain on the Mediterranean: formerly a Moorish kingdom. 3,772,002. **2.** a seaport in E Spain. 738,575. **3.** a city in N Venezuela. 624,113.

Va·len·ci·ennes (və len′sē enz′, -en′), a city in N France, SE of Lille. 43,202.

Val·la·do·lid (val′ə də lid′, -lēd′), a city in N Spain, NW of Madrid: Columbus died here 1506. 341,194.

Val·le d'A·o·sta (vä′lā dä ô′stä), a region in NW Italy. 114,162; 1259 sq. mi. (3260 sq. km).

Val·le·jo (və lā′ō, -yā′hō), a city in W California, on San Pablo Bay, NE of San Francisco. 109,199.

Val·let·ta (və let′ə), the capital of Malta, on the NE coast. 14,049.

Val′ley Forge′, a village in SE Pennsylvania: winter quarters of Washington's army 1777–78.

Val′ley of Ten′ Thou′sand Smokes′, a volcanic area in SW Alaska, in Katmai National Park.

Val·ley of the Kings', a valley on the W bank of the Nile near the site of Thebes: necropolis of many rulers of ancient Egypt. Also called **Val·ley of the Tombs'**.

Va·lois (val wä'), a medieval county and duchy in N France.

Val·pa·rai·so (val'pə ri'zō, -sō), a seaport in central Chile. 278,762. Spanish, **Val·pa·ra·i·so** (bäl'pä Rä ē'sô).

Van (van, vän), **Lake**, a salt lake in E Turkey. 1454 sq. mi. (3766 sq. km).

Van·cou·ver (van kōō'vər), **1.** a large island in SW Canada, off the SW coast of British Columbia. 12,408 sq. mi. (32,135 sq. km). **2.** a seaport in SW British Columbia, on the Strait of Georgia opposite SE Vancouver Island. 431,147; with suburbs 1,226,152. **3. Mount**, a mountain on the boundary between Alaska and Canada, in the St. Elias Mountains. 15,700 ft. (4785 m).

Van Die'men's Land' (van dē'mənz), former name of TASMANIA.

Vä·nern (ven'ərn), a lake in SW Sweden. 2141 sq. mi. (5545 sq. km).

Va·nu·a Le·vu (və nōō'ə lev'ōō), an island in the S Pacific, one of the Fiji Islands. 2145 sq. mi. (5556 sq. km).

Va·nu·a·tu (vä'nōō ä'tōō), a republic consisting of a group of islands in the SW Pacific, W of Fiji: formerly under joint British and French administration; gained independence in 1980. 149,400; ab. 5700 sq. mi. (14,763 sq. km). *Cap.:* Vila. Formerly, **New Hebrides.** —**Va'nu·a'tu·an,** *adj., n.*

Va·ra·na·si (və rä'nə sē), a city in SE Uttar Pradesh, in NE India, on the Ganges River. 794,000. Formerly, **Benares.**

Var·dar (vär'där), a river in S Europe flowing SE from NW Macedonia through N Greece into the Gulf of Salonika. 200 mi. (322 km) long.

Va·re·se (və rä'ze, -se), a city in N Italy, NW of Milan. 90,011.

Var·na (vär'nə), a seaport in NE Bulgaria, on the Black Sea. 305,891. Formerly, **Stalin.**

Väs·ter·ås (ves'tə rôs'), a city in central Sweden. 117,563.

Vat'i·can Cit'y (vat'i kən), an independent state within the city of Rome, on the right bank of the Tiber: established in 1929. 1000; 109 acres (44 hectares). Italian, **Città del Vaticano.**

Vät·tern (vet′ərn), a lake in S Sweden. 733 sq. mi. (1900 sq. km).

Vaud (vō), a canton in W Switzerland. 556,700; 1239 sq. mi. (3210 sq. km). *Cap.:* Lausanne. German, **Waadt.**

Ve·ii (vē′yī, vā′yē), an ancient Etruscan city in central Italy, in Etruria, near Rome: destroyed by the Romans 396 B.C.

Vel·sen (vel′sən), a seaport in W Netherlands. 57,147.

Ven·da (ven′də), a self-governing black homeland in NE South Africa: granted independence in 1979. 513,890; 2510 sq. mi. (6500 sq. km). *Cap.:* Thohoyandou.

Ven·dée (vän dā′), a region in W France, on the Atlantic: royalist revolt 1793–95. —**Ven·de·an** (ven dē′ən, vän dā′-), *n., adj.*

Ve·ne·ti·a (və nē′shē ə, -shə), a historic area in NE Italy, bounded by the Alps, the Po River, and the Adriatic Sea. Italian, **Venezia.**

Ve·ne·to (ven′i tō′, vā′ni-), a region in NE Italy. 4,374,911; 7090 sq. mi. (18,364 sq. km). *Cap.:* Venice.

Ve·ne·zia (ve ne′tsyä), **1.** Italian name of VENETIA. **2.** Italian name of VENICE.

Vene′zia Giu′lia (jōō′lyä), a former region of NE Italy, at the N end of the Adriatic: now mainly in Croatia and Slovenia.

Vene′zia Tri·den·ti′na (trē′den tē′nä), a former department in N Italy, now forming the greater part of the region of Trentino-Alto Adige.

Ven·e·zue·la (ven′ə zwā′lə, -zwē′-), **1.** a republic in N South America. 18,770,000; 352,143 sq. mi. (912,050 sq. km). *Cap.:* Caracas. **2. Gulf of,** a gulf on the NW coast of Venezuela. —**Ven′e·zue′lan,** *adj., n.*

Ven·ice (ven′is), **1.** Italian, **Venezia.** a seaport in NE Italy, built on numerous small islands in the Lagoon of Venice. 361,722. **2. Gulf of,** the N arm of the Adriatic Sea. **3. Lagoon of,** an inlet of the Gulf of Venice. —**Ve·ne·tian** (və nē′shən), *adj., n.*

Ven·lo (ven′lō), a city in the SE Netherlands. 63,820.

Ven·tu·ra (ven tŏŏr′ə, -tyŏŏr′ə), a city in SW California, NW of Los Angeles. 92,575. Official name, **San Buenaventura.**

Ve·nus·berg (vē′nəs bûrg′), a mountain in central Germany in the caverns of which, according to medieval legend, Venus held court.

Ve·ra·cruz (ver′ə krōōz′, -krōōs′), **1.** a state in E Mexico, on the Gulf of Mexico. 6,658,946; 27,759 sq. mi. (71,895 sq. km). *Cap.:* Jalapa. **2.** a seaport in this state: the chief port of Mexico. 305,456.

Verde (vûrd), **Cape,** a cape in Senegal, near Dakar: the westernmost point of Africa.

Ver·dun (vûr dun′, ver-), **1.** a city in NE France, on the Meuse River. 26,927. **2.** a city in S Quebec, in SE Canada. 60,246.

Ve·ree·ni·ging (fə rā′nə ging, -nə ᴋʜəng), a city in the S Transvaal, in NE Republic of South Africa, S of Johannesburg. 196,357.

Ver·khne·u·dinsk (vûrk′nə ōō′dinsk, vârᴋʜ′-); former name of ULAN UDE.

Ver·mont (vər mont′), a state in the NE United States: a part of New England. 562,758; 9609 sq. mi. (24,885 sq. km). *Cap.:* Montpelier. *Abbr.:* VT, Vt. **—Ver·mont′er,** *n.*

Ve·ro·na (və rō′nə), a city in N Italy, on the Adige River. 258,523. **—Ver·o·nese** (ver′ə nēz′, -nēs′), *adj., n., pl.* **-nese.**

Ver·sailles (ver sī′, vər-), a city in N France, near Paris: palace of the French kings; peace treaty between the Allies and Germany 1919. 95,240.

Ves′ter·å·len Is′lands (ves′tə rô′lən), a group of islands, belonging to Norway, in the Norwegian Sea, NE of the Lofoten Islands.

Ve·su·vi·us (və sōō′vē əs), **Mount,** an active volcano in SW Italy, near Naples: its eruption destroyed the ancient cities of Pompeii and Herculaneum A.D. 79. ab. 3900 ft. (1190 m). **—Ve·su′vi·an,** *adj.*

VI or **V.I.,** Virgin Islands.

Vi·cen·za (vi chen′zə), a city in NE Italy. 109,932.

Vi·chy (vish′ē, vē′shē), a city in central France: provisional capital of unoccupied France 1940–42; hot springs. 32,251.

Vicks·burg (viks′bûrg), a city in W Mississippi, on the Mississippi River: Civil War siege and Confederate surrender 1863. 20,908.

Vic·to·ri·a (vik tôr′ē ə, -tōr′-), **1.** Also called **Hong Kong.** the capital of Hong Kong colony, on the N coast of Hong Kong island. 1,100,000. **2.** a state in SE Australia. 4,183,500; 87,884 sq. mi. (227,620 sq. km). *Cap.:* Melbourne. **3.** the capital of British Columbia, on Vancouver Island, in SW Canada. 66,303. **4.** a city in S Texas. 55,076. **5.** the capital of the Seychelles. 23,000. **6.** Lake. Also called **Victo′ria Ny·an′za** (ni an′zə, nē-, nyän′zä). a lake in E central Africa, in Uganda, Tanzania, and Kenya: second largest freshwater lake in the world. 26,828 sq. mi. (69,485 sq. km).

Victo′ria Falls′, falls of the Zambezi River in S Africa, between Zambia and Zimbabwe, near Livingstone. 350 ft. (107 m) high; more than 1 mi. (1.6 km) wide.

Victo′ria Is′land, an island off the coast of N Canada, in the Arctic Ocean. 80,340 sq. mi. (208,081 sq. km).

Victo′ria Land′, a region in Antarctica, bordering on the Ross Sea, mainly in Ross Dependency.

Vi·en·na (vē en′ə), the capital of Austria, in the NE part, on the Danube. 1,482,800. German, **Wien.** —**Vi′en·nese′** (-ə nēz′, -nēs′), *adj., n., pl.* -nese.

Vienne (vyen), a city in SE France, on the Rhone River, S of Lyons: Roman ruins. 28,753.

Vien·tiane (vyen tyän′), the capital of Laos, on the Mekong River, in the NW part. 377,409.

Vi·et·nam or **Vi·et Nam** (vē et′näm′, -nam′, vyet′-, vē′it-), a country in SE Asia, comprising the former states of Annam, Tonkin, and Cochin-China: formerly part of French Indochina; divided into North Vietnam and South Vietnam in 1954 and reunified in 1976. 64,000,000; 126,104 sq. mi. (326,609 sq. km). *Cap.:* Hanoi. Official name, **Socialist Republic of Vietnam.** Compare NORTH VIETNAM, SOUTH VIETNAM. —**Vi·et′nam·ese′** *n., pl.* -ese, *adj.*

Vi·go (vē′gō), **1.** Bay of, an inlet of the Atlantic, in NW Spain. **2.** a seaport on this bay. 263,998.

Vii·pu·ri (vē′pŏŏ Ri), Finnish name of VYBORG.

Vi·ja·ya·wa·da (vē′jə yə wä′də), a city in E Andhra Pradesh, in SE India, on the delta of the Krishna River. 545,000. Formerly, **Bezwada.**

Vi·la (vē′lə), a seaport in and the capital of Vanuatu. 15,000.

Vil·la Cis·ne·ros (vēl′yä sēs när′ōs, vē′yä), former name of **Dakhla** (def. 2).

Vil·la·her·mo·sa (vē′yä er mō′sə), the capital of Tabasco, in E Mexico. 250,903.

Vil·la·vi·cen·cio (vē′yä vi sen′sē ō′), a city in central Colombia. 178,685.

Ville·ur·banne (vēl′öör ban′), a city in E France, near Lyons. 119,438.

Vil·ni·us (vil′nē ŏŏs′), the capital of Lithuania, in the SE part. 582,000. Russian, **Vil·na** (vyēl′nə; *Eng.* vil′nə).

Vim·i·nal (vim′ə nl), one of the seven hills on which ancient Rome was built.

Vi·ña del Mar (vēn′yə del mär′), a city in central Chile, near Valparaiso: seaside resort. 297,294.

Vin′dhya Hills′ (vind′yə), a mountain range in central India, N of the Narbada River.

Vin′dhya Pra·desh′ (prə dāsh′, -desh′), a former state in central India: now part of Madhya Pradesh.

Vine·land (vīn′lənd), a city in S New Jersey. 54,780.

Vin·land (vin′lənd), a region in E North America variously identified as a place between Newfoundland and Virginia: visited and described by Norsemen ab. A.D. 1000.

Vin·ni·tsa (vin′it sə), a city in central Ukraine, on the Bug River. 383,000.

Vin·son Mas·sif (vin′sən ma sēf′, mas′if), a mountain in Antarctica, near the Ronne Ice Shelf: highest point in Antarctica. 16,864 ft. (5140 m).

Vir·gin·ia (vər jin′yə), a state in the E United States, on the Atlantic coast: part of the historical South. 6,187,358; 40,815 sq. mi. (105,710 sq. km). *Cap.:* Richmond. *Abbr.:* VA, Va. —**Vir·gin′·ian,** *n., adj.*

Virgin′ia Beach′, a city in SE Virginia. 393,069.

Virgin′ia Cit′y, a mining town in W Nevada: famous for the discovery of the rich Comstock silver lode 1859.

Vir′gin Is′lands, a group of islands in the West Indies, E of Puerto Rico: comprises the Virgin Islands of the United States and the British Virgin Islands. *Abbr.:* VI, V.I.

Vir′gin Is′lands Na′tional Park′, a national park on St. John Island, in the Virgin Islands. 23 sq. mi. (59 sq. km).

Vir·gin Is·lands of the Unit·ed States·, a group of islands in the West Indies, including St. Thomas, St. John, and St. Croix: purchased from Denmark 1917. 110,000; 133 sq. mi. (345 sq. km). *Cap.:* Charlotte Amalie. Formerly, **Danish West Indies.**

Vi·sa·kha·pat·nam (vi sä´kə put´nəm), a seaport in Andhra Pradesh, in E India, on the Bay of Bengal. 594,000.

Vi·sa·lia (vī säl´yə), a city in central California. 75,636.

Vi·sa·yan Is·lands (vi sī´ən), a group of islands in the central Philippines, including Panay, Negros, Cebú, Bohol, Leyte, Samar, Masbate, and smaller islands. Spanish, **Bisayas.**

Vis·by (viz´bē), a seaport on the Swedish island of Gotland, in the Baltic: an important member of the Hanseatic League. 55,346.

Vis·ta (vis´tə), a town in SW California. 71,872.

Vis·tu·la (vis´chŏŏ lə), a river in Poland, flowing N from the Carpathian Mountains into the Baltic. 677 mi. (1089 km) long. Polish, **Wisła.**

Vi·tebsk (vē´tepsk), a city in NE Belarus, on the Dvina River. 347,000.

Vi·ti Le·vu (vē´tē lev´ōō), the largest of the Fiji Islands, in the S Pacific. 4027 sq. mi. (10,430 sq. km). *Cap.:* Suva.

Vi·to·ria (vi tôr´ē ə, -tōr´-), a city in N Spain. 207,501.

Vi·tó·ri·a (vi tôr´ē ə, -tōr´-), the capital of Espírito Santo, in E Brazil. 144,143.

Vlaar·ding·en (vlär´ding ən), a city in the W Netherlands, at the mouth of the Rhine. 75,023.

Vla·di·kav·kaz (vlad´i käf käz´; *Russ.* vlə dyi kuf käs´), the capital of the North Ossetian Autonomous Republic, in the Russian Federation in SE Europe. 300,000. Formerly (1944–91), **Ordzhonikidze.**

Vlad·i·mir (vlad´ə mēr´, vlə dē´mir), a city in the W Russian Federation, E of Moscow. 343,000.

Vla·di·vos·tok (vlad´ə vos´tok, -və stok´), a seaport in the SE Russian Federation in Asia, on the Sea of Japan: eastern terminus of the Trans-Siberian Railroad. 648,000.

Vlis·sing·en (vlis´ing ən), Dutch name of FLUSHING.

Vlo·rë (vlôr′ə, vlōr′ə), a seaport in SW Albania. 61,000. Formerly, **Avlona.**

Vl·ta·va (vul′tə və), a river in W Czechoslovakia, flowing N to the Elbe. 270 mi. (435 km) long. German, **Moldau.**

Voj·vo·di·na (voi′və din/ə, -dē′nə), an autonomous province within the Serbian republic, in N Yugoslavia. 2,050,000; 8303 sq. mi. (21,506 sq. km). *Cap.:* Novi Sad.

Volca′no Is′lands, three islands in the W Pacific, including Iwo Jima, belonging to Japan: under U.S. administration 1945–68.

Vol·ga (vol′gə, vōl′-), a river flowing from the Valdai Hills in the W Russian Federation E and then S to the Caspian Sea: the longest river in Europe. 2325 mi. (3745 km) long.

Vol·go·grad (vol′gə grad′, vōl′-), a city in the SW Russian Federation, on the Volga River: battles in World War II, 1942–1943. 999,000. Formerly, **Stalingrad, Tsaritsyn.**

Vo·log·da (vô′ləg də, vō′-), a city in the W Russian Federation in Europe, NNE of Moscow. 278,000.

Vol·ta (vōl′tə, vol′-), a river in W Africa, in Ghana, formed by the confluence of the Black Volta and the White Volta and flowing S into the Bight of Benin. ab. 250 mi. (400 km) long; with branches ab. 1240 mi. (1995 km) long.

Vol·ta Re·don·da (vōl′tə ri don′də, vol′-), a city in SE Brazil, NW of Rio de Janeiro. 183,917.

Vol·tur·no (vōl tŏor′nō, vol-), a river in S central Italy, flowing from the Apennines into the Tyrrhenian Sea. 110 mi. (175 km) long.

Volzh·sky or **Volzh·skiy** (vôlsh′skē), a city in the SW Russian Federation in Europe, near Volgograd on the Volga River. 257,000.

Vor·arl·berg (fôr′ärl/bərg, -bârk, fōr′-), a province in W Austria. 314,444; 1004 sq. mi. (2600 sq. km). *Cap.:* Bregenz.

Vo·ro·nezh (və rō′nish), a city in the SW Russian Federation in Europe. 887,000.

Vo·ro·shi·lov·grad (vôr′ə shē′ləf grad′), former name (1935–90) of Lugansk.

Vosges (vōzh), a range of low mountains in NE France: highest peak, 4668 ft. (1423 m).

Voy′a·geurs Na′tional Park′ (voi′ə jərz), a

national park in N Minnesota: lakes and forests. 343 sq. mi. (888 sq. km).

VT or **Vt.,** Vermont.

Vyat·ka (vyät′kə), former name of KIROV.

Vy·borg (vē′bôrg), a seaport in the NW Russian Federation in Europe, on the Gulf of Finland. 79,000. Finnish, **Viipuri.**

W

WA, Washington.

Waadt (vät), German name of VAUD.

Waal (väl), a river in the central Netherlands, flowing W to the Meuse River: the S branch of the lower Rhine. 52 mi. (84 km) long.

Wa·bash (wô′bash), a river flowing from W Ohio through Indiana, into the Ohio River. 475 mi. (765 km) long.

Wa·co (wā′kō), a city in central Texas, on the Brazos River. 103,590.

Wa·gram (vä′gräm), a village in NE Austria: Napoleon defeated the Austrians here in 1809.

Wai·ki·ki (wī′kē kē′, wī′kē kē′), a beach and resort area on SE Oahu, in central Hawaii: part of Honolulu.

Wa·ka·ya·ma (wä′kə yä′mə), a seaport on S Honshu, in S Japan. 402,000.

Wake·field (wāk′fēld′), a city in West Yorkshire, in N England. 310,300.

Wake′ Is′land, an island in the N Pacific, N of the Marshall Islands, belonging to the U.S. 3 sq. mi. (8 sq. km).

Wal′den Pond′ (wôl′dən), a pond in NE Massachusetts, near Concord: site of Thoreau's cottage.

Wales (wālz), a division of the United Kingdom, in SW Great Britain. 2,791,851; 8018 sq. mi. (20,768 sq. km). Medieval, **Cambria. —Welsh** (welsh, welch), *adj., n.*

Wal·la·chi·a or **Wa·la·chi·a** (wo lā′kē ə), a former principality in SE Europe: united with Moldavia to form Romania in 1861. **—Wal·la′chi·an,** *adj., n.*

Wal·la·sey (wol′ə sē), a city in Merseyside, in W England, on the Mersey estuary, opposite Liverpool. 97,061.

Wall′ Street′, a street in New York City, in S Manhattan: the major financial center of the U.S.

Wal′nut Creek′, a town in W California. 60,569.

Wal·sall (wôl′sôl), a city in West Midlands, in central England, near Birmingham. 261,800.

Wal·tham (wôl′thəm *or, locally,* -tham), a city in E Massachusetts. 57,878.

Wal′tham For′est (wôl′təm, -thəm), a borough of Greater London, England. 214,500.

Wal·tham·stow (wôl′təm stō′, -thəm-), a former borough, now part of Waltham Forest, in SE England.

Wal′vis Bay′ (wôl′vis), **1.** an inlet of the S Atlantic Ocean, on the coast of Namibia, in SW Africa. **2.** a seaport on this inlet. **3.** an exclave of the Republic of South Africa around this seaport. 42,234; 434 sq. mi. (1124 sq. km).

Wands·worth (wondz′wərth, -wûrth), a borough of Greater London, England. 258,100.

Wan·ne-Eick·el (vä′nə i′kəl), a city in the Ruhr region in W Germany. 100,300.

Wan·xian (wän′shyän′) also **Wan′hsien′** (-shyen′), a city in E Sichuan province, in S central China, on the Chang Jiang. 269,758.

Wa·ran·gal (wôr′əng gəl), a city in N Andhra Pradesh, in SE India. 336,000.

War·ren (wôr′ən, wor′-), **1.** a city in SE Michigan, near Detroit. 144,864. **2.** a city in NE Ohio, NW of Youngstown. 50,793.

War·ring·ton (wôr′ing tən, wor′-), a city in Cheshire, in NW England, on the Mersey River. 183,700.

War·saw (wôr′sô), the capital of Poland, in the E central part, on the Vistula River. 2,432,000. Polish, **War·sza·wa** (vär shä′vä).

War·ta (vär′tä), a river in Poland, flowing NW and W into the Oder. 445 mi. (715 km) long. German, **War′the** (-tə).

Wart·burg (värt′bŏŏrk′), a castle in Thuringia, Germany: Luther translated the New Testament here 1521–22.

War·wick (wôr′ik, wor′- *or, for 3,* -wik), **1.** a town in Warwickshire in central England. 117,800. **2.** Warwickshire. **3.** a city in E Rhode Island. 85,427.

War·wick·shire (wôr′ik shēr′, -shər, wor′-), a county in central England. 482,200; 765 sq. mi. (1980 sq. km). Also called **Warwick.**

Wa′satch Range′ (wô′sach), a mountain range in N Utah and SE Idaho. Highest peak, Mt. Timpanogos, 12,008 ft. (3660 m).

Wash (wosh, wôsh), **the,** a shallow bay of the North Sea, on the coast of E England. 20 mi. (32 km) long; 15 mi. (24 km) wide.

Wash., Washington.

Wash·ing·ton (wosh′ing tən, wô′shing-), **1.** Also

called **Washington, D.C.** the capital of the United States, on the Potomac: coextensive with the District of Columbia. 606,900. **2.** a state in the NW United States, on the Pacific coast. 4,866,692; 68,192 sq. mi. (176,615 sq. km). *Cap.:* Olympia. *Abbr.:* WA, Wash. **3. Mount,** a mountain in N New Hampshire, in the White Mountains: highest peak in the northeastern U.S., 6293 ft. (1918 m). **4. Lake,** a lake in W Washington, near Seattle. 20 mi. (32 km) long. —**Wash'ing·to'ni·an** (-tō'-nē ən), *n., adj.*

Wash·i·ta (wosh'i tô', wô'shi-), OUACHITA.

Wa·ter·bur·y (wô'tər ber'ē, -bə rē, wot'ər-), a city in W Connecticut. 108,961.

Wa·ter·ee (wô'tə rē', wot'ə-), a river in South Carolina, the lower portion of the Catawba River, joining with the Congaree River to form the Santee River. ab. 300 mi. (480 km) long.

Wa·ter·ford (wô'tər fərd, wot'ər-), **1.** a county in Munster province, in the S Republic of Ireland. 50,190; 710 sq. mi. (1840 sq. km). **2.** its county seat: a seaport. 41,054.

Wa·ter·loo (wô'tər lōō', wot'ər-, wô'tər lōō', wot'ər-), **1.** a village in central Belgium, S of Brussels: Napoleon decisively defeated here on June 18, 1815. **2.** a city in NE central Iowa. 66,467. **3.** a city in SE Ontario, in S Canada. 58,718.

Wa'ter·ton-Gla'cier Interna'tional Peace' Park' (wô'tər tən, wot'ər-), a park in S Alberta and NW Montana, jointly administered by Canada and the U.S., encompassing Waterton Lakes National Park (Canada) and Glacier National Park (U.S.). 1804 sq. mi. (4672 sq. km).

Wa'terton Lakes' Na'tional Park', a national park in W Canada, in S Alberta. 220 sq. mi. (570 sq. km).

Wau·ke·gan (wô kē'gən), a city in NE Illinois, on Lake Michigan, N of Chicago. 69,392.

Wau·ke·sha (wô'ki shô'), a city in SE Wisconsin, W of Milwaukee. 56,958.

Wau·wa·to·sa (wô'wə tō'sə), a city in SE Wisconsin, near Milwaukee. 51,308.

Wa·zir·i·stan (wə zēr'ə stän', -stan'), a mountainous region in NW Pakistan, on the Afghanistan border.

Weald (wēld), **the,** a region in SE England, in

Kent, Surrey, and Essex counties: once a forest area; now an agricultural region.

Wed/dell Sea/ (wed/l, wə del/), an arm of the Atlantic, E of Antarctic Peninsula.

Wei·fang (wā/fäng/), a city in N Shandong province, in NE China. 371,992.

Wei·hai (wā/hī/), a seaport in NE Shandong province, in E China. 175,000. Formerly, **Wei·hai·wei** (wā/hī/wā/).

Wei·mar (vī/mär, wī/-), a city in Thuringia, in central Germany. 64,000. —**Wei·mar/i·an,** *n.*, *adj.*

Wei/mar Repub/lic, the German republic (1919–33), founded at Weimar.

Weiss·horn (vīs/hôrn/), a mountain in S Switzerland, in the Alps. 14,804 ft. (4512 m).

Wel/land Ship/ Canal/ (wel/ənd), a ship canal in S Canada, in Ontario, connecting Lakes Erie and Ontario. 28 mi. (45 km) long.

Wel·ling·ton (wel/ing tən), the capital of New Zealand, on S North Island. 325,200.

Wem·bley (wem/blē), a former borough, now part of Brent, in SE England, near London.

Wen·zhou (wœn/jō/) also **Wen·chou, Wen·chow** (wun/jō/), a seaport in SE Zhejiang province, in E China. 508,611.

We·ser (vā/zər), a river in Germany, flowing N from S Lower Saxony into the North Sea. ab. 300 mi. (485 km) long.

Wes·sex (wes/iks), an ancient Anglo-Saxon kingdom, later an earldom, in S England. *Cap.:* Winchester.

West/ Al/lis (al/is), a city in SE Wisconsin, near Milwaukee. 63,221.

West/ Bank/, a region in the Middle East, between the W bank of the Jordan River and the E frontier of Israel: occupied in 1967 by Israel; formerly held by Jordan.

West/ Bengal/, a state in E India: formerly part of the province of Bengal. 54,485,560; 33,805 sq. mi. (87,555 sq. km). *Cap.:* Calcutta.

West/ Berlin/, See under BERLIN.

West/ Brom/wich (brum/ij, -ich, brom/-), a city in West Midlands, in central England, near Birmingham. 154,930.

West/ Coast/, the region of the U.S. bordering on the Pacific Ocean.

West′ Co·vi·na (kə vē′nə), a city in SW California, E of Los Angeles. 96,086.

West′ern Austral′ia, a state in W Australia. 1,477,700; 975,920 sq. mi. (2,527,635 sq. km). *Cap.:* Perth. —**West′ern Austral′ian,** *n., adj.*

West′ern Dvi′na, Dvina (def. 1).

West′ern Ghats′, a low mountain range in W India, along the W margin of the Deccan plateau and bordering on the Arabian Sea. ab. 1000 mi. (1600 km) long.

West′ern Hem′isphere, the part of the globe west of the Atlantic, including North and South America, their islands, and the surrounding waters.

West′ern Isles′, Hebrides.

West′ern Reserve′, a tract of land in NE Ohio reserved by Connecticut (1786) when its rights to other land in the western U.S. were ceded to the federal government: relinquished in 1800.

West′ern Ro′man Em′pire, the western part of the Roman Empire, esp. after the division in A.D. 395: became extinct A.D. 476.

West′ern Sahar′a, a region in NW Africa on the Atlantic coast, bounded by Morocco, Algeria, and Mauritania: a former Spanish province comprising Río de Oro and Saguia el Hamra 1884–1976; divided between Morocco and Mauritania 1976; claimed entirely by Morocco 1979, but still under dispute. 180,000; ab. 102,700 sq. mi. (266,000 sq. km). Formerly, **Spanish Sahara.**

West′ern Samo′a, an independent country in the S Pacific, comprising the W part of Samoa: formerly a trust territory of New Zealand. 163,000; 1133 sq. mi. (2935 sq. km). —**West′ern Samo′an,** *n., adj.*

West′ern Thrace′, See under Thrace (def. 2).

West·fa·len (vest fä′lən), German name of Westphalia.

West′ Flan′ders, a province in NW Belgium. 1,035,193; 1249 sq. mi. (3235 sq. km). *Cap.:* Bruges.

West′ Fri′sians, See under Frisian Islands.

West′ Ger′many, a former republic in central Europe, created in 1949 by the coalescing of the British, French, and U.S. zones of occupied Germany established in 1945: reunited with East Germany in 1990. *Cap.:* Bonn. Official name,

Federal Republic of Germany. Compare GER-MANY. —**West′ Ger′man,** *n., adj.*

West′ Glamor′gan, a county in SE Wales. 363,200; 315 sq. mi. (815 sq. km).

West′ Ham′ (ham), a former borough, now part of Newham, in SE England, near London.

West′ Hart′ford, a town in central Connecticut. 61,301.

West′ Ha′ven, a town in S Connecticut, near New Haven. 54,021.

West′ In′dies, 1. Also called **the Indies.** an ar chipelago in the N Atlantic between North and South America, comprising the Greater Antilles, the Lesser Antilles, and the Bahamas. **2. Federation of.** Also called **West′ In′dies Federa′tion.** a former federation (1958–62) of the British islands in the Caribbean, comprising Barbados, Jamaica, Trinidad, Tobago, and the Windward and Leeward island colonies. —**West′ In′dian,** *n., adj.*

West′ In′dies Asso′ciated States′, a former group (1967–81) of states associated with the United Kingdom: members included Antigua, Dominica, Grenada, St. Kitts-Nevis-Anguilla, St. Lucia, and St. Vincent.

West′ I′rian, IRIAN JAYA.

West′ Jor′dan, a town in N central Utah. 50,140.

West·land (west′lənd), a city in SE Michigan, near Detroit. 84,724.

West′ Lo′thi·an (lō′thē ən), a historic county in S Scotland. Formerly, **Linlithgow.**

West·meath (west′mēth′, -mēth′), a county in Leinster in the N central Republic of Ireland. 63,306; 681 sq. mi. (1765 sq. km).

West′ Mid′lands, a metropolitan county in central England. 2,624,300; 347 sq. mi. (899 sq. km).

West·min·ster (west′min/stər), **1.** a central borough (officially a city) of Greater London, England: Westminster Abbey, Houses of Parliament, Buckingham Palace. 173,400. **2.** a city in SW California. 78,118. **3.** a city in NE Colorado. 74,625.

West′ Pa′kistan, a former province of Pakistan: now constitutes the country of Pakistan.

West′ Palm′ Beach′, a city in SE Florida. 67,643.

West·pha·li·a (west fā/lē ə, -fāl/yə), a former province in NW Germany, now a part of North Rhine-Westphalia: treaty ending the Thirty Years' War 1648. German, **Westfalen.** —**West·pha/-li·an,** adj., n.

West/ Point/, a military reservation in SE New York, on the Hudson: U.S. Military Academy.

West/ Prus/sia, a former province of Prussia: since 1945 part of Poland. German, **West·preus·sen** (vest/proi/sən). —**West/ Prus/sian,** n., adj.

West/ Ri/ding (ri/ding), a former administrative division of Yorkshire, in N England.

West/ Sen/eca, a city in NW New York, near Buffalo. 51,210.

West/ Suf/folk, a former administrative division of Suffolk, in E England.

West/ Sus/sex, a county in SE England. 700,000; 778 sq. mi. (2015 sq. km).

West/ Val/ley Cit/y, a city in N Utah, SW of Salt Lake City. 86,976.

West/ Virgin/ia, a state in the E United States. 1,793,477; 24,181 sq. mi. (62,629 sq. km). Cap.: Charleston. Abbr.: WV, W.Va. —**West/ Virgin/ian,** n., adj.

West/ York/shire, a metropolitan county in N England. 2,052,400; 787 sq. mi. (2039 sq. km).

Wex·ford (weks/fərd), a county in Leinster province, in the SE Republic of Ireland. 102,456; 908 sq. mi. (2350 sq. km).

Wey·mouth (wā/məth), a town in E Massachusetts, S of Boston. 55,601.

Whales (hwālz, wālz), **Bay of,** an inlet of the Ross Sea, in Antarctica: location of Little America.

Whea·ton (hwēt/n, wēt/n), a city in NE Illinois, W of Chicago. 51,464.

Whee·ling (hwē/ling, wē/-), a city in N West Virginia, on the Ohio River. 34,882.

White·chap·el (hwit/chap/əl, wit/-), a district in E London, England.

White·horse (hwit/hôrs/, wit/-), capital of the Yukon Territory, in NW Canada. 15,199.

White/ Moun/tains, a range of the Appalachian Mountains in N New Hampshire. Highest peak, Mt. Washington, 6293 ft. (1918 m).

White/ Nile/, the part of the Nile that flows NE to Khartoum, Sudan. ab. 500 mi. (800 km) long. Compare NILE.

White/ Pass/, a mountain pass in SE Alaska, near Skagway. 2888 ft. (880 m) high.

White/ Riv/er, 1. a river flowing SE from NW Arkansas into the Mississippi River. 690 mi. (1110 km) long. **2.** a river flowing NE from NW Nebraska to the Missouri River in S South Dakota. 325 mi. (525 km) long.

White/ Sea/, an arm of the Arctic Ocean in the NW Russian Federation in Europe. ab. 36,000 sq. mi. (93,240 sq. km).

White/ Vol/ta, a river in W Africa, in Ghana, a branch of the Volta River. ab. 550 mi. (885 km) long. Compare VOLTA.

Whit·ney (hwit/nē, wit/-), **Mount,** a mountain in E California, in the Sierra Nevada, in Sequoia National Park: highest peak in the U.S. outside Alaska. 14,495 ft. (4418 m).

Whit·ti·er (hwit/ē ər, wit/-), a city in SW California, E of Los Angeles. 77,671.

WI, Wisconsin.

Wich·i·ta (wich/i tô/), a city in S Kansas, on the Arkansas River. 304,011.

Wich/ita Falls/, a city in N Texas. 96,259.

Wick·low (wik/lō), a county in Leinster province, in the E Republic of Ireland. 94,482; 782 sq. mi. (2025 sq. km).

Wien (vēn), German name of VIENNA.

Wies·ba·den (vēs/bäd/n), the capital of Hesse in W Germany: health resort; mineral springs. 251,800.

Wig·an (wig/ən), a borough of Greater Manchester, in W England. 309,600.

Wight (wīt), **Isle of,** an island off the S coast of England, constituting a county of England. 126,900. 147 sq. mi. (381 sq. km).

Wig·town (wig/tən, -toun/), a historic county in SW Scotland. Also called **Wig/town·shire/** (-shēr/, -shər).

Wil·helms·ha·ven (vil/helms hä/fən), a seaport in NW Germany, NW of Bremen, on the North Sea. 95,570.

Wilkes/ Land/ (wilks), a coastal region of Antarctica, S of Australia.

Wil·lam·ette (wi lam/it), a river flowing N through NW Oregon into the Columbia River at Portland. ab. 290 mi. (465 km) long.

Wil·lem·stad (vil/əm stät/), a seaport on the island of Curaçao, in the S West Indies: capital of the Netherlands Antilles. 50,000.

Willes·den (wilz/dən), a former borough, now part of Brent, in SE England, near London.

Wil·liams·burg (wil/yəmz bûrg/), a city in SE Virginia: colonial capital of Virginia; now restored to its original pre-Revolutionary style. 7304.

Wil·ming·ton (wil/ming tən), **1.** a seaport in N Delaware, on the Delaware River. 71,529. **2.** a seaport in SE North Carolina, on the Cape Fear River. 55,530.

Wil·son (wil/sən), **Mount,** a mountain in SW California, near Pasadena: astronomical observatory. 5710 ft. (1740 m).

Wil/son Dam/, a dam on the Tennessee River, in NW Alabama. 4862 ft. (1482 km) long; 137 ft. (42 km) high.

Wilt·shire (wilt/shēr, -shər), a county in S England. 550,900; 1345 sq. mi. (3485 sq. km). Also called **Wilts** (wilts).

Wim·ble·don (wim/bəl dən), a former borough, now part of Merton, in SE England, near London: international tennis tournaments.

Win·ches·ter (win/ches/tər, -chə stər), a city in Hampshire, in S England: cathedral; capital of the early Wessex kingdom and of medieval England. 95,600.

Wind/ Cave/ Na/tional Park/ (wind), a national park in SW South Dakota: limestone caverns. 44 sq. mi. (114 sq. km).

Win·der·mere (win/dər mēr/), **Lake,** a lake in NW England, in Cumbria: the largest lake in England. 10 mi. (16 km) long.

Wind·hoek (vint/hŏok/), the capital of Namibia, in the central part. 114,500.

Wind/ Riv/er Range/ (wind), a mountain range in W Wyoming, part of the Rocky Mountains. Highest peak, Gannett Peak, 13,785 ft. (4202 m).

Wind·sor (win/zər), **1.** Official name, **Wind/sor and Maid/enhead.** a city in E Berkshire, in S England, on the Thames: the site of the residence (**Wind/sor Cas/tle**) of English sovereigns since William the Conqueror. 129,900. **2.** a city in S Ontario, in SE Canada, opposite Detroit, Michigan. 193,111.

Wind/ward Is/lands (wind/wərd), a group of islands in the SE West Indies, consisting of the S part of the Lesser Antilles: includes British, French, and independent territories.

Wind/ward Pas/sage, a strait in the West Indies, between Cuba and Hispaniola. 50 mi. (80 km) wide.

Wind/y Cit/y, the, Chicago, Illinois (used as a nickname).

Win·ne·ba·go (win/ə bā/gō), **Lake,** a lake in E Wisconsin. 30 mi. (48 km) long.

Win·ni·peg (win/ə peg/), **1.** the capital of Manitoba, in S Canada, on the Red River. 594,551. **2. Lake,** a lake in S Canada, in Manitoba. 9465 sq. mi. (24,514 sq. km). **3.** a river in S Canada, flowing NW from the Lake of the Woods to Lake Winnipeg. ab. 200 mi. (320 km) long. —**Win/ni·peg/ger,** n.

Win·ni·pe·go·sis (win/ə pi gō/sis), **Lake,** a lake in S Canada, in W Manitoba, W of Lake Winnipeg. 2086 sq. mi. (5405 sq. km).

Win/ston-Sa/lem (win/stən), a city in N North Carolina. 143,485.

Win·ter·thur (vin/tər toŏr/), a city in Zurich canton, in N Switzerland, NE of Zurich. 84,400.

Wis. or **Wisc.,** Wisconsin.

Wis·con·sin (wis kon/sən), **1.** a state in the N central United States. 4,891,769; 56,154 sq. mi. (145,440 sq. km). *Cap.:* Madison. *Abbr.:* WI, Wis., Wisc. **2.** a river flowing SW from N Wisconsin to the Mississippi. 430 mi. (690 km) long. —**Wis·con/sin·ite/,** n.

Wi·sła (vē/swä), Polish name of VISTULA.

Wit·ten·berg (wit/n bûrg/, vit/-), a city in E central Germany, on the Elbe: Luther taught in the university here; beginnings of the Reformation 1517. 54,190.

Wit·wa·ters·rand (wit/wô/tərz rand/, -wot/ərz-), a rocky ridge in S Transvaal, in the Republic of South Africa, near Johannesburg: gold mining. Also called **The Rand.**

Wolfs·burg (woŏlfs/bûrg; *Ger.* vôlfs/boŏRk), a city in Lower Saxony, in N central Germany, near Brunswick. 124,900.

Wol·las·ton Lake/ (woŏl/ə stən), a lake in NE Saskatchewan, in central Canada. ab. 796 sq. mi. (2062 sq. km).

Wol·lon·gong (woŏl/ən gông/, -gong/), a seaport in E New South Wales, in E Australia. 232,510.

Wol·ver·hamp·ton (wŏŏl′vər hamp′tən), a city in West Midlands, in W England. 250,500.

Wŏn·san (wœn′sän′), a seaport in E North Korea. 350,000.

Wood·bridge (wŏŏd′brij′), a city in NE New Jersey. 90,074.

Woods (wŏŏdz), **Lake of the,** LAKE OF THE WOODS.

Wool·wich (wŏŏl′ij, -ich), a former borough of Greater London, England, now part of Greenwich and Newham.

Worces·ter (wŏŏs′tər), **1.** a city in central Massachusetts. 169,759. **2.** a city in Hereford and Worcester, in W England, on the Severn. 74,300. **3.** WORCESTERSHIRE.

Worces·ter·shire (wŏŏs′tər shēr′, -shər), a former county in W central England, now part of Hereford and Worcester. Also called WORCESTER.

Worms (wûrmz; *Ger.* vôrms), a city in E Rhineland-Palatinate, in SW Germany. 71,827.

Wound′ed Knee′ (wŏŏn′did), a village in SW South Dakota: site of a massacre of about 300 Lakota Indians on Dec. 29, 1890.

Wran·gel (rang′gəl, vrang′-), an island belonging to the Russian Federation in the Arctic Ocean, off the coast of the NE Russian Federation in Asia. ab. 2000 sq. mi. (5180 sq. km).

Wran·gell (rang′gəl), **Mount,** an active volcano in SE Alaska, in the Wrangell Mountains. 14,006 ft. (4269 m).

Wran′gell Moun′tains, a mountain range in SE Alaska. Highest peak, 16,500 ft. (5029 m).

Wrangell-St. E·li·as Na′tional Park (i lī′əs), a national park in E Alaska: mountains, glaciers, and wildlife. 12,730 sq. mi. (32,970 sq. km).

Wroc·ław (vrôts′läf), a city in SW Poland on the Oder River. 1,119,000. German, **Breslau.**

Wu·chang (wŏŏ′chäng′), a former city in E Hubei province, in E China: now part of Wuhan.

Wu·han (wŏŏ′hän′), the capital of Hubei province, in E China, at the junction of the Han and Chang Jiang: comprises the former cities of Hankou, Hanyang, and Wuchang. 3,400,000. Also called **Han Cities.**

Wu·hu (wŏŏ′hŏŏ′), a port in E Anhui province, in E China, on the Chang Jiang. 456,219.

Wup·per·tal (vŏŏp′ər täl′), a city in North Rhine-

Westphalia, in W Germany, in the Ruhr Valley. 365,500.

Würt·tem·berg (wûr′təm bûrg′; *Ger.* vʏʀ′təm-beʀk′), a former state in SW Germany: now part of Baden-Württemberg.

Würz·burg (wûrts′bûrg; *Ger.* vʏʀts′bŏŏʀk′), a city in NW Bavaria, in S Germany, on the Main River. 123,500.

Wu·xi or **Wu·sih** (wy′shœ′), also **Wu·hsi** (wōō′-shē′), a city in S Jiangsu province, in E China. 812,610.

WV or **W.Va.,** West Virginia.

WY or **Wy.** or **Wyo.,** Wyoming.

Wye (wī), a river flowing from central Wales through SW England into the Severn estuary. 130 mi. (210 km) long.

Wy·o·ming (wī ō′ming), **1.** a state in the NW United States. 453,588; 97,914 sq. mi. (253,595 sq. km). *Cap.:* Cheyenne. *Abbr.:* WY, Wyo., Wy. **2.** a city in W Michigan, near Grand Rapids. 63,891. **—Wy·o′-ming·ite′,** *n.*

Wyo′ming Val′ley, a valley in NE Pennsylvania, along the Susquehanna River: Indian massacre 1778.

X

Xan·thus (zan′thəs), an ancient city of Lycia, in SW Asia Minor.

Xe·res (her′ās, -ēz), former name of JEREZ.

Xia·men (shyä′mœn′) also **Hsiamen, 1.** an island near the Chinese mainland in the Taiwan Strait. **2.** a seaport on this island. 507,390. Also called **Amoy.**

Xi·an or **Xi′·an** or **Si·an** (shē′än′), the capital of Shaanxi province, in central China: a capital of China under the Han and T'ang dynasties. 2,330,000. Formerly, **Changan.**

Xiang·tan or **Siang·tan** (shyäng′tän′), a city in E Hunan, in S China. 300,000.

Xi Jiang (shē′ jyäng′) also **Si Kiang,** a river in S China, flowing E from Yunnan province to the South China Sea near Guangzhou. 1250 mi. (2012 km) long.

Xi·kang or **Si·kang** (shē′käng′), a former province in W China, now part of Sichuan.

Xin·gú (shing gōō′), a river flowing N through central Brazil to the Amazon. 1300 mi. (2090 km) long.

Xi·ning or **Hsi·ning** or **Si·ning** (shē′ning′), the capital of Qinghai province, in W central China. 571,545.

Xin·jiang Uy·gur or **Sin·kiang Ui·ghur** (shin′-jyäng′ wē′gər), an autonomous region in NW China, bordering Tibet, Kazakhstan, Kyrgyzstan, and Mongolia. 13,840,000; 635,830 sq. mi. (1,646,800 sq. km). *Cap.:* Ürümqi.

Xin·xiang or **Hsin·hsiang** or **Sin·siang** (shin′-shyäng′), a city in N Henan province, in E China. 508,604.

Xi·zang (shē′zäng′), Chinese name of TIBET.

Xu·zhou (shy′jō′) also **Süchow,** a city in NW Jiangsu province, in E China. 779,289.

Y

Ya·blo·no·vyy (or **Ya·blo·no·vy**) **Range** (yä′blə nə vē′), a mountain range in the SE Russian Federation in Asia, E of Lake Baikal.

Yad·kin (yad′kin), a part of the Pee Dee River that flows SE through central North Carolina.

Ya·fo (yä′fō), JAFFA.

Yak·i·ma (yak′ə mô′, -mə), **1.** a city in S Washington. 54,827. **2.** a river in S central Washington. 203 mi. (327 km) long.

Ya·kut′ Auton′omous Repub′lic (yə kōōt′), an autonomous republic in the NE Russian Federation in Asia. 1,081,000; 1,198,146 sq. mi. (3,103,200 sq. km). *Cap.:* Yakutsk.

Ya·kutsk (yə kōōtsk′), the capital of the Yakut Autonomous Republic in the NE Russian Federation in Asia, on the Lena River. 187,000.

Yal·ta (yôl′tə, yäl′-), a seaport in the Crimea, in S Ukraine, on the Black Sea: site of wartime conference of Roosevelt, Churchill, and Stalin 1945. 83,000.

Ya·lu (yä′lōō), a river in E Asia, forming part of the boundary between Manchuria and North Korea and flowing SW to the Yellow Sea. 300 mi. (483 km) long.

Yang·kü (yäng′ky′), former name of TAIYUAN.

Yan·gon (yang gon′, -gôn′), the capital of Burma, in the S part. 2,459,000. Formerly, **Rangoon.**

Yang·tze (yang′sē′, -tsē′), CHANG JIANG.

Yaoun·dé (youn′dā, youn dā′), the capital of Cameroon, in the SW part. 436,000.

Yap (yäp, yap), one of the Caroline Islands, in the W Pacific: part of the Federated States of Micronesia. 46 sq. mi. (119 sq. km).

Ya·qui (yä′kē), a river in NW Mexico, flowing into the Gulf of California. 420 mi. (676 km) long.

Yar·kand (yär kand′, -känd′), SHACHE.

Yar·mouth (yär′məth), **Great,** GREAT YARMOUTH.

Ya·ro·slavl (yär′ə slä′vəl), a city in the W Russian Federation in Europe, on the Volga. 634,000.

Ya·va·rí (yä′və rē′), Spanish name of JAVARI.

Yaz·oo (yaz′ōō, ya zōō′), a river flowing SW from N Mississippi into the Mississippi River at Vicksburg. 188 mi. (303 km) long.

Ye·do (yed′ō), a former name of Tokyo.

Yel·low·knife (yel′ō nīf′), the capital of the Northwest Territories, in N central Canada, on Great Slave Lake. 11,753.

Yel′low Riv′er, Huang He.

Yel′low Sea′, an arm of the Pacific N of the East China Sea, between China and Korea. Also called **Huang Hai, Hwang Hai.**

Yel·low·stone (yel′ō stōn′), a river flowing from NW Wyoming through Yellowstone Lake and NE through Montana into the Missouri River in W North Dakota. 671 mi. (1080 km) long.

Yel′lowstone Falls′, two waterfalls of the Yellowstone River, in Yellowstone National Park: upper falls, 109 ft. (33 m) high; lower falls, 308 ft. (94 m) high.

Yel′lowstone Lake′, a lake in NW Wyoming, in Yellowstone National Park. 20 mi. (32 km) long; 140 sq. mi. (363 sq. km).

Yel′lowstone Na′tional Park′, a park in NW Wyoming and adjacent parts of Montana and Idaho: geysers, hot springs, falls, canyon. 3458 sq. mi. (8955 sq. km).

Yem·en (yem′ən, yā′mən), **1. Republic of,** a country in S Arabia, formed in 1990 by the merger of the Yemen Arab Republic and the People's Democratic Republic of Yemen. 12,000,000; 207,000 sq. mi. (536,130 sq. km). *Cap.* (political): San'a. *Cap.* (economic): Aden. **2.** Also called **Yem′en Ar′ab Repub′lic, North Yemen.** a former country in SW Arabia: since 1990 a part of the Republic of Yemen. *Cap:* San'a. **3.** Also called **People's Democratic Republic of Yemen, South Yemen.** a former country in S Arabia: since 1990 a part of the Republic of Yemen. *Cap:* Aden. —**Yem′en·ite′, Yem′e·ni,** *n., pl.* **-en·ites, -e·nis.**

Ye·ni·sei (yen′ə sā′), a river in the Russian Federation in Asia, flowing N from the Sayan Mountains to the Kara Sea. 2566 mi. (4080 km) long.

Ye·re·van (yer′ə vän′), the capital of Armenia, in the W part. 1,199,000.

Ye·şil·köy (yesh′ēl koi′), a town in Turkey, near Istanbul. Formerly, **San Stefano.**

Ye·zo (yez′ō), former name of Hokkaido.

Yin·chuan (yin′chwän′), the capital of Ningxia Hui region, in N China. 363,509.

Ying·kou (ying′kō′) also **Ying·kow** (-kou′, -kō′), a port in Liaoning province, in NE China, near the Gulf of Liaodong. 419,640.

Yok·ka·i·chi (yō′kə ē′chē), a city on S Honshu, in central Japan. 266,000.

Yo·ko·ha·ma (yō′kə hä′mə), a seaport on SE Honshu, in central Japan, on Tokyo Bay. 3,072,000.

Yo·ko·su·ka (yō′kə soo′kə, yə koos′kə), a seaport on SE Honshu, in central Japan, on Tokyo Bay. 429,000.

Yon·kers (yong′kərz), a city in SE New York, on the Hudson, N of New York City. 188,082.

Yor′ba Lin′da (yôr′bə lin′də, yōr′-), a city in SW California. 52,422.

York (yôrk), **1.** YORKSHIRE. **2.** Ancient, **Ebora-cum.** a city in North Yorkshire, in NE England, on the Ouse: the capital of Roman Britain. 102,700. **3.** a city in SE Pennsylvania: meeting of the Continental Congress 1777–78. 42,192. **4.** an estuary in E Virginia, flowing SE into Chesapeake Bay. 40 mi. (64 km) long. **5. Cape,** a cape at the NE extremity of Australia.

York·shire (yôrk′shēr, -shər), a former county in N England, now part of Humberside, North Yorkshire, South Yorkshire, Cleveland, and Durham. Also called **York.**

York·town (yôrk′toun′), a village in SE Virginia: surrender in 1781 of Cornwallis to Washington in the American Revolution.

Yo·sem′i·te Na′tional Park′ (yō sem′i tē), a national park in E central California, in the Sierra Nevada: granite peaks, a series of high waterfalls (**Yosem′ite Falls′**), giant sequoias, and a steep-walled valley (**Yosem′ite Val′ley**). 1182 sq. mi. (3060 sq. km).

Youngs·town (yungz′toun′), a city in NE Ohio. 95,732.

Youth (yōōth), **Isle of,** an island in the Caribbean, south of and belonging to Cuba. 68,700; 1182 sq. mi. (3061 sq. km). Formerly, **Isle of Pines.** Spanish, **Isla de la Juventud.**

Y·pres (*Fr.* ē′pR°), a town in W Belgium: battles 1914–18. 34,758. Flemish, **Ieper.**

Y·ser (ē zâr′), a river flowing from N France through NW Belgium into the North Sea. 55 mi. (89 km) long.

Y.T., Yukon Territory.

Yuan Jiang (yōō än′ jyäng′, yyän′), **1.** a river in SE China, flowing NE from Guizhou province to Dongting Lake. 540 mi. (869 km) long. **2.** Chinese name of RED RIVER.

Yu·ca·tán or **Yu·ca·tan** (yōō/kə tan′, -tän′), **1.** a peninsula in SE Mexico and N Central America comprising parts of SE Mexico, N Guatemala, and Belize. **2.** a state in SE Mexico, in N Yucatán Peninsula. 1,302,600; 14,868 sq. mi. (38,510 sq. km). *Cap.:* Mérida.

Yu·go·sla·vi·a or **Ju·go·sla·vi·a** (yōō/gō slä′-vē ə), **1.** a federal republic in SE Europe: since 1992 comprised of Serbia and Montenegro. 10,392,000; 39,449 sq. mi. (102,173 sq. km). *Cap.:* Belgrade. **2.** Formerly (1918–29), **Kingdom of the Serbs, Croats, and Slovenes.** a republic in S Europe on the Adriatic: formed 1918 from the kingdoms of Serbia and Montenegro and part of Austria-Hungary; a federal republic 1945–91 comprised of Bosnia and Herzegovina, Croatia, Macedonia, Montenegro, Serbia, and Slovenia. —**Yu′go·slav′** (-släv′, -slav′), *n.* —**Yu′go·sla′·vi·an,** *adj., n.* —**Yu′go·slav′ic,** *adj.*

Yu·kon (yōō/kon), **1.** a river flowing NW and then SW from NW Canada through Alaska to the Bering Sea. ab. 2000 mi. (3220 km) long. **2.** Also called **Yu′kon Ter′ritory.** a territory in NW Canada. 23,504; 207,076 sq. mi. (536,325 sq. km). *Cap.:* Whitehorse.

Yu·ma (yōō/mə), a city in SW Arizona, on the Colorado River. 54,923.

Yun·nan or **Yün·nan** (yōō nan′, -nän′), **1.** a province in S China. 34,560,000; 168,417 sq. mi. (436,200 sq. km). *Cap.:* Kunming. **2.** former name of KUNMING.

Yu·zov·ka (yōō/zəf kə), a former name of DONETSK.

Zab·rze (zäb′zhä), a city in SW Poland. 198,000.

Za·ca·te·cas (zä′kə tā′kəs, sä′-), **1.** a state in N central Mexico. 1,251,331; 28,125 sq. mi. (72,845 sq. km). **2.** the capital of this state. 56,829.

Za·dar (zä′där), a seaport in W Croatia, on the Adriatic coast. 116,174.

Za·ga·zig (zä′gə zēg′, zag′ə-) also **Zaqaziq**, a city in NE Egypt, on the Nile delta. 202,637.

Za·gorsk (zə gôrsk′), former name (1930–91) of SERGIYEV POSAD.

Za·greb (zä′greb), the capital of Croatia, in the NW part. 1,174,512.

Zag·ros Moun·tains (zag′rəs), a mountain range in S and SW Iran, extending along the borders of Turkey and Iraq. Highest peak, 14,921 ft. (4550 m).

Za·ire or **Za·ïre** (zä ēr′, zä′ēr), **1. Republic of.** Formerly, **Democratic Republic of the Congo, Belgian Congo, Congo Free State.** a republic in central Africa: a former Belgian colony; gained independence 1960. 32,560,000; 905,063 sq. mi. (2,344,113 sq. km). *Cap.:* Kinshasa. **2.** official name within Zaire of the CONGO River. —**Za·ir′i·an, Za·ir′e·an,** *adj., n.*

Za·kin·thos (zä′kēn thôs), Greek name of ZANTE.

Za·ma (zā′mə, zä′mä), an ancient town in N Africa, SW of Carthage: the Romans defeated Hannibal near here in the final battle of the second Punic War, 202 B.C.

Zam·be·zi (zam bē′zē), a river in S Africa, flowing S and E from NW Zambia into the Mozambique Channel of the Indian Ocean. 1650 mi. (2657 km) long. —**Zam·be′zi·an,** *adj.*

Zam·bi·a (zam′bē ə), a republic in S central Africa: formerly a British protectorate; gained independence 1964. 7,384,000; 290,586 sq. mi. (752,614 sq. km). *Cap.:* Lusaka. Formerly, **Northern Rhodesia.** —**Zam′bi·an,** *adj., n.*

Zam·bo·an·ga (zam′bō äng′gə), a seaport on SW Mindanao, in the S Philippines. 449,000.

Zan·te (zän′tē, -tā, zan′-), a Greek island, off the W coast of Greece: southernmost of the Ionian Islands. 157 sq. mi. (407 sq. km). Greek, **Zakin-thos.**

Zan·zi·bar (zan/zə bär/, zan/zə bär/), **1.** an island off the E coast of Africa: with Pemba and adjacent small islands it formerly comprised a British protectorate that became independent in 1963; now part of Tanzania. 640 sq. mi. (1658 sq. km). **2.** a seaport on W Zanzibar. 110,669. —**Zan/zi·ba/ri,** adj., n., pl. **-ris.**

Za·po·ro·zhye (zä/pə rô/zhə), a city in SE Ukraine, on the Dnieper River. 884,000.

Za·qa·ziq (zä/kä zēk/, zə-), ZAGAZIG.

Za·ra·go·za (Sp. thä/ʀä gô/thä, sä/ʀä gô/sä), SARAGOSSA.

Za·ri·a (zär/ē ə), a city in N central Nigeria. 274,000.

Zea·land (zē/lənd), the largest island of Denmark: Copenhagen is located here. 2709 sq. mi. (7015 sq. km). Danish, **Sjaelland.** —**Zea/land·er,** n.

Zee·brug·ge (zē/brŏŏg/ə, zā/-), a seaport in NW Belgium: port for Bruges; German submarine base in World War I.

Zee·land (zē/lənd; Du. zā/länt/), a province in the SW Netherlands, consisting largely of islands. 355,501; 1041 sq. mi. (2695 sq. km). —**Zee/land·er,** n.

Zeist (zist), a city in the central Netherlands. 59,727.

Zer·matt (tser mät/), a village in S Switzerland, near the Matterhorn: resort. 3101.

Zet·land (zet/lənd), former name of SHETLAND.

Zhang·jia·kou (jäng/jyä/kō/) also **Chang-chiak'ou,** a city in NW Hebei province, in NE China. 605,906. Formerly, **Kalgan.**

Zhang·zhou (jäng/jō/) also **Changchou,** a city in S Fujian province, in SE China. 300,000.

Zhda·nov (zhdä/nəf, zhə dä/-), former name (1948–89) of MARIUPOL.

Zhe·jiang (jœ/jyäng/) also **Chekiang,** a province in E China, on the East China Sea. 40,700,000; 39,300 sq. mi. (101,800 sq. km). Cap.: Hangzhou.

Zheng·zhou (jœng/jō/) also **Chengchow,** the capital of Henan province, in E China. 1,590,000.

Zhen·jiang (jœn/jyäng/) also **Chenchiang,**

Chinkiang, a port in S Jiangsu province, in E China, on the Chang Jiang. 346,024.

Zhi·to·mir (zhi tō′mēr), a city in central Ukraine, W of Kiev. 287,000.

Zhu Jiang (jy′ jyäng′) also **Chu Kiang,** a river in SE China, in S Guangdong province, flowing from Guangzhou to the South China Sea and forming an estuary between Macao and Hong Kong. ab. 110 mi. (177 km) long. Also called **Pearl River.**

Zhu·zhou (jy′jō′) also **Chuchow,** a city in NE Hunan province, in SE China. 385,646.

Zi·bo (zœ′bô′) also **Tzepo, Tzupo,** a city in central Shandong province, in NE China. 2,300,000.

Zi·gong (zœ′gông′) also **Tzekung, Tzukung,** a city in S Sichuan province, in central China. 875,337.

Zim·bab·we (zim bäb′wä, -wē), a republic in S Africa: a former British colony; unilaterally declared independence in 1965; gained independence in 1980. 9,174,000; 150,804 sq. mi. (390,580 sq. km). *Cap.:* Harare. Formerly, **Southern Rhodesia** (until 1964), **Rhodesia** (1964–80). —**Zim·bab′we·an,** *adj., n.*

Zi·on (zī′ən), a hill in Jerusalem, on which the Temple was built: used to symbolize the city itself, esp. as a religious or spiritual center.

Zi′on Na′tional Park′, a national park in SW Utah: colorful canyons and mesas. 229 sq. mi. (593 sq. km).

Zla·to·ust (zlä′tə ōōst′), a city in the W Russian Federation in Asia, in the Ural Mountains. 206,000.

Zom·ba (zom′bə), a city in S Malawi: the former capital. 53,000.

Zug (tsōōk), **1.** a canton in N central Switzerland. 83,100. 92 sq. mi. (238 sq. km). **2.** the capital of this canton, on the Lake of Zug. 22,200. **3. Lake of,** a lake in N central Switzerland. 15 sq. mi. (39 sq. km).

Zui·der Zee (zī′dər zā′, zē′), a former shallow inlet of the North Sea in the central Netherlands. Compare IJSSELMEER.

Zuid-Hol·land (zoit′hôl′änt), Dutch name of SOUTH HOLLAND.

Zu·lu·land (zōō′lōō land′), a black homeland in

NE Natal province, in the Republic of South Africa. Also called **Kwazulu.**

Zu·rich (zŏŏr′ik), **1.** a canton in N Switzerland. 1,136,700; 668 sq. mi. (1730 sq. km). **2.** the capital of this canton, on the Lake of Zurich. 840,313. **3. Lake of,** a lake in N Switzerland. 34 sq. mi. (88 sq. km). German, **Zü·rich** (tsY′RiKH) (for defs. 1, 2).

Zwick·au (zwik′ou, swik′-, tsvik′-), a city in W Saxony, in E Germany. 121,749.

Zwol·le (zvôl′ə), a city in the central Netherlands. 90,570.

Member Nations of the United Nations

Afghanistan (1946)
Albania (1955)
Algeria (1962)
Angola (1976)
Antigua and Barbuda (1981)
Argentina*
Armenia (1992)
Australia*
Austria (1955)
Azerbaijan (1992)
Bahamas (1973)
Bahrain (1971)
Bangladesh (1974)
Barbados (1966)
Belarus*
Belgium*
Belize (1981)
Benin (1960)
Bhutan (1971)
Bolivia*
Bosnia and Herzegovina (1992)
Botswana (1966)
Brazil*
Brunei (1984)
Bulgaria (1955)
Burkina Faso (1960)
Burma (1948)
Burundi (1962)
Cambodia (1955)
Cameroon (1960)
Canada*
Cape Verde (1975)
Central African Republic (1960)
Chad (1960)
Chile*
China*¹
Colombia*
Comoros (1975)
Congo (1960)
Costa Rica*
Croatia (1992)

Cuba*
Cyprus (1960)
Czechoslovakia*
Denmark*
Djibouti (1977)
Dominica (1978)
Dominican Republic*
Ecuador*
Egypt*
El Salvador*
Equatorial Guinea (1968)
Estonia (1991)
Ethiopia*
Fiji (1970)
Finland (1955)
France*
Gabon (1960)
Gambia (1965)
Germany (1973)
Ghana (1957)
Greece*
Grenada (1974)
Guatemala*
Guinea (1958)
Guinea-Bissau (1974)
Guyana (1966)
Haiti*
Honduras*
Hungary (1955)
Iceland (1946)
India*
Indonesia (1950)
Iran*
Iraq*
Ireland (1955)
Israel (1949)
Italy (1955)
Ivory Coast (1960)
Jamaica (1962)
Japan (1956)
Jordan (1955)
Kazakhstan (1992)
Kenya (1963)

*Indicates charter member in 1945. (Year in parentheses shows date of admission.)
¹The People's Republic of China replaced the Republic of China in 1971.

Kuwait (1963)
Kyrgyzstan (1992)
Laos (1955)
Latvia (1991)
Lebanon*
Lesotho (1966)
Liberia*
Libya (1955)
Liechtenstein (1990)
Lithuania (1991)
Luxembourg*
Madagascar (1960)
Malawi (1964)
Malaysia (1957)
Maldives (1965)
Mali (1960)
Malta (1964)
Marshall Islands (1991)
Mauritania (1961)
Mauritius (1968)
Mexico*
Micronesia (1991)
Moldova (1992)
Mongolia (1961)
Morocco (1956)
Mozambique (1975)
Nepal (1955)
Netherlands*
New Zealand*
Nicaragua*
Niger (1960)
Nigeria (1960)
North Korea (1991)
Norway*
Oman (1971)
Pakistan (1947)
Panama*
Papua New Guinea (1975)
Paraguay*
Peru*
Philippines*
Poland*
Portugal (1955)
Qatar (1971)
Romania (1955)
Russian Federation*²
Rwanda (1962)

St. Kitts-Nevis (1983)
St. Lucia (1979)
St. Vincent and the
   Grenadines (1980)
São Tomé and Príncipe
   (1975)
Saudi Arabia*
Senegal (1960)
Seychelles (1976)
Sierra Leone (1961)
Singapore (1965)
Slovenia (1992)
Solomon Islands (1978)
Somalia (1960)
South Africa*
South Korea (1991)
Spain (1955)
Sri Lanka (1955)
Sudan (1956)
Suriname (1975)
Swaziland (1968)
Sweden (1946)
Syria*
Tajikistan (1992)
Tanzania (1961)
Thailand (1946)
Togo (1960)
Trinidad and Tobago (1962)
Tunisia (1956)
Turkey*
Turkmenistan (1992)
Uganda (1962)
Ukraine*
United Arab Emirates (1971)
United Kingdom*
United States*
Uruguay*
Uzbekistan (1992)
Vanuatu (1981)
Venezuela*
Vietnam (1977)
Western Samoa (1976)
Yemen (1947)
Yugoslavia*
Zaire (1960)
Zambia (1964)
Zimbabwe (1980)

*Indicates charter member in 1945. (Year in parentheses shows date of admission.)
²The Russian Federation replaced the U.S.S.R. in 1992.

States of the United States

State	Postal Abbr.	Capital	Population (1990 census)	Sq. Mi.	Sq. Km.	State Flower	State Nickname
Alabama	AL	Montgomery	4,062,608	51,609	133,670	Camellia	Heart of Dixie
Alaska	AK	Juneau	551,947	586,400	1,519,000	Forget-me-not	Last Frontier
Arizona	AZ	Phoenix	3,677,985	113,909	295,025	Giant Cactus	Grand Canyon State
Arkansas	AR	Little Rock	2,362,239	53,103	137,537	Apple Blossom	Land of Opportunity
California	CA	Sacramento	29,839,250	158,693	411,015	Golden Poppy	Golden State
Colorado	CO	Denver	3,307,912	104,247	270,000	Columbine	Centennial State
Connecticut	CT	Hartford	3,295,669	5009	12,975	Mountain Laurel	Constitution State
Delaware	DE	Dover	668,696	2057	5330	Peach Blossom	First State
Florida	FL	Tallahassee	13,003,362	58,560	151,670	Orange Blossom	Sunshine State
Georgia	GA	Atlanta	6,508,419	58,876	152,489	Cherokee Rose	Empire State of the South
Hawaii	HI	Honolulu	1,115,274	6424	16,638	Hibiscus	Aloha State
Idaho	ID	Boise	1,011,986	83,557	216,415	Mock Orange	Gem State
Illinois	IL	Springfield	11,466,682	56,400	146,075	Native Violet	Land of Lincoln
Indiana	IN	Indianapolis	5,564,228	36,291	93,995	Peony	Hoosier State
Iowa	IA	Des Moines	2,787,424	56,290	145,790	Wild Rose	Hawkeye State
Kansas	KS	Topeka	2,485,600	82,276	213,094	Sunflower	Sunflower State
Kentucky	KY	Frankfort	3,698,969	40,395	104,625	Goldenrod	Bluegrass State
Louisiana	LA	Baton Rouge	4,238,216	48,522	125,672	Magnolia	Pelican State
Maine	ME	Augusta	1,233,223	33,215	86,027	Pine Cone and Tassel	Pine Tree State
Maryland	MD	Annapolis	4,798,622	10,577	27,395	Black-eyed Susan	Old Line State
Massachusetts	MA	Boston	6,029,051	8257	21,385	Trailing Arbutus	Bay State
Michigan	MI	Lansing	9,328,784	58,216	150,780	Apple Blossom	Wolverine State
Minnesota	MN	St. Paul	4,387,029	84,068	217,735	Lady's-slipper	Gopher State

State	Abbr.	Capital			Flower		Nickname
Mississippi	MS	Jackson	2,586,443	47,716	Magnolia	123,585	Magnolia State
Missouri	MO	Jefferson City	5,137,804	69,674	Hawthorn	180,455	Show Me State
Montana	MT	Helena	803,655	147,138	Bitterroot	381,085	Treasure State
Nebraska	NE	Lincoln	1,584,617	77,237	Goldenrod	200,044	Cornhusker State
Nevada	NV	Carson City	1,206,152	110,540	Sagebrush	286,300	Silver State
New Hampshire	NH	Concord	1,113,915	9304	Purple Lilac	24,100	Granite State
New Jersey	NJ	Trenton	7,748,634	7836	Purple Violet	20,295	Garden State
New Mexico	NM	Santa Fe	1,521,779	121,666	Yucca	315,115	Land of Enchantment
New York	NY	Albany	18,044,505	49,576	Rose	128,400	Empire State
North Carolina	NC	Raleigh	6,657,630	52,586	Dogwood	136,198	Tarheel State
North Dakota	ND	Bismarck	641,364	70,665	Prairie Rose	183,020	Flickertail State
Ohio	OH	Columbus	10,887,325	41,222	Scarlet Carnation	106,765	Buckeye State
Oklahoma	OK	Oklahoma City	3,157,604	69,919	Mistletoe	181,090	Sooner State
Oregon	OR	Salem	2,853,733	96,981	Oregon Grape	251,180	Beaver State
Pennsylvania	PA	Harrisburg	11,924,710	45,333	Mountain Laurel	117,410	Keystone State
Rhode Island	RI	Providence	1,005,984	1214	Violet	3145	Ocean State
South Carolina	SC	Columbia	3,505,707	31,055	Carolina Jessamine	80,430	Palmetto State
South Dakota	SD	Pierre	699,999	77,047	American Pasqueflower	199,550	Sunshine State
Tennessee	TN	Nashville	4,896,641	42,246	Iris	109,415	Volunteer State
Texas	TX	Austin	17,059,805	267,339	Bluebonnet	692,410	Lone Star State
Utah	UT	Salt Lake City	1,727,784	84,916	Sego Lily	219,930	Beehive State
Vermont	VT	Montpelier	564,964	9609	Red Clover	24,885	Green Mountain State
Virginia	VA	Richmond	6,216,568	40,815	American Dogwood	105,870	Old Dominion State
Washington	WA	Olympia	4,887,941	68,192	Rhododendron	176,615	Evergreen State
West Virginia	WV	Charleston	1,801,625	24,181	Rosebay Rhododendron	62,629	Mountain State
Wisconsin	WI	Madison	4,906,745	56,154	Wood Violet	145,440	Badger State
Wyoming	WY	Cheyenne	455,975	97,914	Indian Paintbrush	253,595	Equality State
Dist. of Columbia	DC	Washington	609,909	69	American Beauty Rose	179	